The Monograph

The volumes within the *Tall Buildings and Urban Environment* series correspond to the Council's group and committee structure. The present listing includes all current topical committees. Some are collaborating to produce volumes together, and Groups DM and BSS plan, with only a few exceptions, to combine all topics into one volume.

PLANNING AND ENVIRONMENTAL CRITERIA (PC)
Philosophy of Tall Buildings
History of Tall Buildings
Architecture
Rehabilitation, Renovation, Repair
Urban Planning and Design
External Transportation
Parking
Social Effects of the Environment
Socio-Political Influences
Design for the Disabled and Elderly
Interior Design
Landscape Architecture

DEVELOPMENT AND MANAGEMENT (DM)
Economics
Ownership and Maintenance
Project Management
Tall Buildings in Developing Countries
Decision-Making Parameters
Development and Investment
Legal Aspects

SYSTEMS AND CONCEPTS (SC) ✓
Cladding
Partitions, Walls, and Ceilings
Structural Systems
Foundation Design
Construction Systems
High-Rise Housing
Prefabricated Tall Buildings
Tall Buildings Using Local Technology
Robots and Tall Buildings
Application of Systems Methodology

CRITERIA AND LOADING (CL)
Gravity Loads and Temperature Effects
Earthquake Loading and Response
Wind Loading and Wind Effects
Fire
Accidental Loading
Safety and Quality Assurance
Motion Perception and Tolerance

TALL STEEL BUILDINGS (SB)
Commentary on Structural Standards
Methods of Analysis and Design
Stability
Design Methods Based on Stiffness
Fatigue Assessment & Ductility Assurance
Connections
Cold-Formed Steel
Load and Resistance Factor Design (Limits States Design)
Mixed Construction

TALL CONCRETE AND MASONRY BUILDINGS (CB)
Commentary on Structural Standards
Selection of Structural Systems
Optimization
Elastic Analysis
Nonlinear Analysis and Limit Design
Stability
Stiffness and Crack Control
Precast Panel Structures
Creep, Shrinkage, & Temperature Effects
Cast-in-Place Concrete
Precast-Prestressed Concrete
Masonry Structures

BUILDING SERVICE SYSTEMS (BSS)
HVAC/Energy Conservation
Plumbing and Fire Protection
Electrical Systems
High-Tech Buildings
Vertical & Horizontal Transportation
Environmental Design
Urban Services

The basic objective of the Council's Monograph is to document the most recent developments to the state of the art in the field of tall buildings and their role in the urban habitat. The following volumes can be ordered through the Council.

Planning and Design of Tall Buildings, 5 volumes (1978-1981 by ASCE)

Developments in Tall Buildings–1983 (Van Nostrand Reinhold Company)

Advances in Tall Buildings (1986, Van Nostrand Reinhold Company)

High-Rise Buildings: Recent Progress (1986, Council on Tall Buildings)

Second Century of the Skyscraper (1988 Van Nostrand Reinhold Company)

Tall Buildings: 2000 and Beyond, 2 volumes. (1990 & 1991, Council on Tall Buildings)

Council Headquarters
Lehigh University, Building 13
Bethlehem, Pennsylvania 18015 USA

Fire Safety
in Tall Buildings

Library of Congress Cataloging-in-Publication Data

Fire safety in tall buildings / Council on Tall Buildings and Urban
  Habitat, Committee 8A ; contributors, Cliff Barnett ... [et al.] ;
  editorial group, Duiliu Sfintesco, chairman ; Charles Scawthorn,
  vice-chariman ; Joseph Zicherman, editor.
      p.   cm. — (Tall buildings and urban environment series)
  (Tall building criteria and loading)
    Includes bibliographical references and indexes.
    ISBN 0-07-012531-7
    1. Tall buildings—Fires and fire prevention. 2. Tall buildings—
  Design and construction. I. Barnett, Cliff. II. Sfintesco,
  Duiliu (date). III. Scawthorn, Charles. IV. Zicherman, Joseph
  B. (Joseph Bernard) V. Council on Tall Buildings and Urban Habitat.
  Committee 8A. VI. Series. VII. Series: Tall building criteria and
  loading.
  TH9445.T18F57   1992
  628.9′2—dc20   92–14561
                                                            CIP

1 2 3 4 5 6 7 8 9 0   DOC/DOC   9 8 7 6 5 4 3 2

ISBN   0-07-012531-7

For the Council on Tall Buildings, Lynn S. Beedle is the Editor-in-Chief
and Dolores B. Rice is the Managing Editor.

For McGraw-Hill, the sponsoring editor was Joel Stein, the editing
supervisor was Peggy Lamb, and the production supervisor was Donald
F. Schmidt. This book was set in Times Roman. It was composed by
McGraw-Hill's Professional Book Group composition unit.

Council on Tall Buildings and Urban Habitat

Council on Tall Buildings and Urban Habitat

Contributors

Boundary Layer Wind Tunnel Laboratory (U. Western Ontario), London
H. K. Cheng & Partners Ltd., Hong Kong
Douglas Specialist Contractors Ltd., Aldridge
The George Hyman Construction Co., Bethesda
Johnson Fain and Pereira Assoc., Los Angeles
LeMessurier Consultants Inc., Cambridge
W. L. Meinhardt & Partners Pty. Ltd., Melbourne
Obayashi Corporation, Tokyo
PSM International, Chicago
Tooley & Company, Los Angeles
Nabih Youssef and Associates, Los Angeles

Contributing Participants

Adviesbureau Voor Bouwtechniek bv, Arnhem
American Institute of Steel Construction, Chicago
Anglo American Property Services (Pty.) Ltd., Johannesburg
Artech, Inc., Taipei
Atelier D'Architecture De Genval, Genval
Austin Commercial, Inc., Dallas
Australian Institute of Steel Construction, Milsons Point
B.C.V. Progetti S.r.l., Milano
Bechtel Corporation, San Francisco
W.S. Bellows Construction Corp., Houston
Alfred Benesch & Co., Chicago
BMP Consulting Engineers, Hong Kong
Bornhorst & Ward Pty. Ltd., Spring Hill
Bovis Limited, London
Bramalea Ltd., Dallas
Brandow & Johnston Associates, Los Angeles
Brooke Hillier Parker, Hong Kong
Campeau Corp., Toronto
CBM Engineers, Houston
Cermak Peterka Petersen, Inc., Fort Collins
Connell Wagner (NSW) Pty. Ltd., Sydney
Construction Consulting Laboratory, Dallas
Crane Fulview Door Co., Lake Bluff
Crone & Associates Pty. Ltd., Sydney
Crow Construction Co., New York
Davis Langdon & Everest, London
DeSimone, Chaplin & Dobryn, New York
Dodd Pacific Engineering, Inc., Seattle
Englekirk, Hart, and Sobel, Inc., Los Angeles
Falcon Steel Company, Wilmington
Fujikawa Johnson and Associates, Chicago
Gutteridge Haskins & Davey Pty. Ltd., Sydney
T.R. Hamzah & Yeang Sdn Bhd, Selangor
Hayakawa Associates, Los Angeles
Hellmuth, Obata & Kassabaum, Inc., San Francisco
Honeywell, Inc., Minneapolis
INTEMAC, Madrid
International Iron & Steel Institute, Brussels
Irwin Johnston and Partners, Sydney
Johnson & Nielsen, Irvine
KPFF Consulting Engineers, Seattle
Lend Lease Design Group Ltd., Sydney
Stanley D. Lindsey & Assoc., Nashville
Lohan Associates, Inc., Chicago
Martin & Bravo, Inc., Honolulu
Enrique Martinez-Romero, S.A., Mexico
McWilliam Consulting Engineers, Brisbane

Mitchell McFarlane Brentnall & Partners Intl. Ltd., Hong Kong
Mitsubishi Estate Co., Ltd., Tokyo
Moh and Associates, Inc., Taipei
Mueser Rutledge Consulting Engineers, New York
Multiplex Construction (NSW) Pty. Ltd., Sydney
Nihon Sekkei, U.S.A., Ltd., Los Angeles
Nikken Sekkei Ltd., Tokyo
Norman Disney & Young, Brisbane
O'Brien-Kreitzberg & Associates, Inc., Pennsauken
Ove Arup & Partners, Sydney
Pacific Atlas Development Corp., Los Angeles
Peddle Thorp Australia Pty. Ltd., Australia
Peddle, Thorp & Walker Arch., Sydney
Perkins & Will, Chicago
J. Roger Preston & Partners, Hong Kong
Projest SA Empreendimentos e Servicos Technicos, Rio de Janeiro
Rahulan Zain Associates, Kuala Lumpur
Ranhill Berserkutu Sdn Bhd, Kuala Lumpur
Rankine & Hill, Wellington
RFB Consulting Architects, Johannesburg
Robert Rosenwasser Associates, PC, New York
Emery Roth & Sons Intl., Inc., New York
Rowan Williams Davies & Irwin, Inc., Guelph
Sepakat Setia Perunding (Sdn.) Bhd., Kuala Lumpur
Shimizu Corporation, Tokyo
South African Institute of Steel Construction, Johannesburg
Steel Reinforcement Institute of Australia, Sydney
Steen Consultants Pty. Ltd., Singapore
Stigter Clarey & Partners, Sydney
Studio Finzi, Nova E Castellani, Milano
Taylor Thompson Whitting Pty. Ltd., St. Leonards
BA Vavaroutas & Associates, Athens
Pedro Ramirez Vazquez, Arquitecto, Pedregal de San Angel
VIPAC Engineers & Scientists Ltd., Melbourne
Wargon Chapman Partners, Sydney
Weidlinger Associates, New York
Wimberley, Allison, Tong & Goo, Newport Beach
Wong & Ouyang (HK) Ltd., Hong Kong
Woodward-Clyde Consultants, New York
Yapi Merkezi Inc., Istanbul
Zaldastani Associates, Inc., Boston

Other Books in the Tall Buildings and Urban Environment Series

Fire Safety in Tall Buildings

Council on Tall Buildings and Urban Habitat
Committee 8A

CONTRIBUTORS

Cliff Barnett
Richard W. Bletzacker
Christopher K. Booker
R. Bossart
David Canter
K. P. Cheung
G. M. E. Cooke
Leonard Y. Cooper
Yuh-Chyurn Ding
M. David Egan
Charles M. Fleischman
R. Hass
Yasuo Kajoika

K. Keira
Ezel Kendik
John H. Klote
Karl Kordina
L. Krampf
Patrick R. Kroos
J. Kruppa
Junichiro Maeda
Harold E. Nelson
M. Okamatsu
Nobuhiro Okuyama
M. Oohashi
Jille Powell

Ulrich Quast
E. Richter
K. Rudolph
Y. Sakamoto
Hitoshi Sato
Charles S. E. Scawthorn
J. B. Schleich
Jan A. M. Schreuder
Duiliu Sfintesco
L. Twilt
Takatoshi Ueno
Robert B. Williamson
Joseph Zicherman

Editorial Group

Duiliu Sfintesco, Chairman
Charles Scawthorn, Vice-Chairman
Joseph Zicherman, Editor

McGraw-Hill, Inc.

New York   St. Louis   San Francisco   Auckland   Bogotá
Caracas   Lisbon   London   Madrid   Mexico   Milan
Montreal   New Delhi   Paris   San Juan   São Paulo
Singapore   Sydney   Tokyo   Toronto

AUTHOR ACKNOWLEDGMENT

This Monograph was prepared by Committee 8A (Fire) of the Council on Tall Buildings and Urban Habitat as part of the *Tall Buildings and Urban Environment Series.*

Special acknowledgment is due those individuals whose contributions and papers formed the initial contribution to the chapters in this volume. These individuals are:

Joseph Zicherman, Chapter 1
John H. Klote, Chapter 2
K. P. Cheung, Section 2.14, Chapter 3
Patrick R. Kroos, Section 2.14, Chapter 3
Jan A. M. Schreuder, Section 3.3
Charles S. E. Scawthorn, Chapter 4
Yuh-Chyurn Ding, Chapter 5
Robert B. Williamson, Chapter 5
Charles M. Fleischman, Chapter 5
Christopher K. Booker, Chapter 6
Jille Powell, Chapter 6
David Canter, Chapter 6
G. M. E. Cooke, Chapter 7
Leonard Y. Cooper, Chapter 8
Harold E. Nelson, Chapter 8
Ezel Kendik, Chapter 9
Cliff Barnett, Chapter 10
Richard W. Bletzacker, Chapter 11
J. Kruppa, Section 12.1
Duiliu Sfintesco, Section 12.1

Y. Sakamoto, Section 12.2
M. Okamatsu, Section 12.2
K. Keira, Section 12.2
M. Oohashi, Section 12.2
Yasuo Kajioka, Section 12.3
Hitoshi Sato, Section 12.3
Junichiro Maeda, Section 12.3
Nobuhiro Okuyama, Section, 12.3
Takatoshi Ueno, Section 12.3
J. B. Schleich, Chapter 13
Karl Kordina, Chapter 14
L. Krampf, Chapter 14
R. Hass, Sections 15.1 - 15.4
Ulrich Quast, Sections 15.1 - 15.4
E. Richter, Sections 15.1 - 15.4
K. Rudolph, Sections 15.1 - 15.4
L. Twilt, Sections 15.5 - 15.10
R. Bossart, Section 15.11
M. David Egan, Glossary

CONTRIBUTORS

The following is a complete list of those who have submitted written material for possible use in the Monograph, whether or not that material was used in the final version. The Committee Chairman and Editor were given quite complete latitude. Frequently, length limitations precluded the inclusion of much valuable material. The Bibliography contains all contributions. The contributors are: Cliff Barnett, Richard W. Bletzacker, Christopher K. Booker, R. Bossart, David Canter, K. P. Cheung, G. M. E. Cooke, Leonard Y. Cooper, Yuh-Chyurn Ding, M. David Egan, Charles M. Fleischman, R. Hass, Rolf Jensen, Yasuo Kajoika, K. Keira, Ezel Kendik, John H. Klote, Karl Kordina, L. Krampf, Patrick R. Kroos, J. Kruppa, Junichiro Maeda, Harold E. Nelson, M. Okamatsu, Nobuhiro Okuyama, M. Oohashi, Jille Powell, Ulrich Quast, E. Richter, K. Rudolph, Y. Sakamoto, Hitoshi Sato, Charles S. E. Scawthorn, J. B. Schleich, Jan A. M. Schreuder, Duiliu Sfintesco, L. Twilt, Takatoshi Ueno, Robert B. Williamson, and Joseph Zicherman.

COMMITTEE MEMBERS

David Armston, G. N. Badami, Mohammed Badr, Richard W. Bletzacker, Jack A. Bono, Boris Bresler, Jacques Brozzetti, S. Bryl, Sigge Eggwertz, Herbert Ehm, Raul Estrada, Robert G. Fuller, M. Galbreath, Paul K. Heilstedt, Rolf H. Jensen, Kunio Kawagoe, J. J. Keough, D. Knight, Celal N. Kostem, Patrick R. Kroos, J. Kruppa, Maxwell G. Lay, E. V. Leyendecker, Aluizo Fontana Margarido, Harry W. Marryatt, Murvan M. Maxwell, Charles S. Morgan, Anthony F. Nassetta, Harold E. Nelson, Vladimir Reichel, Leslie E. Robertson, John B. Scalzi, Charles Scawthorn, Lawrence G. Seigel, Duiliu Sfintesco, D. Ajutha Simha, Norman F. Somes, Osami Sugawa, P. S. Symonds, H. D. Taylor, Philip H. Thomas, Jairo Uribe, Erik H. Vanmarcke, Lucien Wahl, R. A. Wheatley, R. Brady Williamson, J. Witeveen, and Joseph Zicherman.

GROUP LEADERS

The committee on Fire is part of Group CL of the Council, "Criteria and Loading." The leaders are:

Alan G. Davenport, Group Chairman
Ben Kato, Group Vice-Chairman

Foreword

This volume is one of a new series of Monographs prepared under the aegis of the Council on Tall Buildings and Urban Habitat, a series that is aimed at updating the documentation of the state-of-the-art of the planning, design, construction, and operation of tall buildings and also their interaction with the urban environment of which they are a part.

The original Monographs contained 52 major topics collected in 5 volumes:

Volume PC: *Planning and Environmental Criteria for Tall Buildings*
Volume SC: *Tall Building Systems and Concepts*
Volume CL: *Tall Building Criteria and Loading*
Volume SB: *Structural Design of Tall Steel Buildings*
Volume CB: *Structural Design of Tall Concrete and Masonry Buildings*

Following the publication of a number of updates to these volumes, the Steering Group of the Council decided to develop a new series. It would be based on the original effort, but would focus more strongly on the individual topical committees rather than on the groups. This would do two things. It would free the Council committees from restraints as to length. Also it would permit material on a given topic to more quickly reach the public.

This particular Monograph was prepared by the Council's Committee 8A, *Fire*. It had its origins in the Fire Chapter in Volume CL, but going beyond that, it has built upon further documentation and recent developments in the field of fire safety. This Monograph focuses on such issues as the performance of tall buildings during fire, fire service operations and safety, occupant behavior and evacuation, and design and construction topics.

The Monograph Concept

The Monograph series *Tall Buildings and the Urban Environment* is prepared for those who plan, design, construct, or operate tall buildings, and who need the latest information as a basis for judgment decisions. It includes a summary and condensation of research findings for design use, it provides a major reference source to recent literature and to recently developed design concepts, and it identifies needed research.

The Monograph series is not intended to serve as a primer. Its function is to communicate to all knowledgeable persons in the various fields of expertise the state of the art and most advanced knowledge in those fields. Our message has more to do with setting policies and general approaches than with detailed applications. It aims to provide adequate information for experienced general practitioners confronted with their first high-rise, as well as to open new vistas to those who have been involved with them in the past. It aims at an international scope and interdisciplinary treatment.

The Monograph series was not designed to cover topics that apply to all buildings in general. However, if a subject has application to *all* buildings, but also is particularly important for a *tall* building, then the objective has been to treat that topic.

Direct contributions to this Monograph have come from many sources. Much of the material has been prepared by those in actual practice as well as by those in the academic sector. The Council has seen considerable benefit accrue from the mix of professions, and this is no less true of the Monograph series itself.

Tall Buildings

A tall building is not defined by its height or number of stories. The important criterion is whether or not the design is influenced by some aspect of "tallness." It is a building in which "tallness" strongly influences planning, design, construction, and use. It is a building whose height creates different conditions from those that exist in "common" buildings of a certain region and period.

The Council

The Council is an activity sponsored by engineering, architectural, construction, and planning professionals throughout the world, an organization that was established to study and report on all aspects of planning, design, construction, and operation of tall buildings.

The sponsoring societies of the Council are the American Institute of Architects (AIA), American Society of Civil Engineers (ASCE), American Planning Association (APA), American Society of Interior Designers (ASID), International Association for Bridge and Structural Engineering (IABSE), International Union of Architects (UIA), Japan Structural Consultants Association (JSCA), and the Urban Land Institute (ULI).

The Council is concerned not only with buildings themselves but also with the role of tall buildings in the urban environment and their impact thereon. Such a concern also involves a systematic study of the whole problem of providing adequate space for life and work, considering not only technological factors, but social and cultural aspects as well.

The Council is not an advocate for tall buildings per se; but in those situations in which they are viable, it seeks to encourage the use of the latest knowledge in their implementation.

Nomenclature

The general guideline was to use SI metric units first, followed by U.S. Customary System units in parentheses, and also "old" metric when necessary. A conversion table for units is supplied at the end of the volume. A glossary of terms also appears at the end of the volume.

The spelling was agreed at the outset to be "American" English.

A condensation of the relevant references and bibliography will be found at the end of each chapter. Full citations are given only in a composite list at the end of the volume.

From the start, the Tall Building Monograph series has been the prime focus of the Council's activity, and it is intended that its periodic revision and the implementation of its ideas and recommendations should be a continuing activity on both national and international levels. Readers who find that a particular topic needs further treatment are invited to bring it to our attention.

Acknowledgment

This work would not have been possible but for the early financial support of the National Science Foundation, which supported the program out of which this Monograph developed. More recently the major financial support has been from the organizational members, identified in earlier pages of this Monograph as well as from many individual members. Their confidence is appreciated.

Professor Le-Wu Lu of Lehigh University served as Group Advisor. Acknowledgment is next due the headquarters staff at Lehigh University with whom it has been our pleasure to be associated, namely, Jean Polzer (secretary) and Elizabeth Easley (student assistant).

All those who had a role in the authorship of the volume are identified in the acknowledgment page that follows the title page. Especially important are the contributors whose papers formed the essential first drafts—the starting point.

The primary conceptual and editing work was in the hands of the leaders of the Council's Committee. The Chairman is Duiliu Sfintesco, Consultant, Paris, France. The Vice-Chairman is Charles Scawthorn of EQE Engineering, San Francisco, California, USA. Comprehensive editing was the responsibility of Joseph Zicherman of IFT Technical Services, Berkeley, California, USA.

Overall guidance was provided by the Group Leaders Alan G. Davenport of the Boundary Layer Wind Tunnel, University of Western Ontario, London, Canada, and Ben Kato of the University of Toyo, Kawagoe City, Japan.

Lynn S. Beedle
Editor-in-Chief

Dolores B. Rice
Managing Editor

Lehigh University
Bethlehem, Pennsylvania
1992

Preface

Fire Safety represents a major aspect of the design, construction, and operation of tall buildings. Because of its interdisciplinary nature, this topic could well differ from the other Monographs in the Tall Buildings series, by virtue of its intrinsic complexity, which includes the wide range of interrelated subjects that must be addressed to properly treat fire safety issues. For example, fire safety must be taken into account with regard to:

- General architectural concepts of buildings, including interior layouts and the distribution and separation of spaces
- Performance of structure and the nonstructural elements
- Types of included occupancies
- Fire performance of interior finish materials and room contents
- Interrelated behavior of materials used in buildings in fire situations whether permanent or temporarily present there
- Performance of mechanical equipment and operational supervision needed for these components under fire situations
- Means of fire detection and alarm, passive and active protection, as well as building characteristics which effect fire fighting activities
- and finally, probable behavior of building occupants in a fire emergency

Much significant research has been conducted since the preparation in the early 1970s of the chapter on "Fire" that was included in the original Volume CL of the Mongraph on Tall Buildings. The purpose of this Mongraph is to present a number of more recent studies which reflect significant advances in knowledge in this subject area for the benefit of practicing architects, engineers, those engaged in research and teaching activities, and most importantly, code writers and enforcement authorities.

The contents of this Monograph are based on contributions by specialists from various parts of the world. As such, it reflects their varied approaches and experiences. While not an exhaustive presentation of the individual subject areas, the volume presents their significant observations and findings and thus represents a valuable contribution to the advancement of practical knowledge. Some of these subjects presented include:

- Fire performance of exterior walls
- Fire performance of tall buildings following earthquake
- Evacuation of disabled people from tall buildings in fire situations
- New approaches to structural fire engineering
- Recent studies on human behavior in fire emergency situations

In considering possible future developments, new construction or materials techniques have been included. These include:

- Use of fire-resistant steel for structural members
- Use of robots for applying fire protective coatings to structural steel members

To introduce the Monograph, thirty-seven (37) recorded fire-cases are briefly presented and followed by general commentaries. While this does not claim to be a systematic statistical survey, it nevertheless represents a significant reference, showing that casualties and property losses in tall building fires are largely due to nonconformity with essential design requirements for fire safety, as well as a lack of proper planning and provision for emergency operation.

Regarding the latter observation, it is striking that general safety provisions and guidelines, which have been well defined for decades and are required by most regulations, are still neglected in some buildings—and not only in old ones! Thus the editors have restated the importance of many of the well-known and time-honored technical and regulatory control measures included here.

In addition to the contributors to this volume (acknowledged elsewhere) the editors would like to express their appreciation to Dolores B. Rice at Lehigh University, and Jerri Holan, AIA, and Judd Amberg, Editorial Assistant, of Berkeley, California, for the invaluable assistance in preparing the Monograph. Without their able assistance its preparation would have been much more difficult. We would like to thank M. David Egan, Clemsen University, South Carolina, USA, for his work on the glossary.

Finally, acknowledgement is due the following individuals and organizations, for their assistance and support given to our contributors and to the project itself:

- Rolf Jensen and his staff at Rolf Jensen, Associates, USA
- Fire Research Station Building Research Establishment, England
- The Institution of Engineers, Australia
- H. Saito, Chiba University, Japan
- H. Yoshida, General Building Research Corp., Japan
- I. Saito, Japan Testing Center for Construction Materials
- Studiengesellschaft fur Anwendungstechnik von Eisen and Stahl e.V., Germany
- European Community for Coal and Steel
- Bundeminister fur Forschung and Technologie, Germany
- Technical Committee 3 of ECCS, and S. Bryl

Duiliu Sfintesco
Chairman

Charles Scawthorn
Vice-Chairman

Joseph Zicherman
Editor

Contents

Fire Safety
in Tall Buildings

1

Introduction

The recognition that the potential for fire-safety problems in high-rise buildings impacts people all over the world led the Council on Tall Buildings' organizers of the 1972 Lehigh Conference on Tall Buildings to dedicate a substantial portion of their work (and the resulting volume) to fire performance characteristics of such structures. Much of the information contained in that First International Conference volume addresses basic fire science and technology issues, which are still relevant 20 years since its publication (Council on Tall Buildings, 1972).

Contemporary fire-related topics are the subject of this Monograph, building upon that original groundwork as well as the subsequent chapter on fire in a 1980 Monograph (Council on Tall Buildings, Group CL, 1980). Thus our objective is to present information in selected areas to update and enhance available information on the subject of tall building fire performance.

In considering the fire performance of tall buildings, it is important to remember that while such structures are found in virtually every country around the world, local expectations as to appropriate levels of fire performance vary widely. For this reason, the international genesis of the contents of this Monograph needs to be considered and recognition given to the fact that a wide variety of technical, economical, and social factors affect the successful achievement of similar generic objectives for tall buildings around the world.

1.1 IMPACT OF FIRE

Overall incident data and statistics show that the percentages of injuries and property damage associated with fires in tall buildings are small. In addition, when such fires do happen, most often they are readily controlled. Large fires in tall buildings, though relatively few, often have consequences which are related to the nature of buildings themselves, that is, construction features may lead to extensive fire or smoke spread, or to reductions in the occupants' ability to exit readily. The small number of tall building fires which do occur usually impact substantially on the urban environment by taxing emergency services, affecting the local economy, and giving rise to legislative changes and legal actions. Because of these potential influences, descriptions of tall building fire performance can be approached in many ways.

It is obvious that technical factors define fire growth patterns, smoke spread,

1

and human response to fires in tall buildings. However, societal concerns define acceptable standards for any given tall building *before* a fire incident occurs. For example, in some locales such requirements as detailed emergency action plans, controversial classes of building products, and sophisticated fire-detection and alarm systems may or may not be mandated.

Sprinklers, too, may or may not be mandatory. Because sprinklers are effective in controlling fires and reducing property loss, in the United States they are required in all new tall buildings and in some older ones as part of retrofit regulations. In most other countries, however, sprinkler protection is not mandated. Rather, it is seen as one of a variety of fire-safety options available when designing tall buildings.

Another aspect of tall building fire safety involves the protection of these structures within the context of their use—for example, residences (permanent and temporary), hospitals, and business occupancies each require specific building safety plans and features.

Even though a tall building may possess a variety of fire-safety features, this does not provide a guarantee of safety. At best, newer tall buildings are likely to have active features such as improved fire detection, sprinklers, or emergency features to control elevator systems. Ironically, when misused or "traded off" incorrectly, these can *contribute* to occupant risk or exacerbate problems of spreading smoke. In-place construction features such as compartmentation, for example, should prevent critical fire spread with appropriate fire suppression. However, if these features are damaged by an earthquake, they may not function. Furthermore, the building might be lost altogether if passive features (such as fire-rated walls and doors) had been eliminated through tradeoff because (damaged) active features (such as sprinklers) were used instead. Such a situation recently occurred in the Loma Prieta earthquake in Northern California. In that case, although no serious high-rise fires occurred, several buildings suffered major sprinkler failures, which released thousands of gallons of water and caused substantial property damage. Had fires occurred in those buildings following the earthquake, these buildings might have been lost, because even though the sprinklers were present, they became severely damaged and only partially functional.

Because of the many differing variables which occur in emergency situations, high-rise fires are difficult to categorize. However, available fire-incident records show that tall-building fire occurrences span extremes in severity and impact. For that reason, Section 1.2 reviews data on high-rise fire incidents and provides synopses of 37 significant tall-building fires. (The objective is not to review all fire incidents in tall buildings.) Besides illustrating a spectrum of incidents and their results, these case studies will help in understanding how fires in tall buildings differ from those in low-rise structures *and* how they may be similar.

1.2 TALL BUILDING FIRE CASE STUDIES

The National Fire Protection Association (NFPA) has maintained an "international" listing of high-rise fires since 1911. NFPA reporting is based on information "submitted and documented" to the NFPA; they caution that, since not all fires are reported to them, such information is not complete. Note, for instance, the dearth of data from Europe (only one fire in France). Are there really so few fires in those countries' high rises? What social factors, customs, and laws might account for such large differences in fire incidence? Despite such questions, Table 1.1 summarizes the data accumulated since 1960. It illustrates that by far the bulk of tall building fires seems to occur in the United States and that, as the number of incidents increase, the distribution of fatalities per incident normalizes.

The following are synopses of specific fires, listed chronologically, with references. These cases illustrate a spectrum of causes, outcomes, and damage trends for high-rise fires.

1. Dale's Penthouse Restaurant (mercantile/apartment complex), Montgomery, Ala., Feb. 7, 1967; 10 stories (+ penthouse), penthouse origin, 25 deaths.
Largest loss of life in U.S. restaurant fire to that point since 1942. No sprinklers; combustible ceiling tiles, wall paneling, and decorations. Only one exit—a stairway (Juillerat and Gaudet, 1967).

2. Hawthorne House (apartments), Chicago, Ill., Jan. 24, 1969; 39 stories, 36th-floor origin, 4 deaths.
Possibly cigarette ignition on furniture. Elevators used by evacuating residents delayed fire fighting. Blowtorch effect in corridor (Watrous, 1969).

3. Apartment building, Philadelphia, Pa., Sept. 6, 1969; 19 stories, 17th-floor origin, 1 death.
Person died from fire, believed started from cigarette, when she fell asleep smoking (Fire Journal, 1970).

4. Conrad Hilton Hotel, Chicago, Ill., Jan. 25, 1970; 25 stories (+6 basement levels), 9th-floor origin, 2 deaths.
World's largest hotel (at the time of fire). No sprinklers nor fire-detection or automatic alarm systems for guests and fire department. Smoking materials ignited foam chairs piled in lobby (fire-code violation). Important issues raised were questions about alarm systems, fire emergency plans for guests, including those with hearing impairment or other disabilities (Grimes, 1970).

5. One New York Plaza, New York, N.Y., Aug. 5, 1970; 50 stories, 33rd-floor origin, 2 deaths.
Questions of whether pull boxes on each floor functioned properly; new office building. Severe, intense fire spread; fairly complete floor destruction; steel-

Table 1.1 Fire incidents and fatalities (since 1960)

Country	Total number of fire incidents I	Number of fatalities F	Average number of fatalities per incident F/I
United States	226	590	2.6
Canada	11	26	2.4
Mexico	2	4	2.0
Puerto Rico	1	96	96.0
Brazil	3	202	67.3
Columbia	1	4	4.0
France	1	2	2.0
Philippines	1	10	10.0
Japan	1	32	32.0
Korea	2	201	100.5
India	1	1	1.0

frame damage. Air-system issues included the need to plan for venting combustion products. Deaths from elevator stopping at fire floor (Ravers, 1971).

6. Office building (carpet showrooms), New York, N.Y., Dec. 4, 1970; 47 stories (+ 2 basement levels), 5th-floor origin, 3 deaths.

No sprinklers; fire and smoke detectors in air system (not called upon). Ignition from cutting torch sparks; prompt fire discovery and report to fire department. Issues of new (1968) New York building code construction: life-safety, venting, smoke-proof stair towers; elevators took passengers to fire floors (fire affected call buttons); (required) fire separations with 1-hour rating helped. Manual use and separation of air-conditioning systems and fans helped (Powers, 1971).

7. Pioneer International Hotel, Tucson, Ariz., Dec. 20, 1970; 11 stories, 4th-floor origin, 28 deaths.

No sprinklers, no fire-detection or evacuation alarm. Simultaneous fires due to arson; spread due to carpet, wall covering, and open stairways. Window escapes, jumping, and ladder rescues. Improperly installed fire door and covered fire escape on 11th floor (Watrous, 1971).

8. Motor hotel complex, New York, N.Y., July 23, 1971; 17 stories, 12th-floor origin, 6 deaths.

Management delayed reporting fire to fire department; staff mistakes allowed fire spread. Victims mistakenly tried elevators versus stairs. Arson suspected. No sprinklers; no automatic or central alarm system (Watrous, 1972a).

9. Tae Yon Kak Hotel, Seoul, Korea, Dec. 25, 1971; 21 stories, 2d-floor origin, 163 deaths.

Manual sprinkler in basement only; heat detectors in each room; pull stations each floor; automatic fire-evacuation alarm. LP-gas fire in coffee shop spread rapidly along interior finish through lobby, cutting off escape through one stairway (alternate stairway locked); smoke, heat, toxic gases, and fire spread through stairway (floors 1 to 5), HVAC shafts to top floors, then to middle floors. (Office/staff section of building closed for the day.) Over 100 escaped by sheet ropes, jumping (versus 38 jumping to death), and aerial fire ladders; 6 rescued from roof by helicopter (versus 2 falling during attempts) (Willey, 1972c).

10. Andraus building (department store and offices), São Paulo, Brazil, Feb. 24, 1972; 31 stories, 4th-floor origin, 16 deaths.

No sprinklers; separate air-conditioning each floor; no lighted exit signs, emergency lights, manual alarms, nor automatic detection systems. Fire started in 4th-floor light and ventilation well, spread through windows in shaft, continued spreading due to draft, winds, combustible ceiling, and finish; spread externally through windows above 7th floor; 3 nearby buildings also damaged. Emergency vehicles stuck in spectator traffic. Crush of panicked building occupants (panic jumping); rescue over ladders from adjacent building to 50th floor; ad hoc helicopter rescue of 350 from roof (Willey, 1972b).

11. YMCA residence, New York, N.Y., March 22, 1972; 14 stories, 7th-floor origin, 4 deaths.

Fire not reported by employee. Delayed alarm (nearby street box); no automatic fire-detection system; no sprinklers. Combustible wood paneling (Watrous, 1972b).

12. Rault Center (offices, restaurant complex), New Orleans, La., Nov. 29, 1972; 16 stories (+ penthouse), 15th-floor origin, 6 deaths.

No sprinklers; no automatic alarm systems. Earlier fire same day may have obscured second, late detection (after flashover); arson possible. Vertical fire spread via 16th-floor windows; then combustible restaurant finish. Helicopter rescue from roof; frustrated and failed window rescues. Door open in stair tower (15th floor) made tower unusable; error by those trying to take elevators up to fire floor in order to help (Watrous, 1973).

13. Avianca building, Bogota, Columbia, July 23, 1973; 36 stories, 13th-floor origin, 4 deaths.

Neither sprinklers, fire-detection, nor alarm systems; only one stairway. Fire started in storage room; spread due to delay in notifying fire department; interior and exterior spread; lack of water in building tanks to fight fire. 250 rescued from roof by helicopters (Sharry, 1974).

14. Caiza Economica building, Rio de Janeiro, Brazil, Jan. 15, 1974; 31 stories, 1st-floor origin, no deaths.

Partial sprinklers. Fire started in lobby; staff tried to fight it. Fire spread due to combustible ceiling, open stairwells with combustible finish such as carpet wall covering; stopped at 18th floor (19th story and up unfinished). Fire department called (late); then called back after leaving. Building unoccupied (Sharry, 1974).

15. Joelana building, São Paulo, Brazil, Feb. 1, 1974; 25 stories, 12th-floor origin, 179 deaths.

No lighted exit signs, manual alarm, auto fire-detection systems, or emergency plans; water tank on roof. Electrical failure (bypassed circuit breaker) ignited window air-conditioning unit; spread by interior finish, and exterior (12th floor up). Traffic delayed fire service. Major devastation. Lack of safe exit paths (one central stairway). Elevator used heavily for rescue; occupants trapped on roof; helicopter rescues failed during fire (Sharry, 1974).

16. Cavalier Beachfront Hotel, Virginia Beach, Va., Sept. 8, 1974; 11 stories, 9th-floor origin, 1 death.

No sprinklers; neither fire-detection nor automatic alarm systems (for guests and fire department alike). Delay (½ hr) reporting fire to fire department while staff tried fighting fire, only to allow its spread. Faulty stairwell door allowed heat and smoke spread (Sharry, 1975).

17. Century City Office Building, Los Angeles, Calif., Nov. 12, 1974; 15 stories, 8th-floor origin.

No sprinklers; no building communication or alarm systems; no evacuation or fire plans. Ignited lacquer vapors (light-switch spark) during renovation; smoke spread throughout building (partly by air-conditioning system); 2000 people evacuated (some by elevators). Helicopters used for equipment and surveys. Recessed windows helped limit vertical fire spread (Fire Journal, 1975).

18. Esplanade apartment building, Chicago, Ill., Feb. 13, 1975; 29 stories, 17th-floor origin, 1 death.

Exterior fire spread; self-closing door checked internal spread. No sprinkler or alarm system (Best, 1975).

19. World Trade Center, New York, N.Y., Feb. 13, 1975; 110 stories, 11th-floor origin.

Open files ignited; fire spread through vertical cableways (telephone) and combustible cable insulation (PVC) as far as 10th- and 16th-floor telephone closets (Lathrop, 1975).

20. World Trade Center (South Tower), New York, N.Y., April 17, 1975; 110 stories, 5th-floor origin.

Various extinguishing systems (such as sprinklers); smoke detectors at return-air ducts, with emergency exhaust system; evacuation plans and drills. Two-way vocal fire-alarm boxes (each floor) for sophisticated central communication system and fire department. Trash ignited (arson). Needless evacuation of floors 9 to 22 (Lathrop, 1976).

21. Squibb Building, New York, N.Y., July 11, 1975; 34 stories, 18th-floor origin.

Smoke detectors at return-air ducts; sprinklers in basement only; neither evacuation alarms nor public-address system; "stand-pipe fire alarm" pull boxes (not to fire department). Smoking material ignited temporary trash area; fire trapped 40 people on fire floor; Life Safety Code violation on length of common paths of travel (Lathrop, 1976).

22. Apartment complex, Dallas, Tex., Dec. 23, 1975; 21 stories, 14th-floor origin, 2 deaths (fire fighters).

Delay in reporting of fire; incorrect address given. No sprinklers above basement level; no automatic fire-detection system; combustible acoustical tile and vinyl wall paper in central corridor (Fire Journal, 1977).

23. International Monetary Fund Building, Washington, D.C., May 13, 1977; 13 stories (+ penthouse), 10th-floor origin.

Sprinklers (top of atrium, not activated); HVAC air system by windows (separate system for atrium); smoke detectors at fans (optional manual control). Four of six smoke vents failed in atrium, after office fire produced smoke which banked down to all levels of atrium; building engineer's long delayed absence for smoke purging. Heavy smoke damage; need for automatic mechanical ventilation of smoke (Lathrop, 1979).

24. Apartment building for elderly, Charleston, W.Va., Aug. 7, 1979; 13 stories, 9th-floor origin.

Smoke detectors in apartments and hallways. Delayed alarm from staff's mistaken instructions for fire department to disregard smoke alarm (Bimonthly Fire Record, 1980a).

25. Apartment building, Los Angeles, Calif., Oct. 29, 1979; 19 stories, 11th-floor origin, 3 deaths.

Disconnected local alarm system (no 24-hr fire watch performed); smoke and heat at fifth-floor stairwell (Bimonthly Fire Record, 1980a).

26. Westvaco office building, New York, N.Y., June 23, 1980; 42 stories, 20th-floor origin.

No sprinklers; smoke detectors had been shut down; fire drills and plans, pull alarms each floor. Smoking materials ignited fire; wide fire spread; wide smoke spread. Elevator malfunctions, a severe problem; compartmentation validated. Increased office fuel load issues; little exterior vertical fire spread (Bell, 1981).

27. Dormitory, Iowa, July 1980; 13 stories, 6th-floor origin, 1 death.

Cigarette or cigar left in upholstered chair. Student fell to death trying to escape through window (Bimonthly Fire Record, 1980b).

28. MGM Grand Hotel, Las Vegas, Nev., Nov. 21, 1980; 23 stories, 1st-floor origin, 85 deaths.

Partial sprinklers; evacuation alarm not activated; no smoke detectors at air vents; no use of emergency plan. Worst continental U.S. hotel fire since 1946; one of largest dollar-loss fires in 1980. Fire from short circuit (electrical ground fault) in restaurant; rapid fire spread from heavy fuel loads, building layout, lack of fire-resistant barriers, unprotected vents, and substandard stairway and tower enclosures; fire stopped at sprinklered areas. Smoke spread through air systems, stairways, and tower's elevator hoistways. Death from smoke in rooms, corridors, stairs, and elevators. Delays in notifying guests and fire department; excessive length of some escape paths; 300 evacuated from roof by helicopter (locked roof doors) (Fire Journal, 1982a).

29. Las Vegas Hilton Hotel, Las Vegas, Nev., Feb. 10, 1981; 30 stories, 8th-floor origin, 8 deaths.

Combination of fire-detection devices and evacuation alarms; some automatic closing (fire) doors; fire-rated elevator vestibules; smoke-proof stair towers, with smoke trap ceilings. Fire started in east tower elevator lobby (arson); rapid exterior vertical spread through 22 floors. Failed to extinguish fire early, and combustible wall and ceiling carpet contributed. Fire dampers in corridor air grills helped limit spread to guests. Helicopter rescue from roof; 1500 safely evacuated from showroom (Fire Journal, 1982b).

30. Westchase Hilton Hotel, Houston, Tex., March 6, 1982; 13 stories, 4th-floor origin, 12 deaths.

Sprinklers (linen chutes only); smoke detectors ineffective or broken. Cigarette ignited furniture; wide smoke spread; deaths from toxic fumes and soot. Ineffective early fire detection; poor notification of fire department. No evacuation alarm for guests. Room door at fire origin left open and allowed smoke spread. Poor staff reaction; exits not marked clearly (Fire Journal, 1983).

31. Alexis Nihon Plaza (apartment and office complex), Montreal, Canada, Oct. 26, 1986; 15 stories, 10th-floor origin.

No sprinklers; electrical room heat detectors; smoke detectors in lobbies and air vents; stairway pull boxes. Fire started in empty rooms and went undetected for a long period. Spread (vertical and horizontal) through communication wire pathways, interim stairway, and air ducts all the way to top floor. Remote/hidden fire annunciator panels, low water pressure (standpipes) hindered fire fighting. Building needed major reconstruction (Isner, 1988a).

32. Dupont Plaza Hotel and Casino, San Juan, Puerto Rico, Dec. 31, 1986; 20 stories, 1st-floor origin, 97 deaths.

No sprinklers, fire-detection system, evacuation plan, nor relevant staff training; manual evacuation alarm (broken); few exits. Arson initiated in new furniture stored in ballroom; fire spread through interior, quickly reached flashover. Smoke spread through elevators, air system, utility passageways, stairways, and building exterior. Some helicopter and ladder rescue (Klem, 1987; Nelson, 1987).

33. Apartment, office, restaurant complex, New York, N.Y., Jan. 11, 1988; 10 stories, 1st-floor origin, 4 deaths.

Neither central alarm system nor sprinklers. History of renovations increased fuel load, limited interior exits. Stairway doors propped open. Electrical wiring ignited couch, wood paneling (Isner, 1988b).

34. Condominium, Wisconsin, 1st half of 1988; 24 stories, 16-floor origin.

Smoke detectors in each unit and hallway. Short circuit ignited wiring, nearby combustibles (Fire Journal, 1988).

35. First Interstate Bank building, Los Angeles, Calif., May 4, 1988; 62 stories, 12th-floor origin, 1 death.

No sprinklers; fire spread in open office settings to flashover (see NBS engineering method and analytical calculations for this fire). Exterior fire spread through floors 13 to 15 after flashover via window walls (Nelson, 1989; Klem, 1989).

36. Union Bank building, Los Angeles, Calif., July 18, 1988; 39 stories, 34th-floor origin.

No sprinklers; pull boxes (evacuation bells only) each floor; smoke detectors at elevator lobbies. Spark from electrical box ignited vapors from refinishing liquids. Compartmentation limited spread; vertical fire spread quickly through windows, but damage limited due to eyebrow construction. Fire fighting hindered by lack of marked keys (Peterson, 1989).

37. John Sevier Retirement Center, Johnson City, Tenn., Dec. 24, 1989; 11 stories, 1st-floor origin, 16 deaths.

No sprinklers. Typifies evacuation problems in elderly housing (ENR, 1990).

1.3 FIRE CHARACTERISTICS

Given that fires such as those cited in the previous section occur in tall buildings, what do we know about their occurrence from a phenomenological perspective? First, in its early stages, fire growth in a tall building is similar to fire growth in a low-rise building. Questions of fire incidence and frequency are more closely related to the use of a structure than its physical size (height). In other words, most fires begin in residences, and most often the fire cause is associated with smoking or cooking activities. Likewise business or industrial occupancies possess their own characteristics and resulting subsets of data relating to frequency of fire cause. These subsets also vary from country to country. Nevertheless, fire begins most often with furnishings in a room or space; in growing it will encounter barriers such as walls or floors and, hopefully, closed doors. In tall buildings (and fire-resistive low-rise buildings as well) the fire will eventually encounter fire-resistive construction assemblies which are designed to inhibit fire spread.

The growth of a fire to where it may threaten a room is divided into two regimes, called pre- and postflashover periods. In the former, boundaries of the affected room (walls, doors, windows, and such) are not threatened and will contain the small, growing fire. In addition (and by definition), occupants of rooms in a preflashover fire will not be at major risk. There is substantial literature dealing with preflashover fires. Common critical factors tend to be the finish materials (such as wall paper), furnishings, and room geometry (size, ventilation, and the like). Crucial variables most often relate to initial and subsequent ignition of individual items in the room: ventilation, rate of flamespread over surfaces, and the ability of furnishings and finish materials to sustain combustion and liberate heat after they ignite. Analyzing whether or not a fire *does* grow to reach flashover lends itself to probabilistic modeling. To do this, one can conduct classical fault-tree analyses or apply state-transition techniques (Williamson and Ling, 1980, 1985, and 1986; Williamson, 1981). Such methods ask a series of "what-if" questions to determine the probability of a small fire growing and spreading to become a serious incident.

The separation between the two (flashover) regimes is characterized by:

1. A sudden, rapid rise in room temperature to over 550°C

2. A change from the fire being a localized or two-dimensional phenomenon (as at a single burning object or a fire moving up a wall) to a three-dimensional one involving the whole room uniformly

3. A marked propensity of the fire to spread from the compartment of origin

The third characteristic is critical because the ability of a room or fire compartment to contain a serious fire and not allow it to spread is strongly influenced by the performance of its component boundaries such as walls, doors, and floor/ceiling assemblies. This is a reflection of its fire endurance. (Fire endurance describes the ability of an assembly to resist a given level of fire threat without collapsing or otherwise allowing excessive heat transfer from the exposed to the unexposed face.) Therefore the practical outcome of a postflashover fire, should it develop in a tall building, depends on the integrity of the fire compartment boundaries. This is because a key factor is whether a fire of given intensity will be contained in the area of origin for the specified design period.

Compared to preflashover fires, postflashover fires are easier to describe analytically, as is the assessment of their impact on building elements. Direct fire testing and computational method tools are both used to accomplish this. Tests to determine fire endurance properties of rated assemblies such as walls or floors and ceilings have a long history. In the United States and Canada the ASTM standard E-119, developed many years ago, provides test conditions and pass-fail criteria for such evaluations (Babrauskas, 1976; Babrauskas and Williamson, 1978). Similar test methods can be found under various British standard, DIN, and ISO designations. Specialized versions of these test protocols exist for door and windows, as well as structural elements such as beams and columns. The ASTM E-814 test, developed in the 1970s, is used to assess the fire performance of elements (pipes, ducts, and cables) which go through ("through-penetrate") fire-rated assemblies comprising compartment boundaries and building perimeters.

1.4 COMPARTMENTATION

Regardless of precisely why some fire incidents become large and some do not, if a fire does progress to flashover, it is a very dangerous fire. Compared to a low-rise structure, when flashover occurs in a tall building, risk levels are greater both for the occupant exiting and for fire-suppression personnel accessing the affected areas. For this reason, emphasis needs to be placed on the control of early fire spread through design and construction features. In this way the initial fire is isolated and fire fighters will have sufficient time to control a growing fire *before* it spreads from the immediate area of origin. Features found in contemporary tall buildings such as fire-rated doors and walls and fire-rated floor-ceiling assemblies are used to prevent this unwanted fire spread.

Historically, older, unsprinklered high rises, with adequate construction features to prevent floor-to-floor fire spread, have not presented unusual threats of fire spread beyond compartments of origin. This changes, however, when the use of such buildings or their fuel loading changes dramatically, or when in-place fire-safety features are removed or otherwise rendered useless.

1.5 EXTERIOR/FACADE DESIGN

Of particular interest to tall-building fire safety is the potential for exterior or perimeter fire spread (see also Council on Tall Buildings, Committee 12A, 1992). Fire-resistive compartmentation has been stressed within and between individual floors. Performance of construction details at floor perimeters, however, has not been stressed as much in fire-resistive design. In these cases, curtain-wall constructions are used frequently, and fires have occasionally spread upward in certain, spectacular cases.

The First Interstate Bank fire in Los Angeles, Calif. (1988), is one such case. There, fire spread did not occur internally within that building beyond the floor of origin, but rather was associated with exterior and perimeter features. Required internal partitions and fire-stopping details (fire walls and doors) functioned largely as expected when the building was constructed in the early 1970s. If not for the (unanticipated) fire spread associated with the building's exterior, fire damage might well have been restricted to the floor of origin, even with a delayed alarm to the Los Angeles Fire Department.

From an exterior design standpoint, the vertical fire spread at the Las Vegas Hilton is also of interest. This fire was localized at elevator lobbies stacked at the end of one wing of that building. The fire spread upward much more rapidly than the one at First Interstate Bank because the finish material on elevator lobby walls (a carpetlike wall covering) became target fuel for fire plumes emanating from lower lobbies. As each of these lobbies reached flashover, its own fire plume contributed to the major plume which "climbed" the side of the building. Given no sprinkler protection, this fire growth dynamic was a foreseeable consequence of the geometry of those lobbies and of the wall coverings and furnishings installed there.

1.6 SMOKE SPREAD

Smoke migration in a high-rise fire is a primary threat to life. In terms of design, this threat is usually addressed through HVAC (heat, ventilation, and air-conditioning) and structural design and smoke-spread modeling (SSM) techniques. The latter techniques attempt to identify aspects of building design which could allow smoke spread to threaten occupants and fire fighters, and thus increase potential for property losses. Examples include the Las Vegas MGM fire, where defects in seismic shaft design led to a majority of fatalities occurring from smoke inhalation 23 stories *above* the fire location. In both this and the Dupont Plaza fire, San Juan, P.R., the actual fire incidents occurred near ground level and could just as well have occurred in low-rise structures. Design features, however, exacerbated losses.

Many contemporary buildings rely on HVAC smoke control and pressurization systems to contain smoke by applying positive pressure above and below a fire floor (or floors). In theory this works well; whether or not such systems (which are not often used in the life of a building) work in practice has not been demonstrated in a major fire. By necessity, simulation and testing of such systems cannot involve the kind of smoke loading that occurs due to burning furnishings and finish materials found in typical modern high rises. Moreover, to operate properly, these systems require training of building and fire service personnel, and development and implementation of effective emergency plans. Local fire services *must* coordinate with building personnel so that smoke control and other emergency systems do work when needed.

1.7 HUMAN RESPONSE

Much has been written about the behavior of occupants in tall building fires, and altruistic behavior between tenants in such situations is an accepted fact. Though threatened individuals do help one another, to meet the challenge of a major fire they *must* depend on sound design features *and* good emergency planning. Areas of refuge and smoke-free stairwell designs are intended to assist occupants in surviving high-rise fires, since immediate evacuation of (fully occupied) buildings is not usually a reasonable option. Helicopter evacuation should not be regarded as a routine option, even though it is occasionally a necessary course and has been used in several major high-rise fires.

Apart from major construction defects or deficiencies, the outcome of a tall-building fire is often governed by human factors. Where tall buildings depend in particular on the functioning of staff personnel for fire-safe performance (such as in hotels, hospitals, and high-rise residences), there is additional potential for disaster. Fuel heat release characteristics and incidence of ignition are higher in such buildings as compared to similar commercial or business occupancies. Such structures (regardless of whether or not they are sprinklered) must have adequate fire-safety plans and be addressed by local fire-service personnel through emergency preplanning. This problem is particularly critical in areas where limited fire-service personnel is available.

Human response to a fire is more critical in a tall building than in a low-rise structure. This is because margins for safe exiting are more limited. Building occupants, building management and operating staff, and fire-service personnel are all affected differently; each faces a different set of threats. The latter two groups have major responsibilities for the safety of the first group.

The actions and reactions of building staff are thus extremely important in high-rise fires. Where high-rise fires have reached major proportions, the performance of staff has rarely been adequate in the early periods of such an incident; usually due to inadequate emergency training. For example, in serious high-rise fires such as the Westchase Hilton (Houston), the Dupont Plaza (San Juan), the MGM Grand (Las Vegas), and the First Interstate Bank (Los Angeles), actions on the part of the staffs, such as repeated silencing of alarms, failure to promptly call emergency service personnel, and inabilities to deal with small, growing fires before they became major threats, all resulted in serious consequences. The importance of delayed alarms is critical and it is not a new concept: time is perhaps the most crucial factor in arresting fire growth and preventing major fires (Wilson, 1962).

The response of fire-service personnel is obviously important, but frequently it is affected by early performance of the building personnel or detection systems in place. In major cities where particular high-rise fires have grown to substantial size, as many as 25 to 40 percent of on-duty fire-service personnel have been needed to control fire incidents, and this has occurred even where high-rise fire and emergency preparedness drills *have* been conducted.For isolated high-rise buildings in less populated areas it is unlikely that either this concentration of personnel and equipment or the training and preplanning needed will be available to suppress such a major fire. Therefore fire control through initial design and construction features becomes even more critical. Tall hotels in remote areas, with limited fire-service personnel (with associated equipment shortages), are increasing in number. These buildings pose serious fire threats, with increased reliance on hotel staff for early response to avert potentially fatal incidents.

For these reasons, emergency action plans are critical, as is maintenance of fire-protection facilities already in place. Besides ensuring that personnel know how to operate and respond to fire-detection systems, building managements need to maintain standpipes and, if they are present, sprinkler systems.

In all cases, high-rise buildings can be made fire-safe by the installation of appropriately designed, installed, and maintained fire sprinkler systems. Even without sprinklers, the proper combination of design features and human factors can produce acceptable levels of fire safety. The latter is common, for example, in Europe, where fewer tall-building fires occur as compared to the United States.

1.8 CONDENSED REFERENCES/BIBLIOGRAPHY

Babrauskas 1976, *Fire Endurance in Buildings*
Babrauskas 1978, *Post-Flashover Compartment Fires: Basis of a Theoretical Model*
Bell 1981, *137 Injured in New York City High-Rise Building Fire*
Belles 1988, *Between the Cracks...*
Best 1975, *High-Rise Apartment Fire in Chicago Leaves One Dead*
Bimonthly Fire Record 1980a, *High-Rise Apartment Building for the Elderly*
Bimonthly Fire Record 1980b, *Dormitory*
Council on Tall Buildings 1972, *Planning and Design of Tall Buildings*
Council on Tall Buildings 1980, Volume CL, *Tall Building Criteria and Loading*
Council on Tall Buildings 1992, *Cladding*
Degenkolb 1981, *Fire Safety Requirements for Existing High-Rise Buildings*
Egan 1978, *Concepts in Building Fire Safety*
ENR 1990, *Fires*
Fahy 1989, *How Being Poor Affects Fire Risk*
Fire Journal 1970, *Bimonthly Fire Record*
Fire Journal 1975, *High-Rise Office Building*
Fire Journal 1977, *Dallas Fire Kills Two Fire Fighters*
Fire Journal 1982a, *Fire at the MGM Grand*
Fire Journal 1982b, *Investigation Report on the Las Vegas Hilton Hotel Fire*
Fire Journal 1983, *Twelve Die in Fire at Westchase Hilton Hotel*
Fire Journal 1988, *Condominium*
Grimes 1970, *Hotel Fire*
Harmathy 1986, *A Suggested Logic for Trading Between Fire-Safety Measures*
Harmathy 1987, *Fire Drainage System*
Harwood 1989, *What Kills in Fires: Smoke Inhalation or Burns?*

Isner 1988a, *$80 Million Fire in Montreal High-Rise*
Isner 1988b, *Smoky Fire Kills Four in New York High-Rise*
Journal of Fire Protection Engineering 1989, *The Capabilities of Smoke Control: Funda-*
Juillerat 1967, *Fire at Dale's Penthouse Restaurant*
Kajima 1986, *Leakage from Doors of a High-Rise Apartment*
Kansai 1983, *Investigation into the Actual Condition of Folding Fire Escape Ladders In-*
Klem 1987, *97 Die in Arson Fire at Dupont Plaza Hotel*
Klem 1989, *Los Angeles High-Rise Bank Fire*
Klote 1986, *Smoke Control and Fire Evacuation by Elevators*
Klote 1987a, *An Overview of Smoke Control Technology*
Klote 1987b, *Experiments of Piston Effect on Elevator Smoke Control*
Las Vegas Fire Department 1982, *Notification and Alarm Systems—The Las Vegas Story*
Lathrop 1975, *World Trade Center Fire*
Lathrop 1976, *Two Fires Demonstrate Evacuation Problems in High-Rise Buildings*
Lathrop 1979, *Atrium Fire Proves Difficult to Ventilate*
National Research Council of Canada 1989, *Audibility of Fire Alarm Systems in High-Rise*
Nelson 1987, *An Engineering Analysis of the Early Stage of Fire Development—The Fire at*
Nelson 1989, *Science in Action*
Peterson 1989, *Construction Helps Limit High-Rise Fire*
Powers 1971, *Office Building Fire*
Quarantelli 1984, *Organizational Behavior in Disasters and Implications for Disaster Plan-*
Ravers 1971, *New York Office Building Fire*
Said 1988, *A Review of Smoke Control Models*
Sharry 1974, *South America Burning*
Sharry 1975, *High-Rise Hotel Fire*
Watrous 1969, *Fire in a High-Rise Apartment Building*
Watrous 1971, *28 Die in Pioneer Hotel, Tucson, Arizona*
Watrous 1972a, *Fatal Hotel Fire*
Watrous 1972b, *Four Die in New York YMCA Fire*
Watrous 1973, *High-Rise Fire in New Orleans*
Weinroth 1988, *Exit: A Simulation Model of Occupant Decisions and Actions in Residential*
Willey 1972c, *Tae Yon Kak Hotel Fire*
Willey 1972b, *High-Rise Building Fire*
Williamson 1980, *The Modelling of Fire through Probabilistic Networks*
Williamson 1981, *Coupling Deterministic and Stochastic Modelling to Unwanted Fire*
Williamson 1985, *Modeling of Fire Spread through Probabilistic Networks*
Williamson 1986, *Use of Probabilistic Networks for Analysis of Smoke Spread*
Wilson 1962, *T-I-M-E! The Yardstick of Fire Control*

Part 1 – Tall-Building Fire Performance

Many common factors exist when fire threats in tall buildings are considered. Factors of primary importance include response of this class of buildings to both smoke threats and fire spread because of products of combustion, considerations related to timely response of fire departments, and fire-department operations for the high rise in general. Postearthquake fire threats are also a consideration in seismic regions of the world. This part considers areas of interest and presents international case studies related to this topic area.

2

Control of Smoke and Combustion Products

Smoke is recognized as the major killer in all fire situations. Fire scenarios involving tall buildings often represent "worst cases" where high potential for smoke-related injuries exists (Berl and Halpin, 1980).

The term *smoke* is defined in accordance with the American Society for Testing and Materials (ASTM, 1980) and the National Fire Protection Association (NFPA, 1981), which state that smoke consists of airborne solid and liquid particulates and gases that evolve when a material undergoes pyrolysis or combustion.

Research in the field of smoke control has been conducted in Australia, Canada, England, France, Japan, the United States, and Germany. This research has consisted of field tests, full-scale fire tests, and computer simulations. In the late 1960s the idea of using pressurization to prevent smoke infiltration of stairwells started to attract attention. This was followed by the idea of the "pressure sandwich," that is, venting or exhausting the fire floor and pressurizing the surrounding floors. Frequently the building's ventilation system is used for this purpose. The term *smoke control* was coined for these systems, which use pressurization produced by mechanical fans to limit smoke movement in fire situations. Many buildings have been built with smoke-control systems, and numerous others have been retrofitted for smoke control.

2.1 SMOKE MOVEMENT

A smoke-control system must be designed so that it is not overpowered by the driving forces that cause smoke movement. For this reason, an understanding of the fundamental concepts of smoke movement and of smoke control is a prerequisite to intelligent smoke-control design. The major driving forces causing smoke movement are stack effect, buoyancy, expansion, wind, and the HVAC system. Gradually, in a fire situation, smoke movement will be caused by a combination of these driving forces. The following is a discussion of each driving force as it would act without the presence of any other driving force.

17

1 Stack Effect

When it is cold outside, there is often an upward movement of air within building shafts, such as stairwells, elevator shafts, dumbwaiter shafts, mechanical shafts, or mail chutes. This phenomenon is referred to as normal stack effect. The air in the building has a buoyant force because it is warmer and less dense than the outside air. This buoyant force causes air to rise within the shafts of buildings. The significance of normal stack effect is greater for low outside temperatures and for tall shafts. However, normal stack effect can exist in a one-story building.

When the outside air is warmer than the building air, a downward airflow frequently exists in shafts. This downward airflow is called reverse stack effect. The pressure difference due to either normal or reverse stack effect is expressed as

$$\Delta P = (\rho_O - \rho_I)gh \qquad (2.1)$$

where ρ_O = air density outside shaft
ρ_I = air density inside shaft
g = gravitational constant
h = distance from neutral plane

The neutral plane is an elevation where the hydrostatic pressure inside the shaft equals the hydrostatic pressure outside the shaft. Using the ideal gas law $(P - \rho RT)$, Eq. 2.1 can be expressed as

$$\Delta P = \frac{gP}{R}\left(\frac{1}{T_O} - \frac{1}{T_I}\right)h \qquad (2.2)$$

where P = absolute atmospheric pressure
R = gas constant of air
T_O = absolute temperature of outside air
T_I = absolute temperature of air inside shaft

This equation is valid for the SI system of units as well as for the U.S. customary system, and unless otherwise noted, this is true of the remainder of the equations.

For standard atmospheric pressure of air Eq. 2.2 becomes

$$\Delta P = K_s\left(\frac{1}{T_O} - \frac{1}{T_I}\right)h \qquad (2.3)$$

where ΔP = pressure difference, Pa (in. H_2O)
T_O = absolute temperature of outside air, K (°R)
T_I = absolute temperature of air inside shaft, K (°R)
h = distance above neutral plane, m (ft)
K_s = coefficient, = 3460 (7.64)

Because the Fahrenheit and Celsius temperature scales are commonly used by design engineers, these scales are used exclusively in the following text and figures. However, the use of absolute temperatures is recommended in calculations where such temperatures are stipulated.

For a building 60 m (200 ft) tall, with a neutral plane at the midheight, an outside temperature of $-18°C$ (0°F), and an inside temperature of 21°C (70°F), the maximum pressure difference due to stack effect would be 55 Pa (0.22 in. H_2O). This means that at the top of the building, a shaft would have a pressure of 55 Pa (0.22 in. H_2O) less than the outside pressure. At the bottom of the shaft, the shaft would have a pressure of 55 Pa (0.22 in. H_2O) less than the outside pressure. Figure 2.1 is a diagram of the pressure difference between a building shaft and the outside. In the diagram, a positive pressure difference indicates that the shaft pressure is higher than the outside pressure and a negative pressure difference indicates the opposite.

Stack effect is usually thought of as existing between a building and its exterior. The air movement in buildings caused by both normal and reverse stack effect is illustrated in Fig. 2.2. In this case, the pressure difference expressed in Eq. 2.3 would actually refer to the pressure difference between the shaft and the outside of the building.

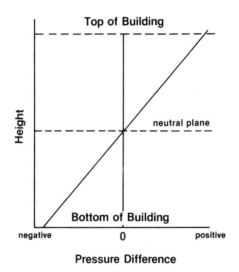

Fig. 2.1 Pressure difference between building shaft and outside due to normal stack effect.

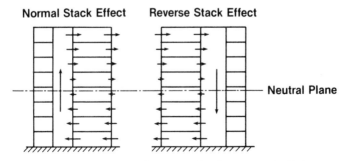

Fig. 2.2 Air movement due to normal and reverse stack effects.

Figure 2.3 can be used to determine the pressure difference due to stack effect. For normal stack effect, the term $\Delta P/h$ is positive, and the pressure difference is positive above the neutral plane and negative below it. For reverse stack effect, the term $\Delta P/h$ is negative, and the pressure difference is negative above the neutral plane and positive below it.

In unusually (air) tight buildings with exterior stairwells, reverse stack effect has been observed even with low outside air temperatures (Klote, 1980). In this situation, the exterior stairwell temperature was considerably lower than the building temperature. The stairwell was the cold column of air, and other shafts within the building were the warm columns of air.

In considering stack effect between a building and the outside, if the leakage paths are fairly uniform in height, the neutral plane will be located near the midheight of the building. However, when the leakage paths are not uniform, the location of the neutral plane can vary considerably, as in the case of vented shafts. McGuire and Tamura (1975) provide methods for calculating the location of the neutral plane for some vented conditions.

Smoke movement from a building fire can be dominated by stack effect, as evidenced in the following descriptions of different types of smoke movement resulting from normal and reverse stack effect.

Normal Stack Effect.   In a building with normal stack effect, existing air currents (as shown in Fig. 2.2) can move smoke considerable distances from a fire's origin. If a fire is below the neutral plane, smoke moves with the building air into

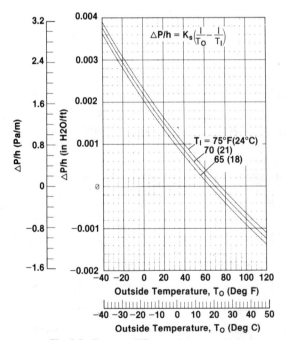

Fig. 2.3   Pressure difference due to stack effect.

and up the shafts. This upward smoke flow is enhanced by buoyancy forces acting on the smoke due to its temperature. Once above the neutral plane, the smoke flows out of the shafts into the upper floors of the building. If the leakage between floors is negligible, the floors below the neutral plane, except the fire floor, will be relatively smoke-free until the quantity of smoke produced is greater than can be handled by stack effect flows.

Smoke from a fire located above the neutral plane is carried by the building airflow to the outside through openings in the exterior of the building. If the leakage between floors is negligible, all floors other than the fire floor will remain relatively smoke-free, again until the quantity of smoke produced is greater than can be handled by stack effect flows. When the leakage between floors is considerable, there is an upward smoke movement to the floor above the fire floor.

Reverse Stack Effect. The air currents caused by reverse effect are also shown in Fig. 2.2. These forces tend to affect the movement of relatively cool smoke in the reverse of normal stack effect. In the case of hot smoke, buoyancy forces can be so great that smoke can flow upward, even during reverse stack effect conditions.

2 Buoyancy

High-temperature smoke from a fire has a buoyancy force due to its reduced density. The pressure difference between a fire compartment and its surroundings can be expressed by an equation of the same form as Eq. 2.3,

$$\Delta P = K_s\left(\frac{1}{T_O} - \frac{1}{T_F}\right)h \qquad (2.4)$$

where ΔP = pressure difference, Pa (in. H_2O)
$\quad T_O$ = absolute temperature of surroundings, K (°R)
$\quad T_F$ = absolute temperature of fire compartment, K (°R)
$\quad h$ = distance above neutral plane, m (ft)
$\quad K_s$ = coefficient, = 3460 (7.64)

The pressure difference due to buoyancy can be obtained from Fig. 2.4 for the surroundings at 20°C (68°F). The neutral plane is the plane of equal hydrostatic pressure between the fire compartment and its surroundings. For a fire with a fire-compartment temperature of 800°C (1470°F), the pressure difference at 1.52 m (5 ft) above the neutral plane is 13 Pa (0.052 in. H_2O). Fang (1980) has studied pressures caused by room fires during a series of full-scale fire tests. During these tests, the maximum pressure differential recorded was 16 Pa (0.064 in. H_2O) between fire exposed and unexposed sides of the burning room's wall at the ceiling level.

Much larger pressure differences are possible for tall fire compartments where the distance h from the neutral plane can be larger. If the fire-compartment temperature is 700°C (1290°F), the pressure difference 10.7 m (35 ft) above the neutral plane is 88 Pa (0.35 in. H_2O). This amounts to an extremely large fire, and the pressures produced by it are beyond state-of-the-art smoke control. However, the example is included here to illustrate the extent to which Eq. 2.4 can be applied.

In a building with leakage paths in the ceiling of the fire room, this buoyancy-induced pressure causes smoke movement to the floor above the fire floor. In addition, this pressure causes smoke to move through any leakage paths in the walls or around the doors of the fire compartment. As smoke travels away from the fire, its temperature drops due to heat transfer and dilution. Therefore the effect of buoyancy generally decreases with the distance from the fire.

3 Expansion

In addition to buoyancy, the energy released by a fire can cause smoke movement due to expansion. In a fire compartment with only one opening to the building, inside air will flow into the fire compartment and hot smoke will flow out of the fire compartment. Neglecting the added mass of the fuel, which is small compared to the airflow, the ratio of volumetric flows can be expressed simply as a ratio of absolute temperatures,

$$\frac{Q_{out}}{Q_{in}} = \frac{T_{out}}{T_{in}} \tag{2.5}$$

where Q_{out} = volumetric flow rate of smoke out of fire compartment, m³/s (ft³/min)

Q_{in} = volumetric flow rate of air into fire compartment, m³/s (ft³/min)

T_{out} = absolute temperature of smoke leaving fire compartment, K (°R)

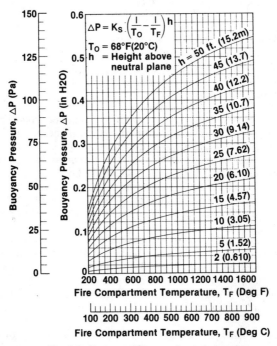

Fig. 2.4 Pressure difference due to buoyancy.

T_{in} = absolute temperature of air into fire compartment, K (°R)

For a smoke temperature of 700°C (1290°F) the ratio of volumetric flows would be 3.32, using absolute temperatures for calculation. In such a case, if the air flowing into the fire compartment is 1.5 m³/s (3180 ft³/min), then the smoke flowing out of the fire compartment would be 4.98 m³/s (10,600 ft³/min). In this case, the gas has expanded to more than three times its original volume. Therefore, for a tightly sealed fire compartment, pressure differences due to expansion may be important.

For a fire compartment with open doors or windows, the pressure difference across these openings is negligible because of the large flow areas involved. The relationship between flow area and pressure difference is discussed in Section 2.3.

4 Wind Effects

In many instances wind can have a pronounced effect on smoke movement within a building. The pressure P_w that the wind exerts on a surface can be expressed as

$$P_w = \frac{1}{2} C_w \rho_o V^2 \qquad (2.6)$$

where C_w = dimensionless pressure coefficient
ρ_o = outside air density
V = wind velocity

For an air density of 1.20 kg/m³ (0.075 lb/ft³) this relationship becomes

$$P_w = C_w K_w V^2 \qquad (2.6a)$$

where P_w = wind pressure, Pa (in. H_2O)
V = wind velocity, m/s (mi/hr)
K_w = coefficient, = 0.600 (4.82×10^{-4})

The pressure coefficients C_w are in the range of ±0.8, with positive values for windward walls and negative values for leeward walls. The pressure coefficient depends on the building geometry and varies locally over the wall surface. In general, the wind velocity increases with height above the surface of the earth. Detailed information concerning wind velocity variations and pressure coefficients is available from a number of sources (Sachs, 1972; Houghton and Carruther, 1976; Simiu and Scanlan, 1978; MacDonald, 1975). Specific information about wind data with respect to air infiltration in buildings has been generated by Tamura and Shaw (1976b).

A 15.6-m/s (35-mi/hr) wind produces a pressure on a structure of 117 Pa (0.47 in. H_2O) with a pressure coefficient of 0.8. The effect of wind on air movement within tightly constructed buildings, and with all their doors and windows closed, is slight. However, the effects of wind can become important for loosely constructed buildings or for buildings with open doors or windows. Usually the resulting airflows are complicated, and, for practical purposes, computer analysis is required.

In fire situations a window frequently breaks in the fire compartment. If the window is on the leeward side of the building, the negative pressure caused by the wind vents the smoke from the fire compartment. This can greatly reduce smoke movement throughout the building. However, if the broken window is on the windward side, the wind forces the smoke throughout the fire floor and even to other floors. This both endangers the lives of building occupants and hampers fire fighting. Pressures induced by the wind in this type of situation can be relatively large and can easily dominate air movement throughout the building.

5 HVAC System

In the early stages of a fire, an HVAC system can serve as an aid to fire detection because it frequently transports smoke during buildings fires. When a fire starts in an unoccupied portion of a building, the HVAC system can transport the smoke to a space where people can smell it and be alerted to the fire. However, as the fire progresses, the HVAC system will transport smoke to every area that it serves, thus endangering life in all those spaces. The HVAC system also supplies air to the fire space, which aids combustion. These are the reasons that HVAC systems, before development of the smoke-control concept, traditionally have been shut down when fires were discovered. Although shutting down the HVAC system prevents it from supplying air to the fire, it does not prevent smoke movement through the supply and return air ducts, air shafts, and other building openings due to stack effect, buoyancy, or wind.

2.2 SMOKE MANAGEMENT

In this discussion, the term *smoke management* includes all methods used independently or in combination to modify smoke movement for the benefit of occupants and fire fighters and for the reduction of property damage. The use of barriers, smoke vents, and smoke shafts are traditional methods of smoke management.

The effectiveness of a barrier in limiting smoke movement depends on the leakage paths in the barrier and on the pressure difference across the barrier. Holes where pipes penetrate walls or floors, cracks where walls meet floors, and cracks around doors are a few possible leakage paths. The pressure difference across these barriers depends on stack effect, buoyancy, wind, and the HVAC system, as discussed in the previous section.

The effectiveness of smoke vents and smoke shafts depends on their proximity to the fire, the buoyancy of the smoke, and the presence of other driving forces. For example, when smoke is cooled due to sprinklers, the effectiveness of smoke vents and smoke shafts is greatly reduced.

Elevator shafts in buildings have been used as smoke shafts. Unfortunately this prevents their use for fire evacuation, and such shafts frequently distribute smoke to floors far from the fire. Specially designed smoke shafts, which have essentially no leakage on floors other than the fire floor, can be used to prevent the distribution of smoke to nonfire floors.

The effectiveness of barriers in a traditional smoke-management system is limited to the extent to which the barriers are free of leakage paths. Use of smoke vents and smoke shafts is thus constrained because the smoke must be sufficiently buoyant to overcome any other driving forces that might be present.

In the last few decades, fans have been used with the intent of overcoming the limitations of traditional systems. Systems with fans are called smoke-control systems. They rely on pressure differences and airflows to limit smoke movement and are discussed in the following section.

2.3 PRINCIPLES OF SMOKE CONTROL

The barriers (such as walls, floors, and doors) used in traditional smoke management are also used for smoke control in conjunction with airflows and pressure differences generated by mechanical fans.

Figure 2.5 illustrates the pressure difference across a barrier acting to control smoke movement. Within the barrier is a door. The high-pressure side of the door can be either a refuge area or an escape route. The low-pressure side is exposed to smoke from a fire. Airflow through the cracks around the door, and through other construction cracks, prevents smoke infiltration to the high-pressure side.

When the door in the barrier is opened, airflow through the open door results. When the air velocity is low, smoke can flow against the airflow into the refuge area or escape route, as shown in Fig. 2.6. This smoke backflow can be prevented if the air velocity is sufficiently large, as illustrated in Fig. 2.7. The magnitude of the velocity necessary to prevent backflow depends on the energy release rate of the fire as discussed in Subsection 1.

The two basic principles of smoke control can be stated as follows:

1. Airflow by itself can control smoke movement if the average air velocity is of sufficient magnitude.

2. Air-pressure differences across barriers can act to control smoke movement.

The use of air-pressure differences across barriers to control smoke is frequently referred to as pressurization. Pressurization results in airflows of high velocity in the small gaps around closed doors and in construction cracks, thereby prevent-

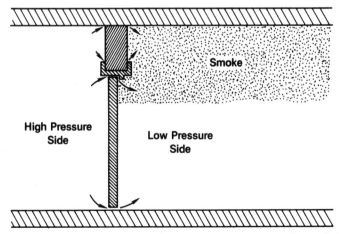

Fig. 2.5 **Pressure difference across barrier of smoke control system preventing smoke infiltration to high-pressure side of barrier.**

ing smoke backflows through these openings. Therefore, in a strict physical sense, the second principle is a special case of the first. However, considering the two principles separately is advantageous for smoke-control design. For a barrier with one or more large openings, air velocity is the appropriate physical quantity for both design considerations and acceptance testing. However, when there are only small cracks, such as around closed doors, designing and measuring air velocities is impractical. In this case, the appropriate physical quantity is pressure difference. Considering the two principles separately has the added advantage of emphasizing the different problems that need to be solved for open and closed doors.

Because smoke control relies on air velocities and pressure differences pro-

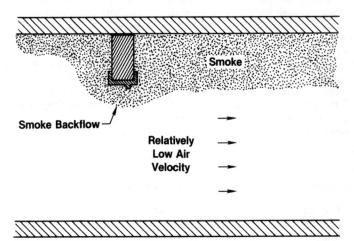

Fig. 2.6   Smoke backflow against low air velocity through open doorway.

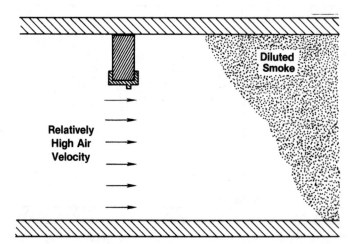

Fig. 2.7   No smoke backflow with high air velocity through open doorway.

duced by fans, it has the following three advantages in comparison to the traditional methods of smoke management:

1. Smoke control is less dependent on tight barriers. Allowance can be made for reasonable leakage through barriers.
2. Stack effect, buoyancy, and wind are less likely to overcome smoke control than passive smoke management. In the absence of smoke control, these driving forces cause smoke movement to the extent that leakage paths allow. However, pressure differences and airflows of a smoke-control system act to oppose these driving forces.
3. Smoke control can be designed to prevent smoke flow through an open doorway in a barrier by the use of airflow. Doors in barriers are opened during evacuation and are sometimes accidentally left open or propped open throughout fires. In the absence of smoke control, smoke flow through these doors is common.

Smoke control systems should be designed so that a path exists for smoke movement to the exterior of buildings; such a path can allow smoke to escape.

The smoke-control designer should be cautioned that dilution of smoke in the fire space is not a means of achieving smoke control. In other words, smoke movement cannot be controlled by simply supplying and exhausting large quantities of air from the space, or zone, in which the fire is located. This supplying and exhausting of air is sometimes referred to as purging the smoke. Because of the large quantities of smoke produced in a fire, purging cannot assure breathable air in the fire space. In addition, purging itself cannot control smoke movement because it does not provide the needed airflows at open doors and the pressure differences across barriers. However, for spaces separated from the fire space by smoke barriers, purging can significantly limit the level of smoke. The following discussion examines the basic principles of smoke control.

1 Airflow

Theoretically, airflow can be used to stop smoke movement through any space. However, two places where air velocity is most commonly used to control smoke movement are open doorways and corridors. The problem of preventing smoke movement through doorways is currently being researched. Thomas (1970) has developed an empirical relation for the critical velocity to prevent smoke from flowing upstream in a corridor:

$$V_k = K\left(\frac{gE}{W\rho cT}\right)^{1/3} \tag{2.7}$$

where V_k = critical air velocity to prevent smoke backflow
    E = energy release rate into corridor
    W = corridor width
    ρ = density of upstream air
    c = specific heat of downstream gases
    T = absolute temperature of downstream mixture of air and smoke
    K = constant, on the order of 1
    g = gravitational constant

The downstream properties are taken at a point sufficiently far downstream of the fire for the properties to be uniform across the corridor section. The critical air velocity can be evaluated at ρ = 1.3 kg/m^3 (0.081 lb/ft^3), c = 1.005 kJ/kg · °C (0.24 Btu/lb · °F, T = 27°C (81°F), and K = 1,

$$V_k = K_v\left(\frac{E}{W}\right)^{1/3}$$ (2.7a)

where V_k = critical air velocity to prevent smoke backflow, m/s (ft/min)
 E = energy release rate into corridor, W (Btu/hr)
 W = corridor width, m (ft)
 K_v = coefficient, = 0.0292 (5.68)

This relation can be used when the fire is located in the corridor or when the smoke enters the corridor through an open door, air transfer grille, or other opening. The critical velocities calculated from Eq. 2.7 are approximate because only an approximate value of K was used. However, critical velocities calculated from this relation are indicative of the kind of air velocities required to prevent smoke backflow from fires of different sizes.

Equation 2.7 can be evaluated from Fig. 2.8. For an energy release rate of 150 kW (0.512 × 10^6 Btu/hr) into a corridor 1.22 m (4.0 ft) wide, Eq. 2.7 yields a critical velocity of 1.45 m/s (286 ft/min). However, for a larger energy release rate of 2.1 MW (7.2 × 10^6 Btu/hr) the relation yields a critical velocity of 3.50 m/s (690 ft/min) for a corridor of the same width.

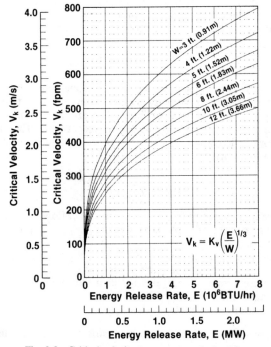

Fig. 2.8   Critical velocity to prevent smoke backflow.

In general, a requirement for high air velocity results in a smoke-control system that is expensive and difficult to design. The use of airflow is most important in preventing smoke backflow through an open doorway that serves as a boundary of a smoke-control system. Thomas (1970) indicated that Eq. 2.7 can be used to obtain a rough estimate of the airflow needed to prevent smoke backflow through a door. Many designers feel that it is prohibitively expensive to design systems that maintain air velocities in doorways greater than 1.5 m/s (300 ft/min). Subsection 3 in Section 2.8 provides a discussion of what constitutes an appropriate design air velocity in a smoke-control system.

Equation 2.7 is not appropriate for sprinklered fires having small temperature differences between the upstream air and downstream gases. Shaw and Whyte (1974) provide an analysis with experimental verification of a method to determine the air velocity needed through an open doorway to prevent backflow of contaminated air. This analysis is specifically for small temperature differences and includes the effects of natural convection. If this method is used for a sprinklered fire where the temperature difference is only 2°C (3.6°F), then an average velocity of 0.25 m/s (50 ft/min) would be the minimum velocity needed through a doorway to prevent smoke backflow. This temperature difference is small, and it is possible that larger values may be appropriate in many situations. Further research is needed in this area.

Even though airflow can be used to control smoke movement, it is not the primary method of smoke control because the quantities of air required are so large. The primary means is by air-pressure differences across partitions, doors, and other building components.

2 Pressurization

The airflow rate through a construction crack, door gap, or other flow path is proportional to the pressure difference across that path raised to the power n. For a flow path of fixed geometry, n is theoretically in the range of 0.5 to 1. However, for all flow paths, except extremely narrow cracks, using $n = 0.5$ is reasonable and the flow can be expressed as

$$Q = CA \sqrt{\frac{2\Delta P}{\rho}} \tag{2.8}$$

where Q = volumetric airflow rate, m³/s (ft³/min)
  C = flow coefficient
  A = flow area (also called leakage area), m² (ft²)
  ΔP = pressure difference across flow path, Pa (in. H$_2$O)
  ρ = density of air entering flow path

The flow coefficient depends on the geometry of the flow path as well as on turbulence and friction. In the present context, the flow coefficient is generally in the range of 0.6 to 0.7. For $\rho = 1.2$ kg/m³ (0.075 lb/ft³) and $C = 0.65$, the flow equation, Eq. 2.8, can be expressed as

$$Q = K_f A \sqrt{\Delta P} \tag{2.8a}$$

where Q = volumetric flow rate, m³/s (ft³/min)
  A = flow area, m² (ft²)
  ΔP = pressure difference across flow path, Pa (in. H$_2$O)
  K_f = coefficient, = 0.839 (2610)

The airflow rate can also be determined from Fig. 2.9. The flow area is frequently the same as the cross-sectional area of the flow path, an exception being the flow area of an open stairwell doorway, as discussed in Section 2.8. A closed door with a crack area of 0.01 m² (0.11 ft²) and with a pressure difference of 2.5 Pa (0.01 in. H_2O) would have an air leakage rate of approximately 0.013 m³/s (29 ft³/min). If the pressure difference across the door was increased to 0.30 75 Pa (in. H_2O), then the flow would be 0.073 m³/s (157 ft³/min).

Frequently in field tests of smoke-control systems, pressure differences across partitions or closed doors have fluctuated by as much as 5 Pa (0.02 in. H_2O). These fluctuations have generally been attributed to the wind, although they could have been due to the HVAC system or some other source. Pressure fluctuations and the resulting smoke movement are current topics of research. To control smoke movement, the pressure differences produced by a smoke-control system must be sufficiently large so that they are not overcome by pressure fluctuations, stack effect, smoke buoyancy, and the forces of the wind. However, the pressure difference produced by a smoke-control system should not be so large that door opening problems result (see Section 2.4 and Section 2.8, Subsection 1).

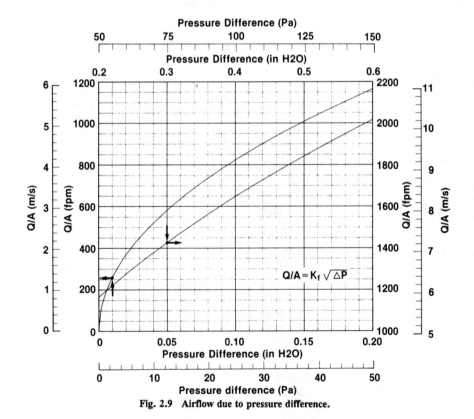

Fig. 2.9 Airflow due to pressure difference.

3 Purging

In general, the systems discussed are based on the two basic principles of smoke control. However, it is not always possible to maintain sufficiently large airflows through open doors to prevent smoke from infiltrating a space that is intended to be protected. Ideally, occurrences of such open doors will only happen for short periods of time during evacuation. Smoke that has entered a protected space can be purged or, in other words, diluted by supplying outside air to the space.

Consider the case where a compartment is isolated from a fire by smoke barriers and self-closing doors so that no smoke enters the compartment when the doors are closed. However, when one or more of the doors are open, there is insufficient airflow to prevent smoke backflow into the compartment from the fire space. In order to facilitate analysis, it is considered that smoke is of uniform concentration throughout the compartment. When all the doors are closed, the concentration of contaminant in the compartment can be expressed as

$$\frac{C}{C_0} = e^{-at} \tag{2.9}$$

where C_0 = initial concentration of contaminant
C = concentration of contaminant at time t
a = purging rate, number of air changes per minute
t = time after doors closed, min
e = constant, ≈ 2.718

The concentrations C_0 and C must both be in the same units, and they can be any units appropriate for the particular contaminant being considered. McGuire et al. (1970) evaluated the maximum levels of smoke obscuration from a number of tests and a number of proposed criteria for tolerable levels of smoke obscuration. Based on this evaluation, they state that maximum levels of smoke obscuration are greater by a factor of 100 than those relating to the limit of tolerance. Thus they indicate that an area can be considered "reasonably safe" for smoke obscuration if its atmosphere will not be contaminated with greater than 1% of the atmosphere prevailing in the immediate fire area. Obviously, this 1% leakage, once diluted with uncontaminated air, would also reduce the concentrations of toxic smoke components. Toxicity is a more complicated problem, and no parallel statement has been made regarding the dilution needed to obtain a safe atmosphere with respect to toxic gases.

Equation 2.9 can be solved for the purging rate,

$$a = \frac{1}{t} \log_e \left(\frac{C_0}{C}\right) \tag{2.10}$$

If, when doors are open, the contaminant in a compartment is 20% of the burning room, and at 6 min after the door is closed, the contaminant concentration is 1% of the burning room, then Eq. 2.10 indicates that the compartment must be purged at a rate of one air change every 2 min.

In reality it is impossible to assure that concentration of the contaminant is uniform throughout the compartment. Because of buoyancy, it is likely that higher concentrations of contaminant would tend to be near the ceiling. Therefore an exhaust location near the ceiling and a supply location near the floor would probably purge the smoke even faster than the calculations indicate. Caution should be exercised in the location of the supply and exhaust points to prevent the supply air from blowing into the exhaust inlet and thus short-circuiting the purging operations.

2.4 DOOR-OPENING FORCES

As mentioned in Subsection 2 in Section 2.3, the door-opening forces resulting from the pressure differences produced by a smoke-control system must be considered in any design. Unreasonably high door-opening forces can result in occupants having difficulty opening, or being unable to open, doors to refuge areas or escape routes. This problem is discussed in more detail in Section 2.8, Subsection 2.

The force required to open a door is the sum of the forces to overcome the pressure difference across the door and to overcome the door closer. This can be expressed as

$$F = F_{dc} + \frac{K_d W A \, \Delta P}{2(W - d)} \qquad (2.11)$$

where F = total door opening force, N (lb)
   F_{dc} = force to overcome door closer, N (lb)
   W = door width, m (ft)
   A = door area, m^2 (ft^2)
   ΔP = pressure difference across door, Pa (in. H$_2$O)
   d = distance from doorknob to edge of knob side of door, m (ft)
   K_d = coefficient, = 1.00 (5.20)

This relation assumes that the door-opening force is applied at the knob. Door-opening forces due to pressure differences can be determined from Fig. 2.10. The force to overcome the door closer is usually greater than 13 N (3 lb) and, in some cases, can be as large as 90 N (20 lb). For a door that is 2.13 m (7 ft) high and 1 m (3 ft) wide, subject to a pressure difference of 62 Pa (0.25 in. H$_2$O), the total door-opening force is 110 N (25 lb) if the force to overcome the door closer is 44 N (10 lb).

2.5 EFFECTIVE FLOW AREAS

The concept of effective flow areas is quite useful for the analysis of smoke-control systems. The paths in the system can be parallel with one another, in series, or a combination of parallel and series paths. The effective area of a system of flow areas is the area that results in the same flow as the system when it is subjected to the same pressure difference over the total system of flow paths. This is analogous to the flow of electric current through a system of electrical resistances.

1 Parallel Paths

Three parallel leakage areas from a pressurized space are illustrated in Fig. 2.11.

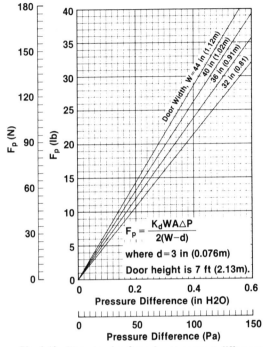

Fig. 2.10 Door-opening force due to pressure difference.

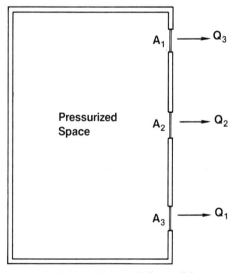

Fig. 2.11 Leakage paths in parallel.

The pressure difference ΔP is the same across each of the leakage areas. The total flow Q_T from the space is the sum of the flows through the leakage paths,

$$Q_T = Q_1 + Q_2 + Q_3 \tag{2.12}$$

The effective area A_e for this situation is that which results in the total flow Q_T. Therefore the total flow can be expressed as

$$Q_T = CA_e \sqrt{\frac{2\Delta P}{\rho}} \tag{2.13}$$

The flow through area A_1 can be expressed as

$$Q_1 = CA_1 \sqrt{\frac{2\Delta P}{\rho}}$$

The flows for Q_2 and Q_3 can be expressed in a similar manner. Substituting the expressions for Q_1, Q_2, and Q_3 into Eq. 2.12 and collecting like terms yields

$$Q_T = C(A_1 + A_2 + A_3) \sqrt{\frac{2\Delta P}{\rho}} \tag{2.14}$$

Comparing this with Eq. 2.13 yields

$$A_e = A_1 + A_2 + A_3 \tag{2.15}$$

In Fig. 2.11, if A_1 is 0.10 m² (1.08 ft²) and A_2 and A_3 are 0.05 m² (0.54 ft²) each, then the effective flow area A_e is 0.20 m² (2.16 ft²).

This logic can be extended to any number of flow paths in parallel. It can be stated that the effective area is the sum of the individual leakage paths, where n is the number of flow areas A_i in parallel,

$$A_e = \sum_{i=1}^{n} A_i \tag{2.16}$$

2 Series Paths

Three leakage areas in series from a pressurized space are illustrated in Fig. 2.12. The flow rate Q is the same through each of the leakage areas. The total pressure difference ΔP_T from the pressurized space to the outside is the sum of pressure differences ΔP_1, ΔP_2, and ΔP_3 across each of the respective flow areas A_1, A_2, and A_3,

$$\Delta P_T = \Delta P_1 + \Delta P_2 + \Delta P_3 \tag{2.17}$$

The effective area for flow paths in series is the flow area that results in the flow Q for a total pressure difference of ΔP_T. Therefore the flow Q can be expressed as

$$Q = CA_e \sqrt{\frac{2\Delta P_T}{\rho}} \tag{2.18}$$

Solving for ΔP_T yields

$$\Delta P_T = \left(\frac{\rho}{2} \frac{Q}{CA_e}\right)^2 \tag{2.19}$$

The pressure difference across A_1 can be expressed as

$$\Delta P_1 = \left(\frac{\rho}{2} \frac{Q}{CA_1}\right)^2 \tag{2.20}$$

The pressure differences ΔP_2 and ΔP_3 can be expressed in a similar manner. Substituting Eq. 2.19 and the expressions for ΔP_1, ΔP_2, and ΔP_3 into Eq. 2.17 and canceling like terms, yields

$$\frac{1}{A_e^2} = \frac{1}{A_1^2} + \frac{1}{A_2^2} + \frac{1}{A_3^2} \tag{2.21}$$

that is,

$$A_e = \left(\frac{1}{A_1^2} + \frac{1}{A_2^2} + \frac{1}{A_3^2}\right)^{-1/2} \tag{2.22}$$

This same reasoning can be extended to any number of leakage areas in series to yield

$$A_e = \left(\sum_{i=1}^{N} \frac{1}{A_i^2}\right)^{-1/2} \tag{2.23}$$

where n is the number of leakage areas A_i in series. In smoke-control analysis there are frequently only two paths in series. For this case, the effective leakage area is

$$A_e = \frac{A_1 A_2}{\sqrt{A_1^2 + A_2^2}} \tag{2.24}$$

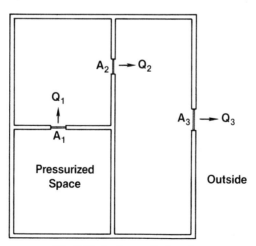

Fig. 2.12 Leakage paths in series.

Example 1. Calculate the effective leakage area of two equal flow paths of 0.02 m^2 (0.2 ft^2) in series. Letting $A = A_1 = A_2 = 0.02$ m^2 (0.22 ft^2) yields

$$A_e = \frac{A^2}{\sqrt{2A^2}} = \frac{A}{\sqrt{2}} = 0.014 \text{ m}^2 \text{ (0.15 ft}^2\text{)}$$

Example 2. Calculate the effective area of two flow paths in series. Letting $A_1 = 0.02$ m^2 (0.22 ft^2) and $A_2 = 0.2$ m^2 (2.2 ft^2),

$$A_e = \frac{A_1 A_2}{\sqrt{A_1^2 + A_2^2}} = 0.0199 \text{ m}^2 \text{ (0.219 ft}^2\text{)}$$

This example illustrates that when two areas are in series and one is much larger than the other, the effective area is approximately equal to the smaller area.

3   Combination of Paths in Parallel and Series

The method of developing an effective area for a system of both parallel and series paths is to combine groups of parallel and series paths systematically. The system illustrated in Fig. 2.13 is analyzed as an example. Figure 2.13 shows that A_2 and A_3 are in parallel. Therefore their effective area is

$$A_{23e} = A_2 + A_3 \tag{2.25}$$

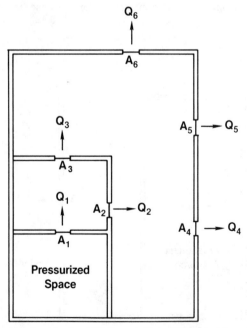

Fig. 2.13   Combination of leakage paths in parallel and in series.

Areas A_4, A_5, and A_6 are also in parallel, so their effective area is

$$A_{456e} = A_4 + A_5 + A_6 \tag{2.26}$$

These two effective areas are in series with A_1. Therefore the effective flow area of the system is given by

$$A_e = \left(\frac{1}{A_1^2} + \frac{1}{A_{23e}^2} + \frac{1}{A_{456e}^2} \right)^{-1/2} \tag{2.27}$$

Example 3. Calculate the effective area of the system in Fig. 2.13 if the leakage areas are $A_1 = A_2 = A_3 = 0.02 \text{ m}^2$ (0.22 ft^2) and $A_4 = A_5 = A_6 = 0.01 \text{ m}^2$ (0.11 ft^2). Then

$$A_{23e} = 0.04 \text{ m}^2 \ (0.44 \text{ ft}^2)$$

$$A_{456e} = 0.03 \text{ m}^2 \ (0.33 \text{ ft}^2)$$

$$A_e = 0.015 \text{ m}^2 \ (0.16 \text{ ft}^2)$$

2.6 SYMMETRY

The concept of symmetry is useful in simplifying problems and thereby easing solutions. Figure 2.14 illustrates the floor plan of a multistory building that can be divided in half by a plane of symmetry. Flow areas on one side of the plane of symmetry are equal to the corresponding flow areas on the other side. For a building to be so treated, every floor of the building must be such that it can be divided in the same manner by the plane of symmetry. If wind effects are not

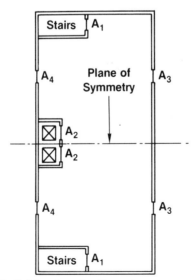

Fig. 2.14 Building floor plan illustrating symmetry concept.

considered in the analysis, or if the wind direction is parallel to the plane of symmetry, then the airflow in only one-half of the building need be analyzed. It is not necessary that the building be geometrically symmetric as in Fig. 2.14; it must be symmetric only with respect to flow.

2.7 FLOW AREAS

In the design of smoke-control systems, airflow paths must be identified and evaluated. Some leakage paths are obvious, such as cracks around closed doors, open doors, elevator doors, windows, and air transfer grilles. Construction cracks in building walls are less obvious but no less important.

The flow area of most large openings, such as open windows, can be calculated easily. However, flow areas of cracks are more difficult to evaluate. The area of these leakage paths is dependent on workmanship—how well a door is fitted or how well weather stripping is installed. A door that is 0.9 by 2.1 m (3 by 7 ft) with an average crack width of 3.2 mm (⅛ in.) has a leakage area of 0.020 m^2 (0.21 ft^2). However, if by accident this door is installed with a 19-mm (¾-in.) undercut, the leakage area is 0.030 m^2 (0.32 ft^2), a significant difference. Leakage areas of elevator doors have been measured in the range of 0.051 to 0.065 m^2 (0.55 to 0.70 ft^2) per door.

For open stairwell doorways Cresci (1973) found that complex flow patterns exist and that the resulting flow through open doorways was considerably below the flow calculated by using the geometric area of the doorway as the flow area in Eq. 2.8a. Based on this research, it is recommended that the flow area of an open stairwell doorway be half that of the geometric area (door height times width) of the doorway. An alternate approach for open stairwell doorways is to use the geometric area as the flow area and use a reduced flow coefficient. Because it does not allow the direct use of Eq. 2.8a, this alternate approach is not used here.

Typical leakage areas for walls and floors of commercial buildings are tabulated as area ratios in Table 2.1. These data are based on a relatively small number of tests performed by the National Research Council of Canada (Tamura and Shaw, 1976a, 1978; Tamura and Wilson, 1966). The area ratios are evaluated at typical airflows of 75 Pa (0.30 in. H$_2$O) for walls and 25 Pa (0.10 in. H$_2$O) for floors. It is believed that actual leakage areas are primarily dependent on workmanship rather than construction materials, and, in some cases, the flow areas in particular buildings may vary from the values listed. Considerable data concerning air leakage through building components is also provided in Chapter 22 of the ASHRAE handbook (1981).

The determination of the flow area of a vent is not always straightforward because the vent surface is usually covered by a louver and screen. Thus the flow area is less than the vent area (vent height times width). Because the slats in louvers are frequently slanted, calculation of the flow area is further complicated. Manufacturers' data should be sought for specific information.

2.8 DESIGN PARAMETERS

Ideally, codes should contain design parameters leading to safe and economical smoke-control systems. Unfortunately, because smoke control is a new field, consensus has not yet been reached as to what constitutes reasonable design parameters. Clearly, the designer has an obligation to adhere to any smoke-control design criteria existing in appropriate codes or standards. However, such criteria should be scrutinized to determine whether or not they will result in an effective system. If necessary, the designer should seek a waiver of the local codes to ensure an effective smoke-control system.

Five areas for which design parameters must be established are (1) leakage areas, (2) weather data, (3) pressure differences, (4) airflow, and (5) number of open doors in smoke-control system.

Leakage areas have already been discussed. An additional consideration affecting pressure differences and airflow is whether or not a window in the fire compartment is broken. In the absence of code requirements for specific parameters, the following discussion may be helpful to the designer.

1 Weather Data

The state of the art of smoke control is such that little consideration has been given to the selection of weather data specifically for the design of smoke-control systems. However, design temperatures for heating and cooling during winter and summer are recommended. In the United States, for example, the *ASHRAE Handbook* (1981) provides 99 and 97.5% winter design temperatures. These values represent the temperatures that are equaled or exceeded during the heating season. The heating season usually consists of three winter months. A more exact definition of these temperatures is available in Chapter 24 of the *ASHRAE Handbook* (1981) or can be found in other similar local publications.

A designer may wish to consider using such temperatures for the design of smoke-control systems. It should be remembered that in a normal winter, if using

Table 2.1 Typical leakage areas for walls and floors of commercial buildings

Construction element	Wall tightness	Area ratio* A/A_w
Exterior building walls (includes construction cracks, cracks around windows and doors)	Tight	0.70×10^{-4}
	Average	0.21×10^{-3}
	Loose	0.42×10^{-3}
	Very loose	0.13×10^{-2}
Stairwell walls (includes construction cracks, but not cracks around windows or doors)	Tight	0.14×10^{-4}
	Average	0.11×10^{-3}
	Loose	0.35×10^{-3}
Elevator shaft walls (includes construction cracks but not cracks around doors)	Tight	0.18×10^{-3}
	Average	0.84×10^{-3}
	Loose	0.18×10^{-2}
		A/A_F
Floors (includes construction cracks and areas around penetrations)	Average	0.52×10^{-4}

*A—leakage area; A_w—wall area; A_F—floor area.

the *ASHRAE Handbook*, there would be approximately 22 hr at or below the 99% design value and approximately 54 hr at or below the 97.5% design value. Furthermore, extreme temperatures can be considerably lower than the winter design temperatures. [For example, the ASHRAE 99% design temperature for Tallahassee, Fla., is −3°C (27°F), but the lowest temperature observed there by the National Oceanic and Atmospheric Administration (1979) was −19°C (− 2°F) on February 13, 1899.]

Temperatures are generally below design values for short periods of time, and because of the thermal lag of building materials, these short intervals of low temperature usually do not result in problems with respect to heating systems. However, the same cannot necessarily be said of a smoke-control system. There is no time lag for a smoke-control system, that is, it is subjected to all the forces of stack effect that exist at the moment it is being operated. If the outside temperature is below the winter design temperature for which a smoke-control system was designed, then problems from stack effect may result. A similar situation can result with respect to summer design temperatures and reverse stack effect.

Wind data are needed for the wind analysis of a smoke-control system. At present no formal method of such an analysis exists, and the approach generally taken is to design the smoke-control system so as to minimize any effects of wind. The development of temperature and wind data for the design of smoke-control systems is an area for future research.

2 Pressure Differences

It is appropriate to consider both the maximum and the minimum allowable pressure differences across the boundaries of smoke-control zones. The maximum allowable pressure difference should be a value that does not result in excessive door-opening forces, although it is difficult to determine what constitutes excessive door-opening forces. Clearly, a person's physical condition is a major factor in determining a reasonable door-opening force for that person. Section 5-2.1.1.4.3 of the National Fire Protection Association (NFPA) *Life Safety Code* (1981) states that the force required to open any door as a means of egress shall not exceed 222 N (50 lb). NFPA is currently evaluating proposals to reduce its maximum door-opening force to 133 N (30 lb).

Many smoke-control designers feel that a value lower than 222 N (50 lb) should be used, especially in occupancies involving the elderly, children, or the handicapped. Also, exposure to smoke during a fire can adversely affect a person's physical capabilities, further complicating the allowable force requirement. In Subsection 4 a method of determining the door-opening force is provided. If, for a particular application, a maximum door-opening force of 178 N (40 lb) is considered appropriate, and the force to overcome the door closer is 49 N (11 lb), then a door 0.91 by 2.13 m (36 by 84 in.) would have a maximum allowable pressure difference of 122 Pa (0.49 in. H_2O).

The criterion used for selecting a minimum allowable pressure difference across a boundary of a smoke-control system is that no smoke leakage shall occur during building evacuation. Other criteria might involve maintaining a number of smoke-free egress routes or preventing smoke infiltration to a refuge area. (Discussion of all possible alternatives is beyond the scope of this Monograph.) In this case, the smoke-control system must produce sufficient pressure differences so that it is not overcome by the forces of wind, stack effect, or buoyancy of hot smoke. The pressure differences due to wind and stack effect can become very large in the event of a broken window in the fire compartment. Evaluation of these pressure differences depends on evacuation time, rate of fire growth, building configuration, and the presence of a fire-suppression system. In the absence of a formal method of analysis, such evaluations must, of necessity, be based on experience and engineering judgment.

A method for determining the pressure difference across a smoke barrier resulting from the buoyancy of hot gases was provided earlier in this chapter. For a particular application it may be necessary to design a smoke-control system that can withstand an intense fire next to a door of a smoke-control zone boundary. It was stated in the section on buoyancy that, in a series of full-scale fire tests, the maximum pressure difference reached was 16 Pa (0.064 in. H_2O) across the burning room's wall at the ceiling. In order to prevent smoke infiltration, the smoke-control system should be designed to maintain a slightly higher pressure in nonfire conditions. A minimum pressure difference in the range of 20 to 25 Pa (0.08 to 0.10 in. H_2O) is suggested.

If a boundary is exposed to hot smoke from a remote fire, a lower pressure difference due to buoyancy will result. For a smoke temperature of 400°C (750°F) the pressure difference caused by the smoke 1.53 m (5.0 ft) above the neutral plane would be 10 Pa (0.04 in. H_2O). In this situation it is suggested that the smoke-control system be designed to maintain a minimum pressure in the range of 15 to 20 Pa (0.06 to 0.08 in. H_2O).

Water spray from fire sprinklers cools smoke from a building fire and reduces the pressure differences due to buoyancy. In such a case it is probably wise to allow for pressure fluctuations (see Subsection 2 in Section 2.3). Accordingly, a minimum pressure difference in the range of 5 to 10 Pa (0.02 to 0.04 in. H_2O) is suggested.

Windows in a fire compartment can break because of exposure to high-temperature gases. In such cases the pressure due to wind on a building's exterior can be determined from Eq. 2.6. If a broken window is the only opening to the outside on the fire floor and the window is oriented into the wind, the boundary of the smoke-control system could be subjected to higher pressures. One possible solution is to vent the fire floor on all sides to relieve such pressures. For a building that is much longer than it is wide, it may be possible to vent only the two longer sides.

In addition to wind effects, stack effect can be increased in the event of a broken fire compartment window. With a fire on a lower floor during cold weather, stack effect will increase pressures found on fire floors above to levels exceeding those of surrounding areas. Even though little research has been done on the subject, the chances of a window breaking in the fire compartment are reduced by the operation of fire sprinklers.

3 Airflow

When the doors in the boundaries of smoke-control systems are open, smoke can flow into refuge areas or escape routes unless there is sufficient airflow through the open door to prevent smoke backflow, as discussed. One criterion for selecting a design velocity through an open door is that no smoke backflow shall occur during building evacuation. (Other criteria might include the allowance of limited smoke leakage into areas to be protected. Under such criteria, the toxicity of the smoke is a factor that must be considered.) Selection of this velocity depends on evacuation time, rate of fire growth, building configuration, and the presence of a fire-suppression system. In the absence of a formal method of analysis, such an evaluation must be based on experience and engineering judgment.

At present there is still much to be learned about the critical velocity needed to stop smoke backflow through an open door. In the absence of specific criteria or formulas for doorways, the method of analysis presented earlier for corridors can be used to yield approximate results. The width of the doorway may be used in place of the width of the corridor. The technique used is based on the assumption that smoke properties are uniform across the corridor width. As previously illustrated for a particular application, it may be necessary to design for an intensive fire, such as one with an energy release rate of 2.4 MW (8×10^6 Btu/hr). A critical velocity of approximately 4 m/s (800 ft/min) would be required to stop smoke.

In another application it may be estimated that the building would be subjected to a much less intense fire with an energy release rate of 125 kW (427,000 Btu/hr). To protect against smoke backflow during evacuation, the critical velocity would be 1.5 m/s (300 ft/min).

In a sprinklered building, the smoke away from the immediate fire area might be cooled to near ambient temperature by water spray from the sprinklers. In such a case a design velocity in the range of 0.25 to 1.25 m/s (50 to 250 ft/min) may be used. Research is needed to fully evaluate the effect of sprinklers on smoke-control design parameters.

4 Number of Open Doors

The need for air velocity through open doors in the perimeter of a smoke-control system is discussed in Subsection 3. Another design consideration is the number of doors that could be opened simultaneously when the smoke-control system is operational. A design that allows for all doors to be opened simultaneously may ensure that the system will always work, but it will probably add to the cost of the system.

Deciding how many doors will be opened simultaneously depends largely on the building occupancy. For example, in a densely populated building it is very likely that all the doors will be opened simultaneously during evacuation. However, if a staged evacuation plan or refuge area is incorporated in the building fire emergency plan or if the building is sparsely occupied, only a few of the doors may be opened simultaneously during a fire.

2.9 STAIRWELLS

1 Pressurized Stairwells

Many pressurized stairwells have been designed and built with the goal of providing a smoke-free escape route in the event of a building fire. A secondary objective is to provide a smoke-free staging area for fire fighters. On the fire floor a pressurized stairwell must maintain a positive pressure difference across a closed stairwell door so that smoke infiltration is prevented.

During building fire situations, some stairwell doors are opened intermittently during evacuation and fire fighting, and some doors may even be blocked open. Ideally, when the stairwell door is opened on the fire floor, there should be sufficient airflow through the door to prevent smoke backflow. Designing such a system is difficult because of the large number of permutations of open stairwell doors and weather conditions that affect the airflow through open doors.

Stairwell pressurization systems are divided into two categories—single- and multiple-injection systems. A single-injection system is one that has pressurized air supplied to the stairwell at one location. The most common injection point is at the top. Associated with this system is the potential for smoke feedback into the pressurized stairwell, such as through the pressurization fan intake. Therefore the capability of automatic shutdown in such an event should be considered.

For tall stairwells, single-injection systems can fail when a few doors are open near the air-supply inject point. All of the pressurized air can be lost through these open doors and the system cannot maintain positive pressures across doors further from the injection point. Such a failure mode is especially likely with bottom injection systems when a ground-level stairwell door is open.

A better system for tall stairwells is supplying air at a number of locations throughout the height of the stairwell. Figures 2.15 and 2.16 are two examples of

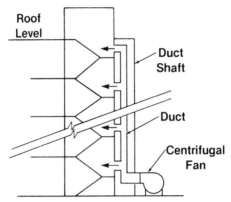

Fig. 2.15   Stairwell pressurization by multiple injection with fan located at ground level.

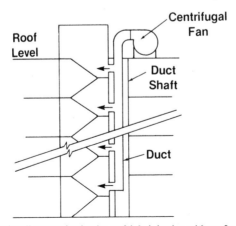

Fig. 2.16   Stairwell pressurization by multiple injection with roof-mounted fan.

many possible multiple-injection systems which can be used to overcome the limitations of single-injection systems. In these figures the supply duct is shown in a separate shaft. However, systems have been built which have eliminated the expense of a separate duct shaft by locating the supply duct in the stairwell itself. Obviously in such a case care must be taken that the duct does not become an obstruction to orderly building evacuation and meets all other code requirements as well.

2 Stairwell Compartmentation

An alternative to multiple injection is compartmentation of the stairwell into a number of sections, as illustrated in Fig. 2.17. When the doors between compartments are open, the effect of compartmentation is lost. For this reason compartmentation is inappropriate for densely populated buildings, where total building evacuation by the stairwell is planned in the event of fire. However, when a staged evacuation plan is used (and given that the smoke-control system is designed to operate successfully when the maximum number of doors between compartments are open), compartmentation can be an effective means of providing stairwell pressurization for tall buildings.

2.10 ZONE SMOKE CONTROL

Pressurized stairwells are intended to prevent smoke infiltration into stairwells. However, in a building with only stairwell pressurization, smoke can flow through cracks in floors and partitions and through shafts to damage property and threaten life at locations remote from the fire. The concept of zone smoke control is intended to limit such smoke movement.

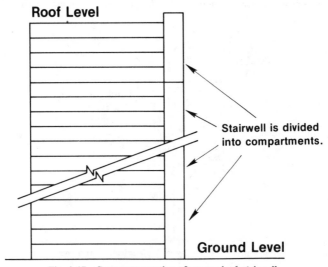

Fig. 2.17   Compartmentation of pressurized stairwell.

A building is divided into a number of smoke-control zones, each zone separated from the others by specially constructed partitions, floors, and doors that can be closed to inhibit smoke movement. In the event of a fire, pressure differences and airflows produced by mechanical fans are used to limit the smoke spread from the zone in which the fire initiated. The concentration of smoke in this zone goes unchecked. Accordingly, in tall buildings using zone smoke-control systems it is intended that the building occupants evacuate the smoke zone where the fire originated as soon as possible after fire detection.

Frequently each floor of a building is chosen to be a separate smoke-control zone. However, a smoke-control zone can consist of more than one floor, or a floor can be divided into more than one smoke-control zone. Some arrangements of smoke-control zones are illustrated in Fig. 2.18. All of the nonsmoke zones in the building may be pressurized. The term *pressure sandwich* is used to describe cases where only adjacent zones to the smoke zone are pressurized, as in Fig. 2.18*b* and *d*.

The intent of zone smoke control is to limit smoke movement to the smoke zone by use of the two principles of smoke control. Pressure differences in the desired direction across the barriers of a smoke zone can be achieved by supplying outside (fresh) air to nonsmoke zones, by venting the smoke zone, or by both methods.

Venting of smoke from a smoke zone is important because it prevents significant overpressures due to the thermal expansion of gases which result from the fire. However, venting results in only slight reduction of smoke concentration in the smoke zone. This venting can be accomplished by exterior wall vents, smoke shafts, and mechanical venting (exhausting).

2.11 COMPUTER ANALYSIS

Some design calculations associated with smoke control are appropriate for hand calculation. However, other calculations involve time-consuming trial-and-error solutions, which are more appropriately left to a digital computer. The National Bureau of Standards has developed a computer program (Klote, 1982) specifically for the analysis of smoke-control systems. A number of other programs applicable to smoke control have been developed. Some calculate steady-state airflow and pressures throughout a building (Sander, 1974; Sander and Tamura, 1973). Other programs go beyond this to calculate the smoke concentrations that would be produced throughout a building in the event of a fire (Yoshia et al., 1979; Evers and Waterhouse, 1978; Wakamatsu, 1977; Rilling, 1978).

Each of these programs is different to some extent. However, the basic concepts are essentially the same. A building is represented by a network of spaces, or nodes, each at a specific pressure and temperature. The stairwells and other shafts are modeled by a vertical series of spaces, one for each floor. Air flows through leakage paths from regions of high pressure to regions of low pressure. These leakage paths are doors and windows that may be opened or closed. Leakage can also occur through partitions, floors, and exterior walls and roofs. The airflow through a flow path is a function of the pressure difference across the path, as presented in Eq. 2.8.

Air from outside the building can be introduced by a pressurization system into any level of a shaft, or even into other building compartments. This allows the simulation of stairwell pressurization. In addition, any building space can be exhausted. This allows simulation of zoned smoke-control systems. The pressures throughout the building and the flow rates through all the flow paths are obtained by solving the airflow network, including the driving forces such as wind, the pressurization system, or an inside-to-outside temperature difference.

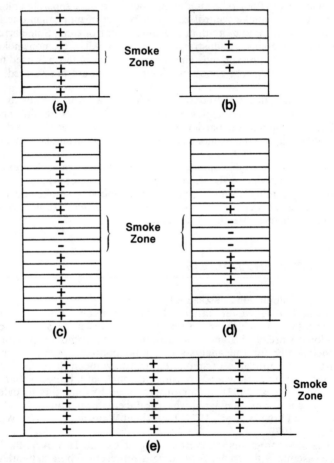

Note: In the Above Figures the Smoke Zone Is Indicated by a Minus Sign and Pressurized Spaces Are Indicated by a Plus Sign. Each Floor Can Be a Smoke Control Zone as in (a) and (b) or a Smoke Zone Can Consist of More Than One Floor as in (c) and (d). All of the Non-Smoke Zones in a Building May Be Pressurized as in (a) and (c) or Only Non-Smoke Zones Adjacent to the Smoke Zone May Be Pressurized as in (b) and (d). A Smoke Zone Can Also Be Limited to a Part of a Floor as in (e).

Fig. 2.18 Some arrangements of smoke-control zones.

2.12 ACCEPTANCE TESTING

Regardless of the care, skill, and attention to detail with which a smoke-control system is designed, an acceptance test is needed as assurance that the system, as built, operates as intended.

An acceptance test should be composed of two levels of testing. The first level is of a functional nature, to determine if everything in the system works as it is supposed to work—in other words, an initial checkout of the system components. The importance of this initial checkout has become apparent because of the many problems that have been encountered during tests of smoke-control systems. These problems include fans operating backward, fans to which no electric power was supplied, and controls that did not work properly.

The second level of testing is of a performance nature, to determine if the system, as a system, performs adequately under all required modes of operation. This testing can consist of measuring pressure differences across barriers and airflows through open doorways under various modes of operation. In addition, real fires, smoke candles (sometimes called smoke bombs), or tracer gas tests can be part of the performance test. At present, no consensus has been reached as to what constitutes a reasonable acceptance test. However, the need for both levels of acceptance testing is obvious.

2.13 IMPACT AND CONTROL OF PRODUCTS OF COMBUSTION

In most cases the existence of leaking elements and substantial temperature differences between the building interior and the exterior (stack effect) make it impossible to apply "confinement" concepts alone for the control of products of combustion. Therefore, as discussed earlier, active design features are introduced in parallel with passive measures.

Such active measures are fire-service installations actuated by electrical or mechanical means, which include fire-venting systems, wet riser (standpipe) and hose-reel systems, sprinkler systems, manual or automatic fire-alarm systems, electromagnetic door closers, halon gas systems, portable fire extinguishers, fire dampers (normally open fire shutters), pressurization systems, surveillance systems, and intercommunication systems.

It had been advocated that fire venting be treated as a basic, active measure (Cheung, 1986). The characteristics and impact of such fire-venting systems are as follows:

1. They can release heat and smoke to open air, minimizing internal migration to other fire compartments.

2. They can provide safe escape routes and improve conditions for fire-fighting and rescue operations.

3. They can prevent flashback or backdrafting, which can severely injure fire fighters.

4. They can help clear smoke and smell after a fire.

Fire venting by means of internal vertical ducts through floor slabs has the disadvantages of providing openings in every floor, a higher possibility of mechanical venting-equipment failure, and uneconomic use of increased floor area.

For these reasons, localized fire venting to atmosphere at each floor periphery merits consideration. This can be achieved in several ways, including spring-loaded electrothermal smoke vents through windows or panel sections, released by either local thermal or remote electrical command, or by means of sprinkler flow detector or manual call-point actuation. Usually localized vents are reliable and provide a low-cost and no-space-consuming option.

The basic principle of local venting has long been adopted by fire fighters. Figures 2.19 and 2.20 demonstrate fire venting using local windows or panels (Cheung, 1986). It is also advocated that elevator shafts should be pressurized to check against the spread of hot gases and smoke to other floors through these shafts (Cheung, 1986). To provide tenable escape routes, air pressurization systems for staircases and other service shafts (BSI 5588, 1978) should be anticipated. To avoid air contamination due to vented fire gases, air pressurization fans should be located at the lowest levels of these shafts to draw in outside air directly. In all cases it is beneficial to provide air-inlet connections to enable fire fighters to boost fresh air to these shafts using portable fans, as illustrated in Fig. 2.21 (Cheung, 1986).

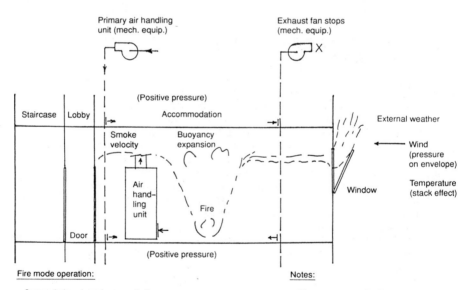

Fire mode operation:

• Some windows/panels open at all sides of the fire floor.
• Grilles for PAU and EF open at floor level (floor level usually at lower pressure and temperature than ceiling level).
• PAU operates.
• EF stops
• Elevator shaft to be pressurized.

Notes:

• The pressures at the floor above and below are at higher pressure than the fire floor.
• Depending on temperatures/pressures of aerodynamic elements, smoke may leak into lobby/staircase/elevator shaft and duct.

Fig. 2.19 Fire venting using local windows or panels.

1 Refuge Floors

As shown in Fig. 2.22, a refuge floor will serve as a relief area for occupants escaping the fire, and as a subbase for fire fighters. The refuge floor must therefore be isolated by construction features to prevent smoke and hot gases from migrating to floors above and below it. According to research on evacuation and sprinkler system design, buildings should have a 45-m (148-ft) maximum-height fire zone, or about 13 floors between any two fire break or refuge floors (Pauls, 1978).

The refuge floor concept enhances evacuation techniques commonly used at ground level. With the fire-fighting command point located on the nearest refuge floor below the fire, it assists in controlling orderly evacuation of the building.

It is important with this concept to provide at least two independent and remotely located staircases between each floor, with a positive air break at the refuge floors, as shown in Fig. 2.23. Passenger elevators not serving the fire floor can also be used for people evacuating from the refuge floors.

In general, each typical floor in a high-rise structure should be designed with a fire-refuge compartment, depending on the number and mobility of the occupants. For facilities such as hospitals, at least two compartments per floor are recommended due to limited mobility. These allow immediate horizontal refuge prior to subsequent vertical escape routes, if further transportation is necessary.

Other essential features should include fire dampers where ductwork passes through fire-compartment boundaries and fire-resistant sealants or fillers to ensure isolation of compartments. The application of intumescent materials to combustible elements such as electrical cable passing through fire-compartment walls contribute to successful fire compartmentation.

The weakest barrier in a compartmentation system is the elevator door since most such doors are not fire-rated. In fact, most elevator doors act as hot absorb-

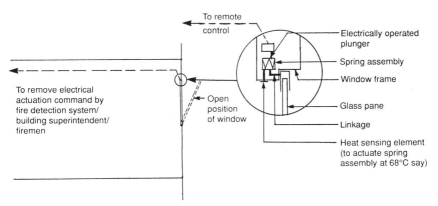

Notes:
· Spring assembly normally holds window/panel closed.
· The linkage of the spring assembly pushes window open when either the electrically operated plunger is actuated by remote means or the local heat sensing element is actuated by rise in local temperature.
· If all remote actuation means fail, local temperature rise will also cause the window/panel to open.

Fig. 2.20 Window or panel opening device for fire venting.

ing panels, radiating heat into the elevator shafts, which can then be transmitted to other floors. The application of intumescent and other lightweight fire-resistant materials on these elevator doors can improve their stability, integrity, and insulation qualities and should make their fire resistance comparable to that of other floor-separation materials.

2.14 CONDENSED REFERENCES/BIBLIOGRAPHY

American Society for Testing and Materials 1980, *Annual Book of ASTM Standards*
American Society of Heating, Refrigerating and Air-Conditioning Engineers 1981,
Berl 1980, *Human Fatalities from Unwanted Fires*
BSI 5588 1978, *Code of Practice for Fire Precautions in the Design of High-Rise*
Cheung 1986, *Staircase Pressurization—The Rationale and the Alternatives*
Cresci 1973, *Smoke and Fire Control in High-Rise Office Buildings—Part II, Analysis of*
Evers 1978, *A Computer Model for Analyzing Smoke Movement in Buildings*

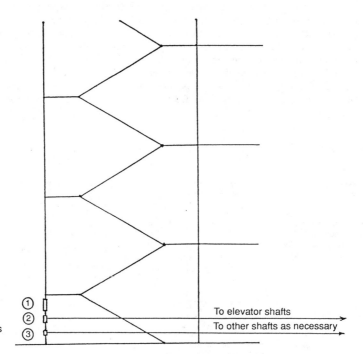

Fixed air inlet connectors

To elevator shafts

To other shafts as necessary

Notes:

• Fixed air inlet connectors ① is for staircase purging/ pressurization protable fan, ② is for pressurization portable fan for lift shafts, ③ is for pressurization fan for other shafts as necessary.

Fig. 2.21 Air inlet connections for fire fighters' portable fans.

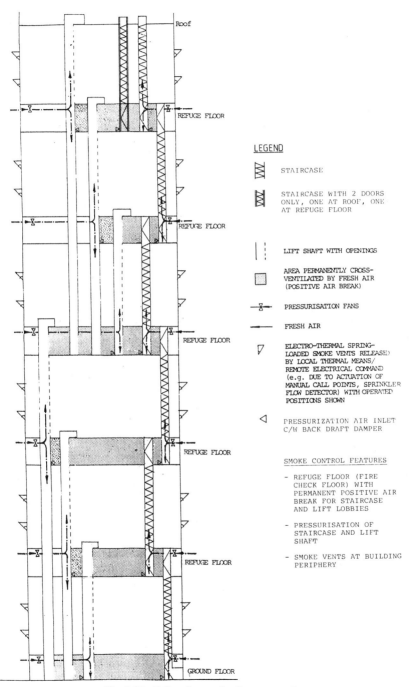

Fig. 2.22   Refuge floor and safe escape route.

Fang 1980, *Static Pressures Produced by Room Fires*
Houghton 1976, *Wind Force on Buildings and Structures*
Klote 1980, *Stairwell Pressurization*
Klote 1982, *A Computer Program for Analysis of Smoke Control Systems*
MacDonald 1975, *Wind Loading on Buildings*
McGuire 1970, *Factors in Controlling Smoke in High Buildings*
McGuire 1975, *Simple Analysis of Smoke Flow Problems in High Buildings*
National Fire Protection Association 1981a, *Code for Safety to Life from Fire in Buildings*
National Fire Protection Association 1981b, *Standard for the Installation of Air Condition-*
National Oceanic and Atmospheric Administration 1979, *Temperature Extremes in the*
Pauls 1978, *Management and Movement of Building Occupants in Emergencies*
Rilling 1978, *Smoke Study, 3rd Phase, Method of Calculating the Smoke Movement be-*
Sachs 1972, *Wind Forces in Engineering*
Sander 1973, *Fortran IV Program to Simulate Air Movement in Multi-Story Buildings*
Sander 1974, *Fortran IV Program to Calculate Air Infiltration in Buildings*
Shaw 1974, *Air Movement through Doorways—The Influence of Temperature and Its Con-*
Simiu 1978, *Wind Effects on Structures: An Introduction to Wind Engineering*
Tamura 1966, *Pressure Differences for a 9-Story Building as a Result of Chimney Effect*
Tamura 1976a, *Air Leakage Data for the Design of Elevator and Stair Shaft Pressurization*
Tamura 1976b, *Studies on Exterior Wall Air Tightness and Air Infiltration of Tall Buildings*
Tamura 1978, *Experimental Studies of Mechanical Venting for Smoke Control in Tall Of-*
Thomas 1970, *Movement of Smoke in Horizontal Corridors against an Air Flow*
Wakamatsu 1977, *Calculation Methods for Predicting Smoke Movement in Building Fires*
Yoshida 1979, *A Fortran IV Program to Calculate Smoke Concentration in a Multi-Story*

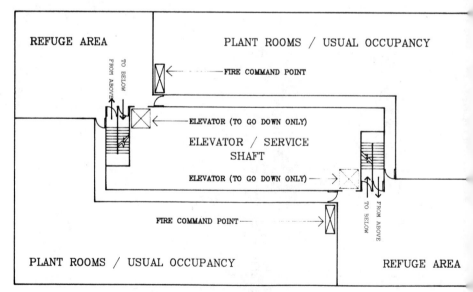

Fig. 2.23 Suggested refuge floor.

3
Fire-Service Operations and Suppression Activities

Primary responsibilities of fire-service personnel involve the control of the fire itself and care and treatment of building occupants. It is obvious from descriptions of tall building fires that the first of these responsibilities can be challenging in itself, regardless of the presence or absence of large numbers of building occupants. However, that scenario is rarely the case, and provision must be made for handling large numbers of people when considering high-rise fire-fighting plans.

Studies on eight vertical escape methods, including rescue by helicopter, conclude that the fire brigade (fire fighter) elevator is ideal as a supplement to primary exit staircases and especially useful for egress by the disabled (Gunter, 1979; Simpson, 1982).

An additional worthwhile feature provides horizontal compartmentation adjacent to the exit staircases to supply transient shelter for the disabled before they can use the elevators or be carried down the staircase by the rescue force (Ministry of Municipal Affairs, Canada, 1984). It would be advantageous to determine the particular requirements for all types of wheelchairs to ensure that staircases can accommodate them, especially in hotels and hospitals.

3.1 FIRE-FIGHTER ACCESS

Figure 3.1 demonstrates two cases of evacuation and fire-fighter access in a building about 270 m (886 ft) in height and with six fire zones. It is assumed that the occupants are given the following instructions:

In case of fire, use the staircases to go down to the nearest refuge floor.

From the refuge floor, use the elevators or staircases to go down to ground level and evacuate the building.

Once people reach a refuge floor and see fire fighters there, they will be psychologically relieved and more confident of their personal safety.

The escape routes for both instructions are downward toward staircases and

53

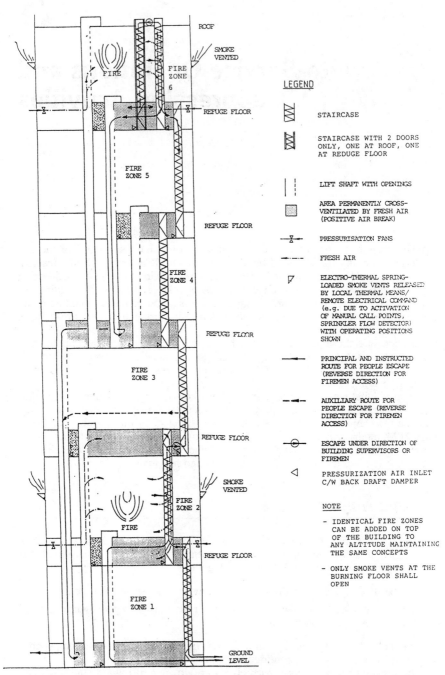

Fig. 3.1   Evacuation and fire fighter access in case of fire (2 cases illustrated).

elevators. However, for fire zone 2, some people may mistakenly go upward to the refuge floor above. In this case they will be able to escape via fire zone 3 elevators under the direction of building supervisors or fire fighters because elevators serving other zones are not given landing openings at the fire-affected zone. For fire zone 6, people who mistakenly go upward to the roof may use a special internal staircase that links the roof and the topmost refuge floor. This staircase has two openings only, one at the roof and the other at the refuge floor itself.

When a fire is detected in a particular zone, separated by two refuge floors, the elevators should descend to the lower refuge floor and stop. Smoke detectors installed in adjacent elevator lobbies can also be controlled to activate this elevator homing concept. It should be noted that Figs. 2.22 and 3.1 also incorporate the fire-venting and pressurization principles mentioned earlier. The routes for access by fire fighters are basically the same as for occupant evacuation, except in reverse direction.

Elevators, if available, are invaluable for carrying fire-fighting equipment (such as portable fire pumps, breathing apparatus, resuscitators, emergency power generators and hoses, and portable pressurization fans) to higher levels, and for the transportation of injured people to ground level. Some elevators should be designed to accept double- and triple-deck stretchers if appropriate.

With the design shown in Fig. 3.1, passenger elevators outside the fire zone are safe for use by both building occupants and fire fighters. However, conventional fire-fighter elevators capable of reaching the fire floor will still be very helpful in fire-fighting operations, provided that these fire-fighter elevators do not serve more than one fire zone and have their lowest point of discharge at ground level or a refuge floor. The design shown in Fig. 3.1 will also enable identical fire zones to be added to the building at any height, maintaining the same concepts of evacuation and fire-fighter access.

3.2 REFUGE-FLOOR COMMAND POINTS

As far as evacuation and fire-fighting operations are concerned, the refuge-floor concept is most advantageous. However, the success of escape and fire fighting is dependent on timely and orderly evacuation to the ground level, and on support from the ground level and other fire command points below the fire floor.

Figure 3.2 illustrates the provision of fire-fighting command points and the building's command center. The ground-level command center acts as a coordination center for other remote fire command points located at positive air breaks adjacent to staircases at refuge floors. Each command point comprises a fire-control subpanel, fire-department connections for sprinkler systems or wet risers, landing valves extended from the lower floors, sprinkler installation control valve set, and connectors for fire fighters' portable air pressurization fans. The fire-control subpanel is equipped with the various communication command facilities indicated in Fig. 3.3. Figure 3.4 suggests a possible arrangement of the fire-fighting and command facilities at a refuge-floor command point.

Evacuation messages and fire-fighting instructions are given to the affected fire zone in the following order:

Before the fire fighters arrive at the foremost command point located at the nearest refuge floor below the fire floor, these messages and instructions are initiated from the ground-level fire command center.

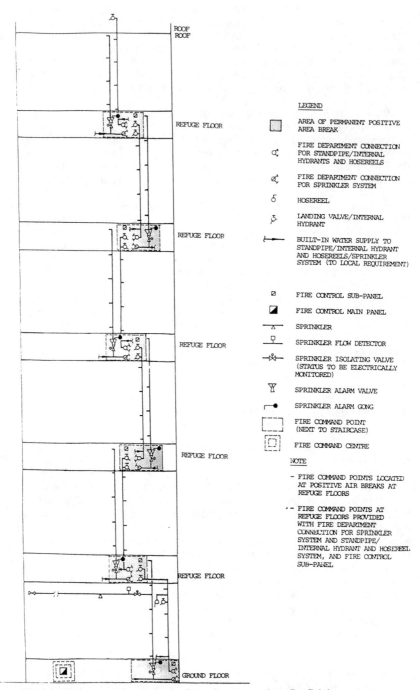

Fig. 3.2   Fire command center and point and water supply to fire-fighting systems.

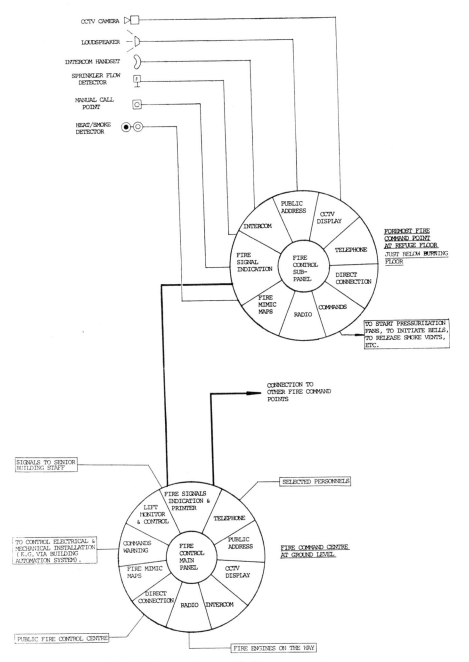

Fig. 3.3   Fire control communication arrangement.

After the fire fighters reach the refuge floor, these messages and instructions will be initiated from the foremost command point, and repeated at the ground-level fire command center, which remains the overall coordinating center.

In terms of the availability of extinguishing agents, water remains the most suitable for high-rise buildings due to its nontoxic characteristic, low relevant cost, good cooling effect, and availability. Wet risers, standpipes, hose-reel cabinets, and sprinkler systems are all essential fire-engineering systems for high-rise buildings. When suitably designed, the wet-riser system will continue to supply water to the sprinklers from a remote source after the building sprinkler pumps fail or the building water tanks are emptied. The sprinkler system is also essential because of its function to protect all parts of the building in the fire's vicinity, regardless of conditions. For example, smoke toxicity, high temperature, oxygen deficiency, and poor visibility, which usually prevent manual fire-fighting operations, do not prevent the sprinkler system from functioning. Other merits which make the sprinkler system essential are reliability and a capacity for raising the alarm (Nash and Young, 1978). Water also contains the fire, thus aiding fire fighters in extinguishing it and gaining time for people to evacuate in an orderly manner.

Fire-department connections provided at ground level and at refuge-floor command points will allow water to be relayed from street hydrants and boosted via fire engines and fire fighters' portable diesel-powered pumps to the highest operational sprinklers, landing valves, and hose reels. These pumps would be carried into place by the fire fighters and would replace either a depleted building water supply or failed building pumps. This water "relay" system, illustrated in Fig. 3.5, would permit fire-fighting techniques used commonly at ground level to be applied from the foremost command point, except that they would be totally internal.

The basic design and operation shown here is effective for all high-rise structures. Note that once the number of floors, fire-department connections, and

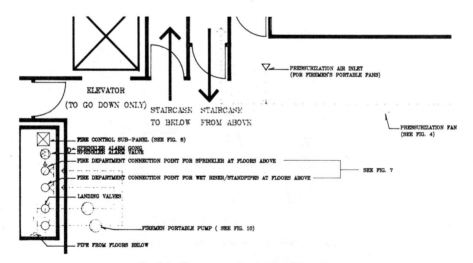

Fig. 3.4 Fire command point at refuge room.

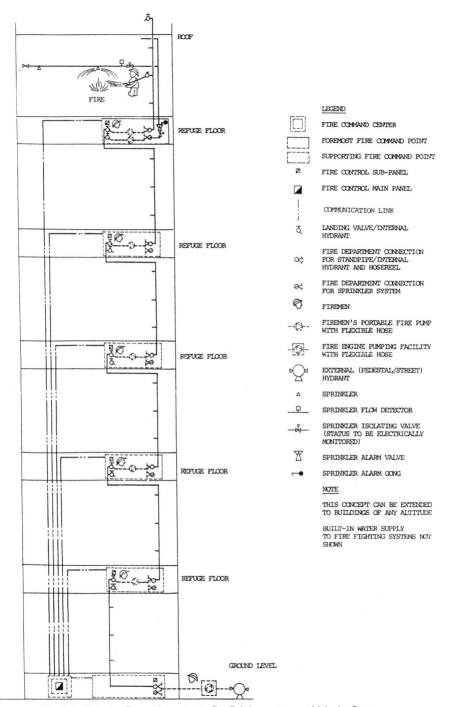

Fig. 3.5 Relaying fire-engine water to fire-fighting systems at high-rise floors.

water-supply requirements are known, the number of portable diesel-powered pumps and fire fighters required to fight the fire can be determined.

The advantage of the refuge floor is that it is adaptable for all building heights. This means that occupants at 500 m (1640 ft) above the ground (the highest structure presently conceived) are only "a little bit more" unsafe than persons at other floors, say at the 30th floor—about 110 m (361 ft) (Engineering News Record, 1986). A 500-m (1640-ft) building would, by adopting the concept, require 10 refuge floors. The concept is also adaptable to existing structures. It allows fire-department connections, pressurization air inlet connections for fire fighters' portable fans, and communication equipment to be added to existing buildings to improve fire safety.

Refuge floors are suitable to being regulated for high-rise buildings due to their identical repetition and associated provisions. Furthermore, the concept enhances fire safety by duplicating features of zoned elevators, wet risers and standpipes, staircases, fire-department connections, pressurization air inlet connections, and portable fire pumps. In addition, this design allows the fire risk to be better quantified.

In common with other life safety techniques, the system must be properly maintained and managed. Staff training and the establishment of contingency planning, fire drills, equipment tests, and inspections are essential. Other safety provisions such as proper design of escape routes and provision of emergency lighting systems are equally important.

The expenses of providing fire-department connections, internal fire communications equipment, air pressurization installations, and related emergency power supply are not significant in comparison with total building costs.

3.3 NETWORK MODELING OF FIRE-SERVICE RESPONSE

It is imperative that a fire department respond to a fire call in an acceptable amount of time. The determination of this response, or attendance time, is a difficult problem when considered in light of questions such as are the attendance times reasonable or within prescribed norms? If not, what is the cause? Are the turn-out times of the brigade fast enough? Is the travel time to the fire address short enough? Are there obstacles on the route or are the distances too long?

To answer such questions, the attendance time for each possible fire location should be known, that is, the quickest routes, under different circumstances, from the fire stations to the possible fire locations should be determined. Obviously one never knows where and when a fire will start, and compounding this, there are too few fires to use statistics for determining the real attendance times.

Operational research and applied mathematics, a science that provides tools with which a decision maker can improve the quality of decisions by analyzing relevant surroundings, is increasingly used to address such questions. The tools used are mathematical models consisting of sets of equations which optimize a particular fire-safety function. The object is to determine the structure of a problem and to generate, compare, and show the consequences of alternative solutions. These calculations are executed primarily by a computer.

A model is a simplified representation of a real situation. Using such a model is valuable because experimentation with real situations can be costly, dangerous, and time-consuming. Furthermore, discussions about a problem with a model are often more efficient and effective.

The research described here began in 1972 with the objective of (re)locating fire stations (calculate the minimal number) in order to satisfy prescribed risk norms, with the resulting models being used in Rotterdam, the Netherlands (Schreuder, 1984). In more recent years, due to economization, this research has shifted to determining stations that could be closed without resulting in unacceptable fire-service response time and coverage (Heijnen and Schreuder, 1982; Schreuder, 1984a). In the city of Hengelo, the model is used to investigate the effects of building a new quarter outside the city and the influences of traffic obstructions on attendance times (Schreuder, 1984b).

It is difficult to apply the results of this research in other countries, however. In the United States, for instance, there is a strong connection between the resources of a fire department (staff and material for fire coverage) and the fire insurance premiums collected in their district. Such a connection does not exist in the Netherlands. Furthermore, the Dutch do not have a rectangular street system, which was one of the main reasons they developed the network model.

1   Starting Points and Assumptions

In the Netherlands, fire coverage is planned in reference to the risk classification of the areas covered in accordance with the directives for fire fighters and materials of the municipal fire departments. Each risk category has standards for the number of fire trucks that should turn out at a fire call and for their attendance times, that is, the first attendance service (see Table 3.1).

Turn-out time is the time between the receipt of a fire call and the departure of the first fire truck from the station. The average turn-out time used for professionals is 1 min, and for volunteers it is 3 min. This should be taken into account when experiments are conducted and assumptions made. Travel time is the time from the station to the fire address. Attendance time is turn-out time plus travel time. The attendance times for other cars, for example, a ladder truck and technical assistance, are longer and therefore not considered.

Table 3.1   Risk norms for first-attendance service

Risk category	District type	Maximum attendance times, min			
		Truck 1	Truck 2	Truck 3	Truck 4
A	Heavy industry	6	6	8	8
B	Industry	6	6	8	—
C	City center	6	8	—	—
D	Apartment	6	—	—	—
E	Suburban areas	8	—	—	—
F	Rural areas	15	—	—	—

The importance of not exceeding the prescribed attendance time is reflected in Fig. 3.6. If the flashover or full-scale involvement point is passed, levels of damage to the structure and the danger to inhabitants increase very rapidly. Likewise, if the fire department arrives after a fire has progressed beyond the department's fire-fighting capabilities and resources, the fire will remain out of control and total destruction of the structure is inevitable. In such cases, fast attendance will be most important for saving lives and preventing the surroundings from catching fire. An increase of the average times for first attendance results in a disproportionate increase of danger and damage. For example, an increase in attendance time of 10% could result in a damage increase of 30%.

Data of turnouts in Rotterdam indicate an average speed of 40 km/hr (25 mi/hr) for fire trucks (Fig. 3.7). Observations in the city of Hengelo confirm this speed. In the daytime the speed varies mainly between 20 and 40 km/hr (12 and 25 mi/hr) and at night between 40 and 60 km/hr (25 and 37 mi/hr).

The most important starting point is that a reasonable statement should be made about the attendance time for each arbitrary point in the area to be covered

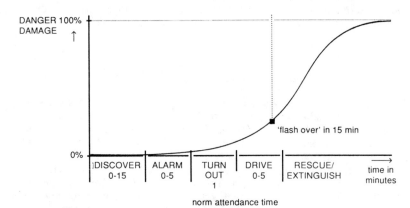

Fig. 3.6 Progress in time of damage and danger after fire starts.

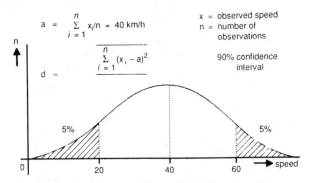

Fig. 3.7 Average travel speed.

by the fire brigade, under every desired condition, such as during traffic rush hours. The districts with risk classification F are usually not taken into account due to sufficient attendance time (see Table 3.1).

2 Network Model

In developing the model, after extensive testing of different models and experimenting with different turnouts, a network was chosen to represent the area covered by the subject fire department. (The model is easy to understand and very useful in practice, which will be demonstrated.) In reality, the problem simply put amounts to measuring the range of effectiveness of the fire department in responding to a fire call. In this context, questions such as the following arise: Must the fire department be strengthened? Is economizing possible? Is a volunteer fire brigade sufficient, or is there a need for professionals (faster turnout versus higher costs)?

The approach often mentioned in the literature is based on economical issues. However, until how, relating fire department coverage to damage levels and danger to human lives has not been very successful in practice. In contrast, the network approach based on attendance and response times has been successful.

The network approach is needed because the distance between two points in a covered area, that is, the fire station and the fire address, cannot simply be calculated with the aid of a formula. The use of distance tables, on the other hand, is hardly practical due to the large number of possible fire locations. Thus the point of the analysis is to reduce this large number to a manageable data set without becoming unrealistic.

A network consists of a number of nodes and edges. The nodes represent road intersections and the edges represent those roads which the fire brigade can use (see Fig. 3.8). Before constructing the network of an area to be covered by the fire brigade, a "norm" map is initially set up, which reflects the risk classifications. Next, the main routes used by the fire brigade are plotted on a city map, and the area to be covered is divided into districts. The risk, the surface area, and the number of inhabitants of these districts should be divided equally and the existing area borders should be used as much as possible. The district borders should also coincide with obstacles for the fire trucks, such as rivers, railroads, and parks, whenever possible.

In each district a center node is chosen. If there is a fire call in a district, the

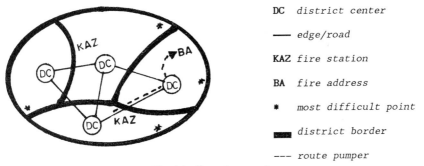

DC   *district center*

——   *edge/road*

KAZ *fire station*

BA   *fire address*

*   *most difficult point*

▬▬ *district border*

--- *route pumper*

Fig. 3.8   Example network.

following is assumed: the truck of the nearest station turns out, it uses the roads of the network to the node that represents the district center, and it travels from there to the fire address (see Fig. 3.8).

To determine the attendance times of the area covered, the shortest routes from each fire station to each district center are calculated with a mathematical algorithm. For this, the travel speed is needed (the distance is measured on the map). The average speed of 40 km/hr (25 mi/hr) is typically used (see Fig. 3.7). Speeds which are outside that range, for example, roads in the city center and freeways, can be calculated easily. A different speed between each pair of nodes can be defined if necessary.

In each district, the most difficult point is selected (see Fig. 3.8). Such a point has the longest travel time in that district with respect to the center node. The attendance time for a district is then calculated: attendance time = turnout time + shortest route to district center + travel time to most difficult point. Of course, the attendance time determined in this way is the largest for a district, but if the attendance time is within accepted norms, the whole district is. Note that the fire department has to cover all points, not the most or the average. This approach reduces all possible fire addresses to a manageable number.

The calculated attendance times can be compared with the prescribed attendance times according to the norms for a district. In this way inconsistencies in the coverage plan can be detected and alternative situations compared.

3 Examples

Upon applying the network model in Rotterdam, a shortage of two fire stations was demonstrated (14 stations exist). Closing one or more existing stations in order to economize would make the fire coverage even worse. In the city of Hengelo, the model showed a sufficient coverage only of the central quarters. The newly built quarter outside the city cannot be covered with an acceptable attendance time and needs a fire station of its own. As a result, economizing in Hengelo is not possible without severe endangerment of the fire coverage.

These examples demonstrate that the network model can be used for strategic as well as tactical planning. In addition, the increasing use of microcomputers and personal computers has made the handling of the model user-friendly and more economical. Another advantage of the model is that it can also be used on an operational level: when a fire call occurs, it is possible, within a few seconds, with the help of a personal computer, to print out a map of the fire-truck route, with specific instructions on the optimum route according to the most recent traffic information.

3.4 CONDENSED REFERENCES/BIBLIOGRAPHY

American Society of Heating, Refrigeration and Air-Conditioning Engineers 1985, Australian Fire Protection Association 1986, *Factory Mutual International ESFR Update* BSI 5588 1978, *Code of Practice for Fire Precautions in the Design of High-Rise Buildings,* Engineering News Record 1986, *The McGraw-Hill Construction Weekly* Factory Mutual Engineering Corp. 1983, *Loss Prevention Data Sheet 1-3* Fire Offices' Committee 1973, *Rules of the Fire Offices' Committee for Automatic Sprin-*

Fire Offices' Committee 1978, *Supplementary High-Rise Sprinkler Draft Rules*
Gunter 1979, *High-Rise Buildings, A German Approach*
Heijnen 1982, *Number, Location and Capacity of Fire Stations in Rotterdam*
Klote 1983, *Design of Smoke Control Systems for Buildings*
Ministry of Municipal Affairs 1984, *The Section 3.7 Handbook*
Nash 1978, *Sprinklers in High-Rise Buildings*
Predtechenskii 1978, *Planning for Foot Traffic Flow in Buildings (1969)*
Public Works Department 1982, *Fire Precautions for Buildings*
Read 1979, *An Investigation of Fire Door Closer Forces*
Schreuder 1981, *Application of a Location Model to Fire Stations in Rotterdam*
Schreuder 1984a, *Risk-Covering Fire Stations in Rotterdam*
Schreuder 1984b, *Risk-Covering Fire Department Hengelo*
Simpson 1982, *The Lift as a Means of Escape for Handicapped Employees*

4

Fire Following Earthquakes

While fire following an earthquake has long existed as a major source of loss potential, the problem has generally gone unrecognized or untreated by most groups, including many who should be specifically concerned with the problem. Fire-protection and structural engineers are examples of groups in which recognition of the problem has generally faded since the 1906 San Francisco and 1923 Tokyo earthquakes and fires. Recognition of the loss potential has remained within the insurance industry, but very little has been done to treat the problem. An exception to the foregoing is Japan, where the problem receives considerable attention.

Aside from Japan, however, this lack of attention is surprising, especially given that the two earthquakes noted have been the sources of the two single greatest (nonmilitary) urban fires of the twentieth century. (The major difference between San Francisco and Tokyo at the time of those earthquakes and the cities of today is the many hundreds of high-rise buildings which were not present in the earlier events.)

Recently attention has been called to the general problem of fire following earthquakes in the United States, although the particular problems of fire following earthquakes in high-rise buildings have not been specifically addressed (Scawthorn, 1987). That the problem of fire following an earthquake in high-rise buildings has largely been overlooked is not too surprising when one considers the relatively small number of earthquakes that have actually caused major damage to high-rise buildings. These would include San Francisco 1906, Tokyo 1923, Alaska 1964, Caracas 1967, Tangshan 1976, Bucharest 1977, and Chile and Mexico 1985—only eight in all, and only six in relatively modern times. (This list neglects several other earthquakes affecting urban areas with only a handful of high-rise buildings, such as Managua 1972 and Guatemala 1976.)

Today, however, we are faced with large and rapidly increasing numbers and densities of high-rise buildings in seismic zones (Council on Tall Buildings, Group CL, 1980) in the United States, Japan, and elsewhere, to the point where it is only a matter of time until a major urban area such as San Francisco, Tokyo, or Los Angeles is strongly shaken with potentially catastrophic results. In a modest attempt to partially avert such a disaster, and fill the gap concerning the problems of fires following earthquakes in high-rise buildings, the following discussion will focus on a general model of fire following earthquake, and then examine ignition, fire spread, and fire prevention specific to the postearthquake high-rise environment.

4.1   THE POSTEARTHQUAKE FIRE PROBLEM

Fires occurring after an earthquake follow the same general process as ordinary fires: ignition, growth, detection, report, response, suppression activities, extinguishment (the latter due to suppression activities or self-extinguishment due to exhaustion of fuel). While the process is similar, fires following earthquakes differ in several significant ways.

Generally speaking, the essence of the postearthquake fire problem is multiple simultaneous ignitions due to numerous disturbances of appliances or earthquake damage. In some cases these ignitions will be undetected for a long period due to the distractions the earthquake creates. These delays will result in large fires which are much more difficult to extinguish. Whether delayed or not, the report of these fire, once detected, will almost assuredly be delayed in the immediate postearthquake period due to the fact that almost all fires are reported by one of two methods:

1. Telephones, which will almost definitely be unusable, due either to direct damage to the system or to system overload because thousands of receivers were shaken off their cradles and because of high customer demand. Even if usable, the fire-emergency number will be overloaded.

2. Automatic detection and alarm systems, which will also overload because of numerous false reports of smoke alarms being activated by dust, motion, and the like. The relatively few true alarms will be indistinguishable from the false alarms.

When finally reported to the fire department, the department's ability to respond is doubtful due to numerous other demands on its resources (other fires, building collapse, hazardous materials releases, and other emergencies). If able to respond, the department will be faced with huge fires (because of these delays), which would normally require large commitments of resources, perhaps in structurally damaged buildings to which entry is not prudent. These additional resources will also be required elsewhere, and will often not be available. The result will be large, uncontrolled fires. The only feasible approach will be to protect exposed buildings and attempt to prevent a conflagration. If the water system is damaged, as it was in San Francisco, even this may not be possible.

This outline is equally true for residential, industrial, and central business district occupancies. A compounding difficulty in the case of high-rise buildings is their typically large demand for fire-fighting personnel. The First Interstate Bank building fire in Los Angeles (May, 1988), for example, required 383 fire fighters (about 40% of the on-duty personnel of one of the largest U.S. fire departments) for extinguishment (Los Angeles City Fire Department, 1988). Because of this, it is likely that most departments will not aggressively pursue high-rise fire fighting following a strong earthquake, but will only seek to ensure evacuation of endangered personnel.

4.2   IGNITIONS

As outlined, the first element in the postearthquake fire problem is ignition. The typical high-rise building offers many sources of ignition. Figure 4.1 depicts several of these, which are primarily electrical or gas-related in nature.

Electrical sources are probably the major hazard in the typical office or residential high rise. There are numerous electrical appliances on each floor, which will be overturned or otherwise disturbed, exposing heating elements and causing possible short circuits. Room contents will be greatly disturbed under strong earthquake shaking, throwing paper and other combustibles onto hot elements of copiers, cooking elements, or similar appliances. Even if the electric power system fails, residual heat in many appliances may still cause ignitions. In addition, even though power may be initially cut off, if it is restored quickly in an unsupervised manner, ignitions in the manner described are almost assured. Lastly, beyond simple appliances, most high-rise buildings contain network transformers, often on each floor, which may short-circuit or otherwise malfunction under strong shaking, creating another source of ignition.

If the high rise has gas service, damage to gas appliances is another source of ignition. This is especially true in residential occupancies, but may also occur in office high rises with restaurants, coffee shops, or cafeterias in the lower floors, which are common. Sloshing deep fryers and other cooking-oil-related appliances and operations are additional ignition sources in these locations.

The mechanical rooms of high rises offer numerous sources of ignition as well. It is ironic that even the emergency power fuel supply in many high rises constitutes a source of ignition; in fact, many of the fuel "day tanks" in older high rises are secured inadequately and may fall and rupture, spewing their diesel or other

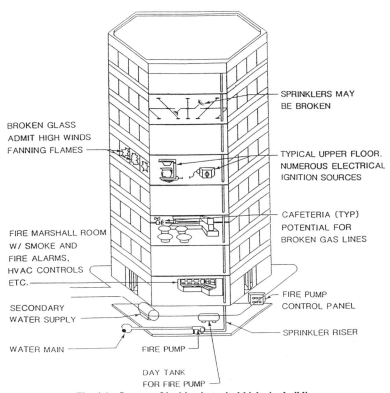

Fig. 4.1 Sources of ignition in typical high-rise building.

petroleum fuel on or near the emergency generators and their batteries just as they are activated.

Lastly, chemical sources of ignition occur in high rises even though their presence may often be less than in low-rise commercial occupancies. Typical locations for chemicals would include photographic, pharmaceutical, or other laboratories and clinics, and storage closets for some older office equipment, specialized cleaning supplies, and such.

At the present time there are insufficient data to differentiate high-rise ignition rates from those of low- and mid-rise buildings. Other studies (Scawthorn, 1987) indicate typical postearthquake serious ignition rates, as listed in Table 4.1. Here MMI refers to the modified Mercalli intensity (Richter, 1958), and serious ignitions refers to ignitions that cannot be suppressed by ordinary citizens but rather require fire-department response.

These ignition rates would indicate, for a typical high-rise building of, say, 46,451 m^2 (500,000 ft^2), probabilities of ignition per building as listed in Table 4.2. These probabilities are not negligible. Since the central business district of San Francisco contains several hundred high rises and would be subjected to intensity levels varying from VII to IX in a repeat of the 1906 earthquake, the rates given in Table 4.2 would indicate several dozen serious high-rise ignitions. The City of Los Angeles contains approximately 1000 high rises.

4.3 POSTEARTHQUAKE HIGH-RISE FIRE SPREAD

High-rise fire spread under ordinary conditions is a complex subject of much discussion (Belles and Beitel, 1988; NFPA, 1982). Earthquake damage acts to degrade active and passive fire-protection features and to enhance fire and smoke

Table 4.1 Postearthquake serious ignition rates by modified Mercalli intensity (MMI)

MMI	One ignition per
VI	Negligible
VII	1,021,934 m^2 (11 $\times$ 10^6 ft^2)
VIII	464,515 m^2 (5 $\times$ 10^6 ft^2)
IX	371,612 m^2 (4 $\times$ 10^6 ft^2)

Source: From Richter (1958).

Table 4.2 Probability of postearthquake ignition for typical high rise

MMI	Probability of ignition
VI	Negligible
VII	0.05
VIII	0.10
IX	0.12

spread, that is, strongly shaken high-rise buildings can be expected to sustain fairly large interstory drifts, resulting in damage such as broken windows, cracked walls, buckled fire and elevator doors, and broken HVAC ducting. The openings thus created more readily permit the entry of hot gases and the spread of fire. The basic point is not that the earthquake will cause such massive damage to a high-rise building as to open it like a sieve, but rather that some small breach in compartmentation will occur. Relatively small openings are all that is required to permit large fire spreads in high rises, and fire departments will probably not pursue high-rise fires for the reasons outlined earlier.

Beyond fire spread within one building is the spread of fire from one building to another. The plume of smoke and hot gases rising from a nearby fire, even if from a low-rise building, can cause ignition in upper stories of an adjacent high rise. This does not usually occur under ordinary circumstances due to active fire-department exposure protection, but because of the absence of the fire department it is likely to happen if fire approaches a high-rise building following a strong earthquake. Even sprinklering of the high rise may not prevent this, since sprinklers may be set off on several floors, overwhelming the hydraulic capacity of the system and permitting ignition.

Under a worst-case scenario, which is entirely possible in the dense concentrations of high rises in the central business portions of major cities, this ignition of upper stories can proceed from one high rise to another, resulting in a phenomenon akin to a forest "crown" fire. This process may be compounded by the fall of burning debris from an involved high rise onto the roofs, skylights, and other portions of intervening lower buildings, igniting them, their fire then spreading to the next high rise, and so on. This, of course, is predicated on the basis of the virtual absence of the fire department, which is not unreasonable. On-duty fire-fighting personnel in San Francisco, for example, with a population of 700,000 and over 500 high rises in the city, is approximately 300, organized into 41 engine companies and 18 truck companies.

4.4 FIRE DETECTION AND SUPPRESSION

As discussed, active suppression on the part of fire-service personnel cannot be relied upon following an earthquake. In many cases the installed detection, alarm, and suppression systems may be unreliable at the same time, that is, fire and smoke alarm detectors may or may not function properly, following strong shaking. Dust raised by shaking will often activate smoke detectors so that dozens of alarms may be received simultaneously over a number of floors of a building. Alternatively, shaking may dislodge detectors or otherwise damage their circuitry so that they cannot function. This problem extends to central fire-alarm panels located in mechanical rooms or in the fire marshall's room. (Central panels are required in most modern high rises, usually located in or near the lobby.) These panels are usually not seismically safe or even properly mounted, that is, the panels may be freestanding or inadequately secured so that they will overturn or fall during strong shaking. In some older high rises, for example, alarm panels are secured to hollow clay tile partitions, the most fragile of building materials, which is almost certain to sustain major cracking and failure. Beyond the adequacy of panel mounting, the seismic qualifications of the panel's interiors need to be addressed. Circuit boards, if not restrained, may become dislodged under strong shaking.

A typical high-rise water and sprinkler system is shown schematically in Fig.

4.1. It consists of the following elements: the water service connection to the building, a fire pump with (sometimes without) emergency power (and associated fuel supply), a fire-pump control panel, and secondary water supply (a tank often located in the basement), and sprinkler risers, mains, branches, and heads. Each of these elements can sustain damage, rendering the system inoperative. Since many urban central business districts are located in, or adjacent to, old waterfronts, the municipal water supply system in the vicinity of the central business district will often sustain a significant number of water-main breaks. This is caused by gross soil failure in soft waterfront soils, filled-in areas, and such, resulting in temporary loss of municipal water pressure immediately following the earthquake. Water supply for high-rise fire protection thus reverts to the secondary supply. In older high rises the tanks may not be adequately secured for strong shaking. Even relatively small rocking or sliding of an older tank can break pipe connections, eliminating the secondary water supply.

Fire pumps will usually be adequately secured to their foundations for ordinary mechanical reasons. Depending on piping configurations and anchorage displacement, however, pumps may be badly damaged if the differential displacement between the piping anchorage and the pump location is more than a few centimeters. Since the municipal electrical supply will often fail toward the end of the ground shaking, emergency power is critical to the functioning of the fire pumps.

Emergency power usually takes the form of a diesel engine or a generator set. These are spring-mounted for ordinary, vibration-isolation purposes. Under strong ground shaking, these spring mounts may resonate, resulting in excessive displacements and tearing of connections. Seismic snubbers are required to restrain excessive displacements. The diesel sets will be started by power from a battery set. These batteries are often not adequately restrained so that they will be tossed about, breaking their circuitry. As discussed, the emergency power fuel, the day tank, may also be inadequately restrained, resulting in an ignition hazard itself, beyond the simple problem of loss of fuel. For all these reasons, emergency power, if it has not been detailed for seismic forces by experienced designers, is not reliable. Modern building codes in high-seismicity zones require higher reliability of life-safety-related systems, such as the emergency power and fire-water system, but the requirement is a general one, and experienced designers are necessary for satisfactory detailing and performance. These building code requirements have only been in place for the last decade or so, and older high rises often lack these features.

The last element in the fire-water system is the sprinkler piping. Building code requirements for sway bracing (both lateral and longitudinal) at nominal intervals are probably satisfactory. Anchorage of pipe hangers into certain types of floors (precast planking, for example) may not be entirely safe; for example, failures occurred in the October 1987 Whittier Narrows earthquake in the Los Angeles region. In addition, simple pounding of sprinkler heads with suspended ceilings may be a problem; breakage of heads due to this was also observed in the Whittier earthquake. The breakage is serious because it directly results in water damage, and, more importantly, if it occurs in more than a few heads, it will exhaust the hydraulic capacity of the sprinkler piping.

4.5 PREVENTIVE MEASURES FOR POSTEARTHQUAKE FIRES

The preceding presents a qualitative discussion of the postearthquake high-rise fire problem. Reduction of this problem often follows the same lines as ordinary high-rise fire protection. Passive compartmentation and proper curtain-wall detailing will aid the problem. Beyond these measures, however, there are a few additional points specific to the postearthquake problem that will aid mitigation. Prevention of ignition is probably the first. Strong shaking and disturbing of contents are difficult to prevent, so ignitions will probably occur. Training occupants for immediate postearthquake reconnaissance of their immediate areas for small fires or potential ignitions (when power is restored) and training for small-fire suppression is recommended. Following this check, evacuation of upper-floor occupants to lower-floor places of refuge is also recommended so that occupants will not be trapped above a fire, should one occur.

As mentioned, the typical high-rise fire-water supply system has a number of potential vulnerabilities. These are well known to experienced engineers and can be reduced substantially through proper design or retrofitting. Building codes in areas of moderate and high seismicity should require proper seismic detailing, and retrofitting should be mandated for older high rises (Council on Tall Buildings, Group CL, 1980). The costs are minuscule compared to the values at risk, not to mention the lives.

In larger, urban central business districts the sheer scale of the problem is enormous, and the fire department may be unavoidably overwhelmed. The measures discussed, if properly instituted, may eliminate some high-rise fires and go a long way toward easing the postearthquake burden on local fire personnel.

4.6 CONDENSED REFERENCES/BIBLIOGRAPHY

Belles 1988, *Between the Cracks...*
Council on Tall Buildings 1980, *Tall Building Criteria and Loading*
Los Angeles City Fire Department 1988, *Executive Summary, Report on First Interstate*
NFPA 1982, *Investigation Report on the Las Vegas Hilton Hotel Fire*
Richter 1958, *Elementary Seismology*
Scawthorn 1987, *Fire Following Earthquake, Estimates of the Conflagration Risk to In-*

5

High-Rise Fire Safety in Taiwan—Case Studies

Taiwan is an island in the western Pacific Ocean which is situated on the Pacific earthquake belt. It covers an area of 36,000 km² (13,900 mi²), two-thirds of which are mountains and hills. The population, which is over 20 million as of mid-1989, is concentrated in the 404 recorded urban districts (Ministry of Interior, ROC, 1989a). High-rise construction has grown rapidly in major urban areas of Taiwan. Taipei, the capital, is situated in the north east corner of the island. Metropolitan Taipei has a population of 2.7 million within an area of 272.14 km² (105.07 mi²) (Ministry of Interior, ROC, 1989b). This high population density led to available land being developed vertically. There are many high-rise buildings, in the range of 7 to 30 stories, in Taipei, and this stock of older high-rise buildings has been a resource for understanding the complex issues of customs and regulations which govern high-rise building fire safety in this case study.

5.1 BACKGROUND

Until 1988 Taiwan experienced about 502 casualties in building fires in an average year. In that year, however, the number of victims rose to 606 (Ministry of Interior, ROC, 1989c). This rapid increase can be attributed to a corresponding increase in the frequency of high-rise fires. The background for these increases, including building regulations, legal aspects, and the traditions and customs of the inhabitants, along with an analysis of 514 fires, is informative, and is expected to improve the fire-safety aspects of planned construction in the 1990s. To better understand the ensuing discussion of high-rise fires in Taiwan, it will be helpful to explain how the government system relates to the fire-safety operations.

1 Legal System

The legal system is divided into three jurisdictions: the county (city), the province (metropolis), and the central government. Until 1987 marshal law prevailed in Taiwan, and democratic mechanisms did not exist for making changes in the building or fire codes.

Each level of government has jurisdiction over specific areas, which are assigned to it by law. Local matters are regulated by local governments, but matters of general concern are enacted into laws by the central government. The laws concerning buildings and fires are on the same level as urban laws. Under the Taiwanese legal system, despite the fact that the building law and the fire law rank on the same level with the urban law, the guidelines and regulations promulgated under the urban law outrank those of the former two. In practice, this means that building plans, design, and occupancy have to meet all these regulations simultaneously. The current management of fire safety in Taiwan is shown schematically in Fig. 5.1.

Before 1990 all of the building regulations for fire safety came under the fire departments, which are administered by the national police administration of the Ministry of the Interior (MOI). After 1990 the building technical regulations (BTR) were switched to the construction and planning administration (CPA), as shown schematically in Fig. 5.1, which can be thought of as a national "building department" with duties similar to those of a U.S. city's building department. In many ways the BTR is a national building code, but there are some aspects of fire safety which were not switched to the CPA from the fire departments. The most important portion is probably the sprinkler requirements, which are still administered by the fire departments.

2 Land Use in Urban Areas

The urban areas are divided into different zoning districts such as residential, commercial, industrial, and agricultural. In turn, each district may be subdivided. In Taipei the land-use and zoning regulations are stipulated by the Taipei metropolitan government in accordance with the prevailing urban laws.

Furthermore, special districts are designated (industrial, agricultural, and the like), which must be used for one purpose only, but most other districts are permitted to consider uses of a similar nature, which inevitably leads to mixed land use. In Taipei there are hundreds of mixed land uses under 44 use divisions. It is quite feasible to encounter public facilities such as garages, markets, and public transport stations to be situated under parks and sport fields. In other words, it is quite acceptable to have a mixed use of urban land, which might lead to shorten the distance between various social activities and decrease the commute time for people.

3 Building Occupancy

In accordance with urban planning principles, the BTR also allows multipurpose occupancies within buildings. To illustrate this, the following examples are given:

Within a single building in a residential district there might be a mixture of occupancies such as apartments, a hotel, a clinic, offices, retail stores, a restaurant, and light industry.

Within a single building in a commercial district there might be a mixture of occupancies such as offices, a theater, a hotel, and a department store.

Within a single building in a public facilities zone there might be a mixture of occupancies such as garages, shops, markets, and mass rapid transportation (MRT) stations, which might be constructed under a park or a sports or school field.

It should also be mentioned here that in most Taiwanese cities there is a high incidence of squatters on the roofs of buildings. They often construct one- or two-story structures which are clearly visible from the street below. This has become such a common problem that is it beyond the powers of the government to

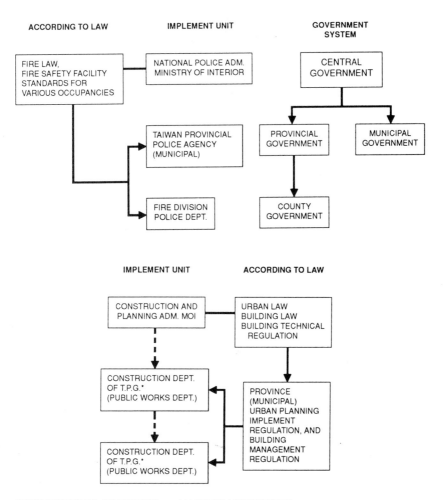

Fig. 5.1 **Schematic diagram of current management of fire safety in Taiwan, ROC. TPG—Taiwan provincial government.**

correct until more housing is constructed. The presence of these squatters means that almost any building planned or constructed at this time has to consider the possible addition of living units on the roof. Some architects plan their roof layout either to accommodate the squatters or to exclude the construction of additional structures.

4 Building Materials

The eastern part of Taiwan has an abundance of cement, and therefore many building frames are constructed of reinforced concrete. But the structural building materials (used for beams, columns, slabs, partitions, and exterior sidings) vary according to the height of buildings, as shown in Table 5.1. Generally speaking, these building materials are fire-resistant, in contrast to the building decorations and furnishings, which are often made of plywood or wood. It should be pointed out here that in Taiwan, to save on costs a number of people use untreated wood or plywood for decorations. Furthermore, to save space (a plywood wall takes up less room than a masonry wall) and money, it is quite common for people to replace a masonry wall with a plywood wall.

5 Heating Sources

Natural gas is the major heating source in the prevalently urban districts. While the more modern buildings have a centrally distributed gas supply system and their individual main gas meters and valves are located on balconies, in the older districts, many structures still exist where gas cylinders, most likely housed in a kitchen cabinet, are being used.

5.2 EXPERIENCE AND CHALLENGE FOR THE FUTURE

There were approximately 514 fires in high-rise buildings in Taipei during the time period of 1977 to 1986. Though records exist for 82 of these high-rise fires, detailed records are only available for 23 big fires in which heavy losses, both human and economic, were sustained. The fires were classified as residential, ho-

Table 5.1 Common materials* used in buildings of various heights

	Story			
	< 5	6–16	16–20	> 20
Frame	RC	RC	RC	SF
Slab	RC	RC	RC	RC
Partition wall	BR	BR, RC	RC	RC, BL
Exterior wall	BR	RC, GC	GLC, PC	PC, GC

*BL—block; GC—glass curtain; SF—steel frame; PC—precast concrete; BR—brick; RC—reinforced concrete.

tel, office, commercial, and mixed-use. The causes of these fires, including inherent building design defects, are discussed in the following sections. A summary of these 23 big fires in Taipei is given in Table 5.2, in which basic data about the buildings and the fire are presented. There are many ways to discuss the fires, but perhaps the most effective way is to present the various aspects of how building fire safety is currently managed in Taiwan, and to give an overview of the current fire-safety regulations. These fires illustrate some of the cultural and regulatory factors which have made the Taiwanese fires unique. It should be remembered, however, that a new way of designing and regulating fire safety is evolving in Taiwan, and there is a growing awareness that the fire departments and the other governmental agencies can work together and create a new era of regulating fire safety.

1 Past and Current Regulations

Sprinklers.   Probably the most unique fire-safety regulation in Taiwan is the one for automatic sprinkler systems. For a building of six or less floors, or for the first six floors of a high-rise building, sprinkler systems are not required except under special conditions, which are described later. This means that the first six floors of most high-rise buildings do not have sprinklers, but sprinklers are usually installed above that height.

In the 82 high-rise fires that had occurred, 27 fires happened in the basement; 34 more fires occurred between the ground floor and fifth floor. Focusing specifically on the 23 big high-rise fires, in which five fires occurred in the basement and 12 fires happened below the sixth floor, these fires resulted in 31 deaths and 77 injuries. Generally speaking, most high-risk occupancies such as restaurants, and theaters; Karaoke, MTV, and KTV (small theaters, sometimes divided by plywood cubicles, in which up to 150 people watch special television programs and interact); and some other high-density occupancies are usually located below the sixth floor. From these cases and data we learn that the lower floors are not only more prone to fire, they also have a greater propensity for flashover.

Exits.   One of the fundamental principles of building fire safety in the United States is for the exits to be wide enough to accommodate the occupants of the

Table 5.2   Summary of 23 large fires in Taipei

	OCCUPANCIES																							
	RESIDENTIAL									HOTEL			OFFICE							COMMERCIAL MIX				
	1	2	3	4	5	6	7	8	9	10	11	12	13	14	15	16	17	18	19	20	21	22	23	24
1. BUILDING COMPLETED	70	66	70	70	69	66	65	66	70	60	68	59	67	66	68	65	70	71	62	66	66	58	67	26
2. HEIGHT (STORIES)	7	7	12	7	12	7	12	7	16	7	14	10	12	12	9	7	10	14	6	12	11	11	14	13
3. BASEMENT (STORIES)	1	1	2	1	2	1	1	1	2	1	2	1	2	1	1	1	1	2	1	1	2	2	1	
4. STORY OF FIRE ORIGIN	B1	7	B1	1	B1	1	B1	B1	15	3	2	7	11	9	9	3	2	3	2	4	4	1	2	2
5. FIRE DURATION (HRS.)	56	16	18	15		29	18	115	48	40	95	19	61	21	13	13	35	30	17	15	16	187	95	243
6. NUMBER OF DEATHS	0	1	0	0	0	7	0	0	2	2	19	0	0	0	0	0	0	0	0	2	0	1	0	103
7. NUMBER OF INJURIES	0	0	0	0	0	3	0	0	0	21	51	3	8	1	1	0	0	0	0	0	0	12	0	120
8. PROPERTY DAMAGE (TSM)	0.5	7	20	100	200	150	20	100	20	3.5		50	8	20	20	150	50	60	28	1	20	1000	100	180

Residential: 1. Central Service-man Housing; 2. Fifth Street Mansion; 3. Ta-Fu Mansion; 4. Hony-Mu Apartment; 5. San-Pu, Jaing-Ai Mansion; 6. Ji-Tai Mansion; 7. Suei-Shan Mansion; 8. Kung-Ming Mansion; 9. San-Pu, An-Ho Mansion.

Hotel: 10. Shih-Men, Diamond Building; 11. Time Hotel; 12. Yun-An Hotel.

Office: 13. Ray-Huang Building; 14. Tien Shian Building; 15. Ta-Ching Building; 16. Tien-Fu Building; 17. Bai-Lo Building; 18. Asian Building; 19. U-Chi Building.

Commercial mix: 20. Chang-Du Mansion; 21. Tai-Young Building; 22. Asian Union Building; 23. Sino Department Store; 24. Japan Department Store.

building. In Taiwan there are few specific regulations in the BTR regarding occupant density in a building and the width that must be provided for exits. This often results in an excessive occupant density of a building, which in turn causes the width of both corridors and stairways to be inadequate for the safe exiting of occupants in case of fire.

To avoid overpopulation in a particular urban area, there is an attempt in Taiwan to control the ratio between the size of a vacant lot and the building density through the urban law, but this has not prevented overcrowding in many residential buildings.

The fire in the Time Hotel (fire no. 11, Table 5.2) is an example of a fire in which the exits were inadequate to accommodate the occupants of the building. More comments about that fire are given later under the discussion of unprotected shafts which spread the smoke to upper stories. As in most high-rise fires, if there had not been pathways for the smoke to travel upward in the building, the exit system would have been adequate. The main reason for the slow occupant evacuation in fires is that the occupants are blocked within corridors and in front of fire-safety doors, which is especially true in the case of high-rise fires.

Three types of staircases are allowed in Taiwan by the BTR: (1) interior, (2) exterior, and (3) emergency. The first two staircases are for use by both the occupants and the fire service, while the emergency staircase is for use by the fire service only. Elevators are not considered part of the exit system of a building, and in new buildings they are automatically recalled to the ground floor.

The type and number of staircases that should be constructed depend on the occupancy, travel distance, door and window openings, building height, and the floor area of each occupancy. The following two types of stairs are most frequently used in Taiwanese high rises:

1. *Piercing stairs.* To save space, staircases were frequently designed to break straight through a building from the basement to the top floor. Besides housing a garage in the basement, buildings often have restaurants, coffee shops, Karaokes, bars, and such in commercial and even residential districts, albeit these occupancies violate the BTR. Because the fire-emergency doors are used by customers, they are frequently left open. In case of fire, the smoke spreads through the open doors and upward through the piercing staircase from the basement to the top floor of the building. There were five fire incidents in which the fire originated in the basement and a stack effect occurred in the piercing staircase.

2. *Scissor Stairs.* Basically this is a staircase that should have access from opposite directions in a building to provide the occupants with a wider selection for exiting in case one stairway is blocked by the fire. In the Taiwanese high rises, to design a lobby or corridor space economically, two of these interior stairs, or two emergency stairs, are designed as one combination set. Because of its crosslike X section, it is called scissors stairs, as illustrated in Fig. 5.2.

In the United States the National Fire Protection Association *Life Safety Code* specifically does not qualify scissor stairs as separate exits unless the entrances to each side of the scissor are separated by a sufficient distance (Lathrop, 1986).

In Taiwan, in order to save space, these scissor stairs are usually designed as part of the lobby, elevator, and vent shaft to form the "core" of a building. On some floors, the emergency doors in these staircases are often left open for use by occupants. These areas are not provided with any automatic detection systems, and there is no warning of smoke spread in case of a fire. There were three residential, two office, and one mixed-use fire incidents in which two people died and one was seriously injured (see fires 9 and 14, Table 5.2).

2 Mixed-Use Occupancy and Change in Occupancy

Mixed-use occupancy in buildings is allowed because of both limited land resources and traditional Taiwanese customs, yet it is also the cause for many problems in building and fire management. Mixed-use occupancies in buildings, except for the parking areas, are considered as a single type occupancy in the BTR. While this does not present problems in the design stage, problems do arise when parts of an existing building undergo a change in occupancy. The building then often meets the building regulation requirements only nominally. In many instances the owners, after having received an occupancy permit for a building, change the partitions in an existing building to meet the needs of the new occupants of the building. This often causes problems to both occupants and management.

When buildings are first constructed, they usually occupy the site to its fullest limits, which means that there is no extra land available for any additional stairs or emergency stairs when the occupancy of a building changes. Such buildings are often called violation occupancy, since they have not applied for a change-of-occupancy permit. When remodeling buildings to suit the new occupancy, the original fire and smoke compartments are often destroyed, and the safety of the new partitions becomes questionable. The current regulations have no requirement for "surface flame spread" or "rate of heat release" (RHR) for any compartment less than 200 m² (2150 ft²). In addition, thin hardwood plywood is

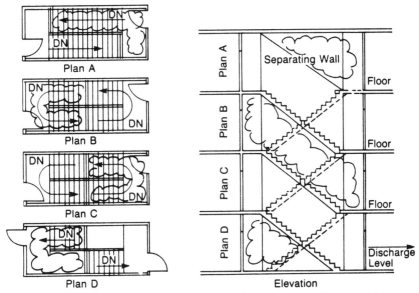

Fig. 5.2 Scissor stairs—two stairways in the same enclosure but completely separated from each other. This results in space saving—two exits are provided with one enclosure. With this arrangement, two entirely independent escape paths are possible even though the two stairs are not remote. Note the continuity of all walls, providing a complete separation at all points. Follow arrows for path of travel.

the most commonly used interior finish material, and it produces extremely fast fire growth with large amounts of excess pyrolysates that can quickly spread the fire beyond the room of origin. Thomas (1986) showed that the burning of room linings produced much larger flame plumes at the doors and windows of a burning compartment than wood cribs. In most cases the new type occupancy has an inherently higher fire risk than the earlier one, and should thus have additional fire-protection facilities. Yet, quite to the contrary, the remodeled fire-safety system often does not fit in with the original design, and instead of being a strengthened system, the new system is less adequate than the original one.

3 Ducts, Shafts, and Penetrations

Because of some design weaknesses in the ducts and shafts, these areas might become another major smoke escape route, especially in high-rise fires.

Vent Ducts.   Taiwan is located in the Asian subtropical zone, and in Taipei the hot summer lasts several months with average temperatures above 30°C (86°F). During the past decade the installation of window air-conditioners has become quite common in residential, office, and commercial buildings, though with the development of glass-curtain walls, central air-conditioning systems have replaced these window-type units in high-rise buildings. Despite air-conditioning regulations in the BTR, vents still remain a source for smoke spread in high-rise buildings in case of fire, as the older type central air-conditioning system does not have automatic dampers. These details are not spelled out in the BTR and the regulations should be updated.

Pipe Shafts.   In buildings of less than seven floors the pipes for water, drainage, and waste as well as the electricity conduits are usually housed in concrete columns, in beams, or under concrete floors. This so-called hidden pipe design is used to save space and money. However, it sometimes makes it difficult to check or repair faulty piping.

In buildings of over seven floors such pipes are always concentrated in shafts, but this does not mean that all such shafts are fire-resistant. There are no regulations in the BTR which directly specify shaft properties, that is, shafts are not regulated. A number of violations can be found in buildings where the walls of the shaft are constructed of plywood. These shafts constitute one of the major reasons for the heavy losses in high-rise building fires in Taipei. For instance, unprotected shafts played a major role in the Time Hotel fire on May 19, 1984, which resulted in 19 deaths and 51 injuries. Most of the deaths and injuries occurred above the second story, the floor where the fire had started. A schematic diagram of how the fire and smoke spread is shown in Fig. 5.3.

4 Fire-Prevention Features of Buildings

Fire departments are administered by the national police administration, Ministry of Interior. In fact, before 1990 all the regulations on building fire prevention within the BTR were administered by the national police administration. As of 1990, these regulations are independent and come under the fire law.

Under the fire law there are the fire safety facility standards, which specify the requirements for the various classified occupancies. However, these standards

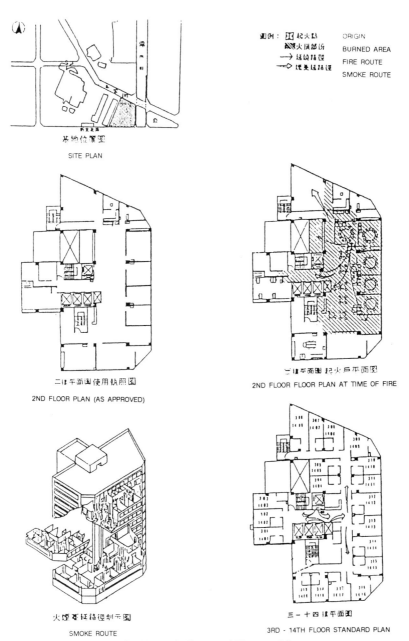

例: 起火点   ORIGIN
火烧部份   BURNED AREA
→ 逃生路径   FIRE ROUTE
⇨ 烟气逃生路径   SMOKE ROUTE

基地位置图
SITE PLAN

二层平面与使用执照图
2ND FLOOR PLAN (AS APPROVED)

二层平面图起火后平面图
2ND FLOOR FLOOR PLAN AT TIME OF FIRE

火烧蔓延情况透视图
SMOKE ROUTE

三一十四层平面图
3RD - 14TH FLOOR STANDARD PLAN

Fig. 5.3 Schematic diagram of Time Hotel fire.

do not take into account mixed-use occupancies or changes in occupancies where part of a building occupancy has changed.

Sprinkler Systems. For a building of six or less floors or, as mentioned, for the first six floors of a high-rise building, sprinkler systems need not be installed under the following conditions:

1. If the stage area of a theater or an assembly hall does not exceed 300 m^2 (32,000 ft^2)
2. If the floor area of a dance hall, night club, bar, or interior playground does not exceed 1000 m^2 (10,800 ft^2)
3. If the floor area of a residence, hotel, restaurant, supermarket, hostel, dormitory, or hospital does not exceed 1500 m^2 (16,000 ft^2)
4. If the open area of a basement does not exceed 50%, or the basement has fewer than 20 parking spaces

Sprinklers are not required in mixed occupancies of less than six floors if the occupancy in each part of a floor meets the preceding conditions. In other words, there may be two or more such occupancies on one floor, and yet it is only necessary for each one of these occupancies to meet the standards.

Exiting Facilities. In the Time Hotel fire (no. 11, Table 5.2) the fire started in a restaurant on the second floor at 9:00 A.M. The Time Hotel was a 14-story building and, as mentioned, the smoke spread to the upper stories through shafts and other openings. The fire caused heavy losses, as 19 were killed and 51 injured. Three of the casualties died by jumping from the upper floors. (The road in front of this site is very steep and the fire fighters were unable to use ladders.) In two other high-rise fires nine occupants died due to the lack of proper exits; emergency doors, windows, and openings were blocked.

Because of the problems of mixed-use occupancies in high-rise buildings and the excessively high density of occupants prevailing in Taipei, the stairs, as presently designed, are not adequate in providing safe exit. At present there are no provisions in the BTR to regulate the staircases for buildings when changes have been made in the occupancy of the structure.

Slab Openings for Escalators. Architects often design escalators for hotels, department stores, financial institutions, and similar occupancies in order to get a fancy interior open space, but these installations cause slab openings wherever they are installed. At present there are no regulations regarding detectors, sprinklers, smoke-control systems, or fire walls around escalator openings. The Asia Union Building (no. 22, Table 5.2), a 12-story commercial high rise, sustained an arson fire (at the basement staircase) at 4:30 A.M. on November 26, 1985. Both flames and smoke propagated through the escalator opening to the upper floors and caused an attendant to die on the 11th floor. A schematic diagram of the Asia Union Building fire is shown in Fig. 5.4.

Another feature of modern buildings in many countries is the atrium, which provides daylight and an interior open space. The fire safety of atria is treated differently in the various countries, and the codes and regulations in most countries are changing with respect to atria. In Taiwan no mechanism has been formulated for incorporating new concepts regarding atria into the BTR. For instance, the roofs above many atria in Taiwan are often fixed glass, which causes many fire-protection experts to fear that in case of a fire, smoke would be accumulated under such an enclosure unless proper ventilation is provided.

5 Customs

There are a number of aspects to the Chinese culture and customs that affect the use of buildings. A number of fires in Taiwan can be attributed to some of these traditional customs.

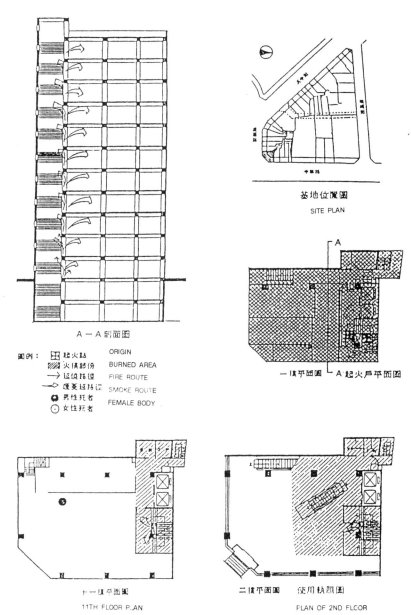

Fig. 5.4 Schematic diagram of Asia Union Building fire.

Installation of New Partitions.   Before moving into their apartments, residents are likely to install new partitions to provide more enclosed or larger spaces. Plywood, instead of the original wall material, usually masonry, is the common material used for these repartitions. This is also true for repartitions of restaurants, massage parlors, barbershops, MTV, KTV, and Karaoke, as it is often necessary to make more independent rooms for the sake of privacy.

Out of the 23 high-rise fires, 16 were within structures where some installation of new partitions had taken place. The problems involved can be divided into two parts:

1. *A decrease in room size.* Rooms repartitioned with untreated plywood are highly combustible and prone to smoke generation. There is also the effect of fire spread as discussed. The addition of new partitions is also likely to lengthen the travel distance between stairs and fire exits. This may be a special problem in quasi-public occupancies, such as MTV and KTV, where the new escape path may be tortuous and difficult to follow under smoky conditions. Another part of this problem is that the addition of new partitions may cause the width of the corridors to be more narrow than required for safe exiting. Finally in some cases, in the process of putting up new partitions, old fire exits were demolished or were no longer properly functioning with the existing structure.

2. *An increase in room size.* This usually means that the density of the new occupancy will be higher than before, and the exit stairs are no longer adequate for the evacuation of the occupants. In addition if there is a smoke-control system, it will no longer be adequate to take care of the smoke accumulated in these larger areas.

Installation of Billboards.   The display of billboards on the facades of buildings is another Chinese tradition. Today this custom translates into another serious problem for the fire-protection engineer. Such signs are usually fixed to the building by a steel truss or a timber frame. The signs are placed on a board made of sheet steel, wood, or plywood, and it often occupies most of the facade of a building. This in turn causes windows to be blocked as these billboards virtually make a facade into a blank wall. Such conditions are often encountered in the commercial districts. From the point of view of fire prevention the following problems arise:

In case of emergency evacuation of occupants through windows by the fire-service personnel, the windows are blocked and make escape impossible.

Natural ventilation is blocked, and the fire is much more difficult to fight. In western countries the addition of this kind of structure would trigger a requirement to add sprinkler protection.

The Ji-Tai Mansion (no. 6, Table 5.2), a seven-story apartment house, underwent remodeling and many new plywood partitions were erected to convert it into a massage and barbershop. During this process a huge billboard was put up, which occupied the whole exterior wall and blocked all the windows. On November 12, 1984, a fire broke out on the ground floor, which caused seven fatalities and three injuries on the second floor. A schematic diagram of the Ji-Tai building fire is shown in Fig. 5.5.

Safety Doors and Barred Windows.   The blocking of evacuation routes is another important reason for the great number of casualties in a fire. One of the major

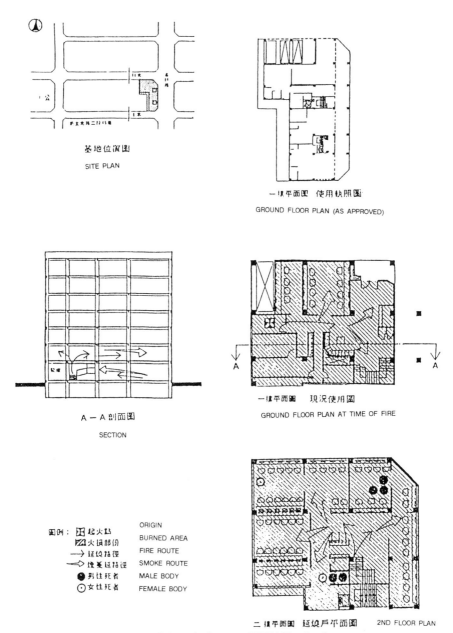

Fig. 5.5   Schematic diagram of Ji-Tai Mansion fire.

reasons for the blocking of escape routes can be attributed to the locking of emergency doors and barring of windows.

Often emergency doors are left open for the convenience of occupants, for going between floors, but they are locked at the ground floor and roof levels for security reasons. In case of a fire, occupants will use these open staircases only to find themselves trapped by the locked doors on the roof and ground floors. This occurred in the Time Hotel fire, in which four occupants lost their lives. In this fire the situation was made even worse by the fact that smoke accumulated and rose to the top floor, where the occupants found themselves trapped.

It is not unusual for residents to install steel bars outside their windows for security reasons. This barring of windows is especially popular in residential districts, and it is often the cause of tragedies, especially for children and disabled people. This problem is more serious in apartment houses than in high-rise buildings, but as more high-rise apartment buildings are constructed, the practice may spread to these buildings. Part of the reason for the need for window bars is the presence of squatters on the roof areas of buildings.

5.3 CONCLUSIONS

1 Reviewing Overall Urban Development

Many of the problems found in the building fires described in this chapter are caused by population pressures and cultural factors which are beyond the scope of fire-safety regulations. Before fire safety can be designed and regulated in existing and new construction it will be necessary to review and change various aspects of urban laws which govern land use and zoning practices in Taiwan. First of all, based on the environmental impact factors and a balanced application of land-use concepts, the urban problems need to be analyzed, and new regulations should be enacted which reflect a sensitivity to the overall challenges. It is very important for urban laws to be revised in order to approach mixed-use occupancies with a new perspective. For instance, the land-use categories could henceforth consider the density of occupants as well as the human behavior of a mixed-use occupancy in high-rise buildings. Since the density and risk of occupancy vary considerably within high-rise buildings, they should be regulated to meet high standards of safety. Land use should be categorized according to its usage, and then should be categorized and graded again according to its inherent dangers. This would eliminate different categories representing the different levels of risk that exist within one building. These regulations should address the building height, the occupancy of each area within a building, and all related safety devices in order to ensure the life safety of all occupants. There would be a direct linkage between the urban laws, the BTR, and the fire law.

2 Improve the Practice of Fire-Safety Engineering

There is little or no recognition in Taiwan for the professional practice of fire-safety engineering. The structural engineering and geological engineering professions in Taiwan are highly developed because of the dangers of earthquakes. If approaches similar to those taken for earthquake safety were instituted to fire safety, the fire-safety picture would improve in many ways. More expert resources would be available for architects and other fields of engineering to utilize before they approached the authority having jurisdiction. Fire-resistant materials and fire-safety systems would also become more available, since it is usually the fire-safety engineer who specifies such items. Furthermore, there would be parallel improvements in industrial fire safety, another area of concern.

3 Building Regulations

Because of the problems of mixed-use occupancies in high-rise buildings and the excessively high density of occupants which prevails in Taipei, the stairs, as presently designed, are not adequate in providing safe exiting. The BTR should regulate the staircases for buildings where modifications in the building occupancies have been made, and it should ensure that there are sufficient staircases for safe exiting. Furthermore, the BTR should also be very specific about regulating the design of staircases in new high-rise buildings.

A method to revise the regulations governing fire safety should be developed so that the regulations reflect:

1. Changed social circumstances
2. Experience from past fires
3. Reliable statistics, records, and fire analyses
4. Ongoing fire research results

5.4 CONDENSED REFERENCES/BIBLIOGRAPHY

Lathrop 1986, *Life Safety Code Handbook*
Ministry of Interior, ROC 1989a, *Construction and Planning Administration*
Ministry of Interior, ROC 1989b, *Statistic Abstract of Interior of the Republic of China,*
Ministry of Interior, ROC 1989c, *Statistical Data Book of the Ministry of the Interior—1989*
Thomas 1986, *The Role of Flammable Linings in Fire Spread*

Part 2 – Occupant Behavior and Performance in Tall-Building Fire Incidents

In all cases, a primary objective in designing a structure is its safe occupancy by the people who use it and consideration of its support of their performance under emergency conditions. Part 2 presents generalized observations on human behavior in fires, as well as contemporary mathematical treatments which describe occupant movements under fire conditions. Disabled personnel and their potential for safe escape in fire situations is also considered. (See also Council on Tall Buildings, Committee 56, 1992.)

6

Understanding
Human Behavior
during Fire Evacuation

Over the last decade or so there has been a proliferation of research on the topic of building fires. The areas of major concern have been the chemistry of combustion, the characteristics of materials under direct heat, flame and smoke behavior and their control, and the development of automatic fire-detection systems. Literature pertaining to human behavior in fire, however, is unfortunately rather stark in comparison. Nevertheless, the insights that have been gained about behavior during fires over the entire range of occupancy types have important implications for design, legislation, and personnel training.

The first problem encountered by fire-safety personnel, when considering the large-scale evacuation of a building, is how best to alert and inform occupants of the impending danger in a way that will facilitate rapid and safe egress. Evacuation of tall buildings poses additional problems, not only because the size of the building's population may be vast and their means of egress limited to horizontal evacuation, but also because the nature of occupancy can be quite varied, even within the same building. Therefore an understanding of how particular individuals are likely to behave under the threat of fire may be helpful in designing fire-safety procedures.

6.1 SOURCES OF KNOWLEDGE ON HUMAN BEHAVIOR IN FIRES

As discussed by Canter (1985), the current basis of knowledge about human behavior in fires is derived from a variety of sources. In the United Kingdom these include special commissions of enquiry, press reports, physical science studies, and scientific psychological research. The efficacy of each source as a valid and objective account, which may lead to a greater understanding of the "human factor," is briefly reviewed in this section.

1 Special Commissions of Enquiry

Commissions of enquiry are *reactive* investigations into major disasters and are based on judicial procedures, with the objective of apportioning responsibility and proposing remedial measures in an attempt to reduce the probability of a similar disaster recurring. These characteristics, then, tend to result in an emphasis on technical aspects of the fire event, close scrutiny of building structures and materials, and the provision of fire-fighting "hardware." More importantly, from the behavioral research point of view, enquiries within a rationale of identifying cause and effect are an inadequate means of describing human actions leading up to, and including, the early stages of awareness and recognition of the fire cues, and the behaviors which may or may not lead to safe egress. Moreover, the minor behavioral events which *are* recorded, and which may give way to an issue of cumulative significance, are often buried or only covertly referred to among a plethora of technical detail.

2 The Physical Science Approach

Technical studies with the prime objective of exploring the physical aspects of fire and the behavior of smoke and flames have resulted in changes which have greatly reduced the danger of fatalities. Physical scientists have also provided, indirectly, complementary information on human behavior in fires. Unfortunately, however, given the absence of scientific psychological research during the first 60 years of this century, architects and engineers have had the responsibility not only of producing buildings which were aesthetically pleasing, functional, and structurally sound, but also of deciding how best to ensure the safety of their buildings.

In general, the building design codes and regulations which resulted from early recommendations are based on the physical science model of pedestrian movement. This rationale is characterized by formulas which predict the flow capacity of corridors, staircases, and exit doors.

The criticism of the physical science approach is that it envisages human movement as analogous to the movement of "granular mass" (Peschl, 1971). Specifically, Pauls (1980a), a Canadian architect, has argued that building regulations are based on the implicit assumption that particular aspects of a building's structure (for example, exit width and maximum travel distance) serve to modify human behavior in the way predicted by "common sense and the first principles of engineering."

To take the example of exit width, which is based on the principle of unit width (in this instance, shoulder width), it is assumed that 45 people can pass through one unit width [559 mm (22 in.)] per minute and that the provision for two people discharging is simply a question of doubling the provision for one. In his observation of evacuation from tall office buildings, however, Pauls discovered that people do not move down staircases shoulder to shoulder as previously assumed. Rather, the evacuating population tends to leave a space of 150 mm (6 in.) at each side of the staircase (a body boundary zone). Data led him to conclude that the conventionally accepted flow of 45 persons through a conventional stairway width appears to be overly optimistic by 50 to 100%.

Evidence also points to the fact that exit routes which carry high normal use loads are used most frequently for emergency egress (Pauls, 1980a), resulting in a 14% increase in the mean evacuation flow and a 9% increase in flow when the exit leads to the main reception area, which is often the center of communications.

This is a fine example of how behavioral research can combine with physical science data to enhance the design and management aspects of occupant safety during emergency egress. Another approach worthy of mention is a dynamic stochastic computer simulation of emergency egress behavior during fires, called BFIRES (Stahl, 1982).

According to Stahl, the BFIRES program can simulate a wide variety of emergency scenarios. BFIRES is based on an "information-processing" explanation of human behavior (Shannon and Weaver, 1949). The basic unit of occupant behavior generated is described as the "individual momentary response" to the state of the environment at a discrete point in time. Thus the sequential presentation of a chain of discrete "time frames" can be seen to generate an "animated film" of the simultaneous egress performance of all occupants in response to a migrating fire threat.

One advantage gained from the consideration of the BFIRES program is that design programmers are encouraged to think more deeply about the consequences of their designs. But such skeletal modeling of person-behavior-environmental interactions also has its limitations, that is, computer simulations must at worst be based on assumptions about the behavioral and cognitive implications of changes in the environment, and at best they must follow empirically proven, theory-driven behavioral research guidelines. With regard to the former case, we might find ourselves using a model based on myths about human behavior in emergencies, which have subsequently been discredited by empirical psychological research [as in the case of Wolpert and Zillman's simulation (1969), which was programmed to cope with panic behavior].

Of course, in the design the architect can extend a considerable influence over the person's choice of escape route. But whether the exit route taken is that which would be predicted is another matter. Psychologists who have studied the role of the environment admit that while the physical environment is an important influence on cognition and behavior, they reject outright any notion of absolute "environmental determinism." Lee (1971) has stressed that individuals actively seek to make sense of and impose meaning upon the environment in ways which deny a control by the environment.

Evidence that the mere provision of fire escapes is not enough to guarantee their use originated from research carried out at Surrey University Fire Research Unit (SUFRU). Canter et al. (1980) found that of the 85 victims of four fires, only 11 attempted to use the fire escape, and of these only five were successful. (The occupancies included hotel, hospital, public bar, and high-rise block.)

3 Press Reports

Journalists and editorial personnel, entrusted as they are with the reporting of issues of public interest, find themselves in a position to influence public opinion and decision making. However, aside from the considerable commercial constraints on news coverage, the reporting of newsworthy items is itself influenced by journalistic style and editorial policy. Such factors have the effect of encouraging the presentation of information (often sparse in detail and accuracy) in the form of a readable story rather than an exacting account of what actually happened. Indeed the most cursory reference is made to minor and domestic fires. The tendency to refer to human behavior in terms of acts of "heroism" or "panic" further suggests that as informants on human behavior under threat, the popular press is sadly misinformed.

Thus far it has been argued that none of the traditional sources of knowledge referred to can accurately describe what people actually do when first confronted with (ambiguous) cues of fire and are faced with a variety of options in the decision to evacuate. In contrast, and despite popular myths, a scientific psychological approach provides more insight into the problem of human behavior in fire.

6.2 THE SCIENTIFIC PSYCHOLOGICAL APPROACH

One of the difficulties frequently encountered when discussing human responses to emergency evacuation is overcoming the widely held belief that "panic" is a sufficiently accurate description of what people do during a fire emergency. Indeed as Sime (1983) points out, even legislative bodies labor under this popular misconception. Panic is assumed to be the automatic response to changes in the immediate social or physical environment such as smell of smoke, smoke itself, absence of light, a cry of alarm or "fire," or a sudden fire alarm, particularly when people are asleep or in a crowded area of a building (Ministry of Works, 1952).

The underlying assumption is that, as a behavioral response, panic is the inevitable, emotional, and uncontrollable response to a fire emergency. But there is little evidence to suggest that human behavior is indeed unpredictable or governed by such stimulus-response, or animalistic, rules.

Careful analyses of witness statements following major fire disasters characterize the human response to be largely altruistic (Bryan, 1981) and affiliative (Sime, 1983). Panic appears to be an exception to the rule, occurring only when the occupants perceive their route of escape to be closing rapidly and their time to search for an alternative means of escape to be inadequate. Too often there is a tendency to describe as panic any response which, in hindsight, and given the full details of more appropriate courses of action, would seem irrational. Thus while escape behavior may in some instances be neither efficient nor effective, it may be rational given the amount of information available to the individual at that particular time. It is essential, therefore, to distinguish between the options which actually are available and those which the individual may *perceive* to be available.

A description of a number of studies follows, each of which has broadened our knowledge of evacuation in an important way, and is helpful in understanding the human reaction toward fire.

1 Questionnaire Study

Wood (1972) carried out the first exploration of behavior in fires with research based on questionnaire interviews of fire victims. His results illustrate that motivation to evacuate is related to the following variables.

Gender. Women are more likely to evacuate immediately than men, who initially tend to fight the fire. This implies that decisions relate to social behavior and gender roles.

Knowledge of an Escape Route. If people are aware that an escape route exists, then they are less likely to leave because they feel less threatened by the fire.

Thus the motivation to escape is only dominant when other objectives, such as extinguishing the fire, are perceived as unattainable.

Intensity and Spread of Smoke. The presence and density of smoke is directly related to the level of perceived threat, so that smoke encourages people to leave. The perceptional relationship cited is believed to outweigh the physiological and spatial disorientation difficulties.

Previous Experience of Fire. People are less likely to leave if they have experienced a fire previously. It would appear that people who have learned that they can cope with a fire threat believe that they can pursue objectives other than evacuation.

Training. The more training an individual has received, the more the person is likely to attempt to control the threat and thus less likely to leave. Again, the learning factor is important, but it is likely that fire training in occupancies such as hospitals will give individuals a set of organizational responsibilities to which they are responding, independently of the intensity of the threat.

Direct Threat Perception. If a fire is judged to be extremely serious, then those facing the threat are most likely to leave.

These findings are useful in a field where previously very little was known. Wood has found that behavior in fires is influenced by social roles and that different groups of people display distinctive patterns of response.

Despite these contributions, Wood's work has a number of shortcomings, which leave many patterns of behavior unexplored. For example, the manner in which the factors listed above operate in combination, or the relative importance of each variable in the behavioral response strategy, are issues that have not been explored. Wood also failed to produce an account of how responses developed with respect to the changing fire conditions. Finally, Wood took as the starting point of his investigation the moment when a person knew that a fire existed. Subsequent research has shown that this overlooks one of the critical periods of response, which is before the fire's existence is known.

2 Affiliative Model of Escape and Exit Choice Behavior

Sime's study (1983) using witness statements from victims of the Summerland Leisure Complex fire (Isle of Man, U.K., 1973) provided strong evidence of affiliative behavior during evacuation. The affiliative model summarizes the strong tendency exhibited by building occupants to move toward familiar people, such as family and friends, and familiar places, such as their usual entrance route. He noted that affiliative behavior had other consequences in that separated individuals responded quickly to ambiguous cues, whereas intact family groups did not begin to evacuate until there was a clear sign of the fire threat. A possible explanation may be that the social pressures of conformity operating within a group make the individual reluctant to evacuate, thinking that they may appear foolish should the cues represent a "false" alarm. There may also be a feeling of security gained from being part of a group.

The behavior of victims in the Summerland fire can be contrasted with the findings of an earlier study of the Kentucky Supper Club fire (Best, 1977). In the Summerland tragedy, staff almost exclusively left by the fire exits, which they used on a daily basis. A considerable proportion of customers, however, who were not familiar with the fire exits, left by the main entrance, often traveling considerable distances and passing fire exits to do so. There was found to be a distant, uninvolved relationship between staff and customers which resulted in different evacuation behavior between the two groups.

The relationship which existed between patrons and waitresses in the dining room of the Kentucky Supper Club was, however, remarkably close. When fire broke out, waitresses guided to the exits exactly those patrons for whom they had had responsibility prior to the fire. Thus equivalent role groups will not always display identical patterns of response. Rather it is the nature of the relationship between two different role groups that sometimes shapes behavior.

3   General Model of Human Behavior in Fires

Canter et al. (1980) conducted open-ended interviews with fire victims which allowed them to follow a response strategy from the point at which an initial alerting cue was perceived to the time when the sequence of behavior was concluded, by rescue or evacuation, for example. An important theme which emerged from the data was the attempt to "make sense" of what was actually happening throughout the various stages of the fire, that is, fire victims were engaged in behavior which sought to reduce the prevailing state of uncertainty arising from ambiguous perceptual cues (such as strange noises and unaccustomed behavior of others).

A second major theme has arisen from the work of Canter et al., which has argued that behavior in fire related to a "role-rule" model. This model postulates that people's conduct is guided by a set of expectations they have formed about their purpose in a particular context. The general framework formed by these expectations is known as their role. The activities in which they engage to fulfill their role are influenced by guiding principles, or rules. Canter et al. argue that when faced with a fire threat, an individual's behavior continued to be guided by the role-rule influences, which had been operating prior to the emergence of the threat.

The main themes can best be understood in the context of the general behavioral model produced by Canter et al. and illustrated in Fig. 6.1. As the general

Stages

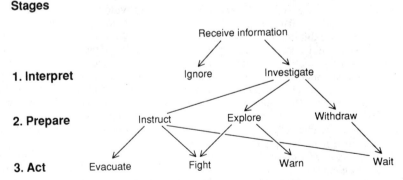

Fig. 6.1   General model of human behavior in fires.

model shows, upon receiving the initial fire cues, the individual may use this information to assess the situation and to further determine the decision to evacuate. Having investigated the initial cues, the individual may be confronted with an actual fire and behave in the ways outlined in Fig. 6.1. The two major decisions to be made, therefore, are first, should I investigate the reason for the fire cues, and second, if I am confronted with an actual fire, what should I do?

The following two behavioral studies highlight the dilemma facing an individual in these situations.

4 Fire Alarms and Fire Growth

Canter (1985) conducted a study which evaluated the effectiveness of conventional alarm systems in motivating egress and found them to be associated with the following three failures:

1. A failure of people to differentiate fire alarms from other types of alarms

2. A failure of people to regard fire alarms as authentic warning of a genuine fire

3. A failure of alarms to present information which will assist fire victims in their attempt to cope with the fire

Clearly, then, conventional alarm systems do little to reduce situational ambiguities and their credibility erodes as a result.

Another psychological factor which hinders the speed of egress from a fire situation is the perceived severity of the fire and the rate with which it is likely to spread. Such judgments may be very difficult to make for nonexperts who rarely encounter fire situations. People thus have difficulty in defining the severity of a fire situation, particularly in its early stages. The initial delay in response to the Bradford City Football Ground fire (1985) clearly illustrated this point.

A small-scale experimental study, which used photographs from the reconstruction of the Stardust Club tragedy (which occurred in Dublin in 1981), was conducted to clarify this issues (Canter et al., 1990). This study assessed the accuracy of the individual's judgments between the stages of the actual fire.

The results showed that, in general, people were rather unclear about the behavior of a fire until approximately 1 minute after it had started. People were far better at judging the rates of simple changes in fire growth such as flame height, but were less able to predict and estimate changes involving lateral spread of flames or smoke production. Further research using a similar methodology would help to clarify this issue.

The standard of fire-safety knowledge among the staff of hospital and commercial premises is, however, very high. The major findings of a fire-safety questionnaire indicate that although most questions were answered correctly, around 40% of respondents did not know what constitutes the essential ingredients of fire or how to fight a small fire in a safe or effective way. Thus although the standard of awareness may be high at a theoretical level, it is difficult to ascertain the extent to which knowledge translates into efficiency of action during a real fire.

Tasks such as the assessment of fire growth rate, as used in the experimental study, might, with more research, be developed into useful teaching devices, which serve to bridge the gap between theoretical and practical training.

6.3 ROLE OF INFORMATIVE FIRE WARNING SYSTEMS IN EFFECTIVE EVACUATION

Traditional fire-alarm systems are unable to provide information beyond the initial stage of warning people that there is (or may be) a fire somewhere in the building. Informative fire-warning (IFW) systems, on the other hand, have the potential to provide building occupants with a variety of different messages, such as enabling them to determine the location, severity, and spread of the fire; whether they need to evacuate or not; and if so, which exit route they should use. The possible influence of IFW systems at various stages of a simple evacuation model is shown in Fig. 6.2.

A number of fire drills have been organized and conducted in various types of buildings to clarify and expand the evacuation model presented, particularly with respect to the input of IFW systems to the egress process. Health care and commercial establishments participated in these studies (Canter et al., 1990).

Drills were monitored and recorded using timed video techniques. Analysis of the video recordings showed that there was a remarkable consistency in the stages of activity occurring after the sounding of the IFW fire alarm. The occupancy nature of the organization itself was the major determinant of which stages occurred and the length of time that these stages took to accomplish. The general four-stage model presented in Fig. 6.2 may thus be elaborated on the basis of the drill investigation (see Fig. 6.3).

It was interesting to find that in all occupancies, the staff "congregated" in either large or small groups following the sounding of the fire alarm, and discussed what they should do. This activity was not written into the official evacuation procedures for any of the occupancies, and would appear to be some kind of social-confirming response. The function of this behavior may be for individuals to seek acknowledgment that what they do is supported by others. If further

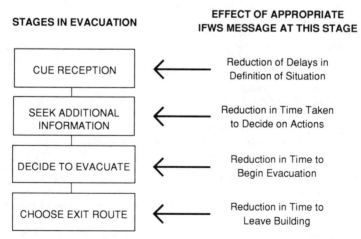

Fig. 6.2 Outline of potential contributions of IFW systems in effective evacuation.

research confirms this finding, it may well have consequences for predicting the total evacuation times for particular buildings.

1 IFW Message Display Characteristics

Although commercially available IFW systems are limited to a 16 to 28 character display, it is envisaged that in the near future the technical and economic constraints (which currently restrict the introduction of more sophisticated displays) will be overcome. With this in mind, a study was performed to assess the individual's ability to recall and comprehend IFW messages as displayed by means of photographic slide simulations (Canter et al., 1990). A number of factors which limit accurate comprehension of visually displayed information were discovered. The results of this study follow in the form of recommendations.

Abbreviations.   If at all possible, the use of abbreviations should be avoided. The inclusion of abbreviations in the midst of unabbreviated text leads to confusion. It was demonstrated that accurate comprehension of various abbreviations fell between the 34 and 50% level of accuracy for messages with more than five message units.

Message Format.   The strategies which appear to facilitate the most accurate recall are those that impose internal consistency upon the message format. The strategy used in one establishment that was studied successfully aided accurate recall by shortening the message length while at the same time preserving precise locational information. This "general/specific" convention dictated that IFW displays on the *affected* floor would present only the precise location of the fire (such as the linen store), while displays on nonaffected floors would present a moderately specific locational address (such as the floor and ward number). Recall of these locational messages exceeded 90%, while recall of messages giving updated information of fire or smoke spread exceed 80%.

Alternatively, the predictable format used in simulated IFW messages for a multistory office tower aided recall by reducing locational information to three units (floor number, the side of the building as dictated by the street onto which

*An asterisk denotes those acts not carried out by office staff.

Fig. 6.3   Elaboration of evacuation model on basis of drills.* Acts not carried out by office staff.

it faced, and the designated use of building space). Recall of these unabbreviated messages in no instance fell below 90%.

If information is to be displayed on a video display screen, the use of grouping text to emphasize specific information is preferable to organizing text in a long string. In accordance with standard recommendations for video display unit text format, both upper- and lowercase characters should be used.

Message Formulation.   Consistency of message format throughout the listings is essential. Location addresses should first refer to the most general area which can be physically demarcated (such as block or wing), a slightly more specific address (like floor number) can follow, and the precise addressable compartment (ward or room) should be given last. However, if the inclusion of the precise addressable area requires that the text be abbreviated, or that the text be longer than has been recommended, it would be wiser to exclude this address in order to ensure a greater degree of accurate comprehension.

Message Length.   Messages containing abbreviations should be no longer than four, and preferably three, units if a 75% level of accuracy is required. Unabbreviated messages should not exceed six units in length if a 75% level of accuracy is required.

Message Specificity.   Individuals responsible for message formulation should resist the temptation to be as specific as possible about the exact location of a fire if this leads to a conflict with the recommendations pertaining to message length. One way to utilize the precision afforded by IFW technology while fulfilling these recommendations is to adopt the "general/specific" convention.

Message Units.   Message units were derived by dividing each message up into its obvious constituents. Given that the messages were all simple directions, there was little difficulty in deciding what the components of the message were. A particular corridor, for example, would be one unit. An end of the corridor would be a separate unit. So a message such as *west entrance* would be regarded as two units.

Naming Conventions.   It is vital that consensus be reached as to the names given to particular locations within a building. Wherever possible, the location names to be used in the IFW address listings should be displayed at the entrance to that area.

Type of Information.   Location addresses should take priority over the inclusion of any other type of information when planning the use of available display facilities. Once the requirements for accurate comprehension of location addresses have been met, the provision of updated information on fire and smoke spread should be displayed along with, but separate from, the address of the fire's source. Instructions should only be targeted at the general population of building occupants or staff; instructions to individuals can be communicated by alternative means.

2   Attitudes Toward IFW Systems and Fire Safety

The study by Canter et al. (1990) concentrated on the individual's ability to make use of an IFW display. However, acceptance of such a system would depend in large part on building occupants holding favorable attitudes toward IFW systems. To assess the degree of acceptance of IFW systems, the individuals who participated in the simulation study were asked to state the extent to which they found the various types of information helpful.

The general response was very favorable. However, while approximately 80% of hospital staff either agreed or strongly agreed that such information was helpful, fewer management staff in commercial office buildings stated likewise (67%).

This less favorable attitude of people working in commercial premises may mean that it will be more difficult to train people to accept the utility of IFW systems and to use them appropriately. In hospitals and similar establishments, where staff typically assume responsibility for the safety and well-being of patients (in other words a more active role in fire safety), it should be easier to train people to respond appropriately.

The attitudes which people have toward various issues related to fire-safety procedures and practices are important since they may influence reactions to different types of alarm systems. These attitudes also provide information relevant to the implementation of IFW systems and training procedures associated with their use. Careful attention must be given to ensure that people are sufficiently informed of the advantages of IFW systems over traditional systems, and there must be a strong commitment by management and fire-safety personnel to ensure that any doubts are erased.

The outcome of a survey which assessed attitudes to fire safety of 597 people who worked in 11 different buildings showed that people tend to conceptualize fire safety in terms of four stages:

1. Fire prevention and general fire safety
2. Alarm systems
3. Fire fighting
4. Evacuation

Given that each stage falls into a rather distinct conceptual category, it would thus seem appropriate to develop training schemes which take these stages into account. If the information were presented in this way, it would relate more closely to people's existing framework about fire safety and would be more likely to be assimilated.

One factor which seemed to affect both knowledge of and attitude toward fire safety was that of occupancy type. In this respect, hospital staff displayed a greater degree of awareness and a more positive attitude toward fire safety than did commercial staff. Since the standard of training in all occupancies was equally high, the differing perspectives may be attributed to the influence of role-related responsibilities.

6.4 AREAS FOR FUTURE RESEARCH

A number of possibilities for future research are suggested by the studies presented. The following is a brief summation of pertinent topics.

1 Speech-Generated Information

The potential confusions in short and abbreviated displays do mean that other forms of display or speech-generated information may be useful in certain situations. A series of studies exploring these possibilities could be of great value.

2 Fire Growth

The identification of specific confusions in the comprehension of fire growth does carry implications for training and the development of automatic sensing devices. The study discussed earlier provides a cost-effective mechanism for a variety of such studies.

3 Fire Preparedness

The attitude and knowledge surveys supported the possibility of developing a standard questionnaire procedure for assessing the general status of any organization with regard to human aspects of fire safety. Such a procedure could have widespread didactic as well as prescriptive value.

4 Case Studies of IFW Systems "In Situ"

Although to date only one extant IFW installation has been examined, it proved most informative. As more IFW systems are installed, further monitoring of them would be worthwhile. In an important sense, all IFW systems installed at present are experimental. The full scientific benefit of these "experiments" is not, as yet, being obtained.

6.5 CONDENSED REFERENCES/BIBLIOGRAPHY

Best 1977, *Reconstruction of a Tragedy: The Beverly Hills Supper Club Fire*
Bryan 1981, *An Examination and Analysis of the Dynamics of the Human Behavior in the*
Canter 1980, *Domestic, Multiple-Occupancy, and Hospital Fires*
Canter 1985, *Studies of Human Behavior in Fire: Empirical Results and Their Implications*
Canter 1990, *The Psychology of Informative Fire Warning Systems*
Lee 1971, *Psychology and Architectural Determinism*
Ministry of Works 1952, *Fire Gradings of Buildings, Means of Escape, Part 3: Personal*
Pauls 1980a, *Building Evacuation: Research and Recommendations*
Peschl 1971, *Flow Capacity of Door Opening in Panic Situation*
Pigott 1979, *Outline Specification for a System of Automatic Fire Detection Offering Re-*
Shannon 1949, *The Mathematical Theory of Communication*
Sime 1983, *Affiliative Behaviour during Escape to Building Exits*
Stahl 1982a, *BFIRES II—A Behavior-Based Computer Simulation of Emergency Egress*
Wolpert 1969, *The Sequential Expansion of a Decision Model in Spatial Context*
Wood 1972, *The Behaviour of People in Fires*

7

Evacuation of the Disabled: A United Kingdom Perspective

An important issue related to evacuation and safe refuge considerations for tall-building design is the treatment of disabled occupants. Because access to buildings by disabled people is being made easier, quite properly, it follows that special provisions now must be made to address their situation when fire occurs (Council on Tall Buildings, Committee 56, 1992).

Consideration of the needs of such building occupants means that two long-standing principles for egress in times of fire threat have to be modified: it is no longer tenable to say that people should be able to get out of a building by their own unaided efforts, or that elevators should not be used for evacuation purposes. The reasons for these changes are discussed here in the context of recent regulations and codes adopted in the United Kingdom. These are probably the most elaborate set of national specifications with provisions for the evacuation of disabled people.

The proposition that all buildings should be accessible to disabled people is now well founded. As a result, recent history in the development of regulations and supporting codes dealing with *access* for the disabled must be considered. In 1967, the British Standards Institution (BSI) published the Code of Practice (CP) 96 entitled *Access for the Disabled to Buildings*. In 1970 the Chronically Sick and Disabled Persons Act was enacted, and it was amended in 1976. Section 4 of the act stipulated that buildings to which the public, or some sections of the public, have access should, whenever practicable and reasonable, be accessible. In 1979, CP 96 was revised as a British standard, and became BS 5810. A further act, the Disabled Persons Act of 1981, had a requirement in Section 6 that access in public buildings should comply with the recommendations of BS 5810. The British government later decided that better progress in the implementation of access requirements would be made by using building regulations as the control instrument. (In England and Wales, building regulations cover fire precautions and means of escape in new buildings and alterations and extensions to existing buildings, whereas other legislation applies once the building is in use. For example, in England and Wales the Fire Precautions Act, 1971 as amended by the Health and Safety at Work, etc., Act, 1974 and the Fire Safety at Places of Sport Act, 1987, applies to certain occupancies of buildings once in use.) Thus, beginning in 1984, access regulations went into force throughout the United Kingdom.

On the other hand, regulations for means of escape for the disabled are not yet enacted for new buildings or alterations and extensions to existing buildings. This is because the relevant code of practice (BS 5588 [1988], hereafter referred to as Part 8) was developed and published after the above-mentioned building regulations.

There are, however, clauses contained in the codes dealing with means of escape for people with no disability, such as BS 5588 Part 2 for shops and BS 5588 Part 3 for offices, which ask for ad hoc escape arrangements for disabled people. Since these two codes are given mandatory status under the building regulations 1985, via the Mandatory Rules for Means of Escape, it follows that there is only a legal requirement to provide means of escape for the disabled in offices and shops whether or not they are high-rise buildings. However, it is expected that the new code (Part 8) will govern the design and management of egress for the disabled in new buildings.

For existing buildings, the Fire Precautions Act of 1971 could be interpreted by fire authorities as requiring a means of escape in buildings to which the disabled have access, and that these should be designed and managed in accordance with Part 8. All new buildings become existing buildings once they are occupied, and thus it could be argued that any temporary inadequacies in building regulations are not important if the Fire Precautions Act of 1971 is interpreted in this way (assuming that the use of the building in question is one that has been designated under the Fire Precautions Act).

7.1 SCOPE OF PART 8

Part 8 gives guidance to designers and the building construction team on measures that should be included in new buildings, and in alterations to existing buildings, to facilitate safe evacuation of disabled people if fire occurs. Part 8 does not cover single- or multifamily dwellings. High-rise blocks of apartments, for instance, should be divided and compartmented so that only those people in the fire compartment need to leave if fire occurs, while their neighbors may remain in their apartments until the fire brigade arrives and decides what, if any, evacuation is needed.

The detailed recommendations of Part 8 are not intended to apply to existing buildings, although such a requirement might be useful. Despite this qualification, Part 8 includes an appendix which gives guidance and suggests where the recommendations in the main text of the code can be slightly relaxed, possibly because of the presence of compensating features. Other parts of the code provide recommendations for people who are not disabled, while buildings which are specifically built for disabled people will usually be provided with means of escape that are more extensive than those recommended in Part 8.

7.2 PRINCIPLES OF PART 8

Some disabled people, for instance wheelchair-bound people, will find it difficult to use an escape stairway without assistance from another person. It is therefore necessary to provide them with a refuge where they can wait for a short time before receiving assistance to get to a place of safety, usually the exterior of the building. The principles of evacuation are shown in Fig. 7.1.

The refuge has to be separated from a fire by fire-resistive construction having at least 30-min fire resistance. Refuge doors should also limit the passage of smoke, for instance, by using flexible edge seals. The refuge should be capable of accommodating at least one wheelchair-bound person and should be placed next to an escape stairway of an evacuation elevator. The refuge can be one of the following:

A room with two separate doors separating it from the fire (Fig. 7.2*a*). [Note that (1) persons in the left-hand compartment would not reach a refuge until they had entered the right-hand compartment; and (2) two door sets in the partition are required in case access to one of the door sets is blocked by fire.]

A landing of a protected stairway large enough to accommodate a parked wheelchair without obstructing the stairway (Fig. 7.2*b*).

A protected lobby to a protected stairway having landings too small to accommodate a wheelchair without causing obstruction (Fig. 7.2*c*).

A protected lobby to a protected evacuation elevator (Fig. 7.3) in which the protected lobby is capable of accommodating at least one wheelchair.

In these examples it is clear that the wheelchair user can safely get to a refuge and, once there, can safely wait for assistance without obstructing other people making their escape. In many buildings, suitable refuges will automatically be provided as a result of meeting the general requirements for means of escape. Where this does not occur, provisions for them should be possible without radically altering the internal layout and without substantial extra costs.

A protected lobby of an evacuation elevator or protected stairway should be provided at every story served by the elevator or stair. At the final exit level, the stairway or elevator should be provided with a protected route to the building's exterior. This ensures continuous protection from the time of entering a refuge to the time of arriving at a place of safety outside the building.

A fire-fighting elevator (as specified in BS 5588 [1988], Part 5) can be used as an evacuation elevator because it is always associated with a protected lobby and a protected stairway. Clause 10 of Part 8 states that:

> . . . a fire fighting elevator may be used for the evacuation of disabled persons prior to the arrival of the fire service, which will then assume responsibility for the evacuation of any remaining persons. Liaison with the fire authority to coordinate procedures for the use of such a fire fighting elevator for evacuation of disabled persons in case of fire is essential.

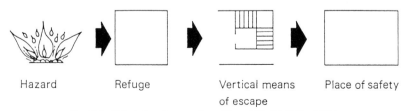

| Hazard | Refuge | Vertical means of escape | Place of safety |

Fig. 7.1 Principles of evacuation. (*From BSI, BS 5588, 1988.*)

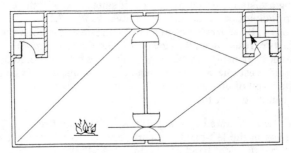

(a) Open plan storey divided into two refuges;
    (stairways not provided with wheelchair space)

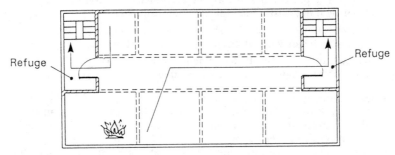

Refuge

Refuge

(b) Protected stairways used as refuges

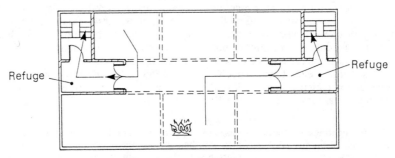

Refuge

Refuge

(c) Protected lobbies used as refuges

Key:

══════ 30 mm (minimum) fire-resisting separation

Fire door having 30 minute fire resistance
and ambient temperature smoke resistance
Note: The doorset may have one or two leaves and,
      dependent on its location, may be single or
      double action (swing)

------- Partitioning for cellular planning

Fig. 7.2   Examples of refuges in buildings not provided with evacuation elevators. (*From BSI, BS 5588, 1988.*)

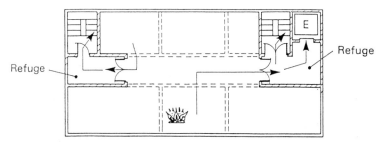

(a) Evacuation lift adjacent to a protected stairway
    protected lobbies used as refuges

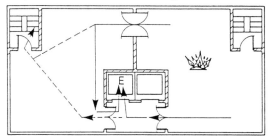

(b) Evacuated lift separated from stairways; storey
    divided into two refuges

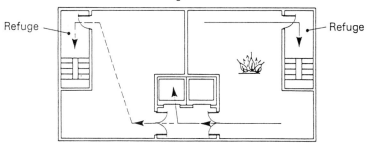

(c) Evacuation lift separated from stairways; protected
    stairways used as refuges

Key:

≡≡≡≡≡ 30 mm (minimum) fire-resisting separation

⌐╙ Fire door having 30 minute fire resistance
    and ambient temperature smoke resistance
    Note: The doorset may have one or two leaves and,
          dependent on its location, may be single or
          double action (swing)

≡≡≡≡≡≡ Partitioning for cellular planning

E   Evacuation lift having doors with 30 minute
       fire resistance

**Fig. 7.3   Examples of refuges in buildings not provided with evacuation elevators. (*From BSI BS
5588, 1988.*)**

It is important to recognize that there is a difference between the use of a fire-fighting elevator for its designed purpose and its use for evacuation purposes. This needs to be fully appreciated by both the management of the building and the fire service.

An acceptable alternative to a fire-fighting elevator for evacuating disabled people is an ordinary passenger elevator which has the following features:

A *duplicated power supply,* comprised of a primary electrical supply obtained from a submain circuit exclusive to the elevator, and an alternative power supply such as an automatically started generator. Preferably, the cable from the alternative supply should be separated from that of the primary supply so that a breakdown on one cable (perhaps caused by fire attack) does not lead to simultaneous failure of the other.

A *switch labeled "evacuation elevator"* situated next to the landing door at the final exit story, which enables an authorized person to control the elevator car. Operation of the switch should isolate the elevator landing call controls (so that others cannot call the elevator) and return the elevator immediately to the final exit story whereupon the elevator can only operate in response to the elevator car control panel. Such a switch is not needed in two-story buildings.

A *communications system,* provided (except in two-story buildings) to relay to the elevator operator rapid identification of those stories with disabled people requiring evacuation. Part 8 gives details of the various ways this can be achieved.

A *fire-resistive shaft* having the same structural protection against fire as a protected exit stairway. (A protected stairway is enclosed in construction, having a fire resistance that is prescribed under building regulations, or recommended in the Approved Document B2/3/4 in England and Wales. This resistance is greater than the half-hour minimum level usually adopted for construction bordering an escape route.)

Part 8 does not recommend that evacuation elevators should be provided in all buildings. Such an elevator simply reduces the need to provide physical assistance in stairwells to evacuating disabled people.

An elevator used for providing access to disabled people should not be used as a fire escape unless the safeguards described in the preceding are built into it and the building structure. Even then there will remain the risk that the elevator may be, or may become, defective (for example, elevator motor failure or smoke gets into the elevator well). Hence, as shown in Fig. 7.3c, it is essential that a disabled person in a refuge have access to a stairway as a last resort.

Wheelchair stair elevators should not be used as a means of escape and should not be installed in exit stairways unless they are the only practical option for providing access for disabled people to upper floors. If such an elevator is used, the stairway width should be sufficient to prevent obstruction of other exiting occupants.

In summary, Part 8 covers the following egress options in high-rise buildings: protected stairways provided for the egress of able-bodied people; ordinary passenger elevators not intended for means of escape; fire-fighting elevators which can be used as a means of escape for disabled people and for priority use by the fire brigade for search and rescue purposes; and evacuation elevators for use by the disabled.

Part 8 also covers other related matters such as full details of the construction of refuges and evacuation elevator enclosures, stairways, ramps, and fire-warn-ing systems. In addition, appendices give information on evacuation procedures, including techniques for evacuating people through stairways, management of evacuation elevators, examples of fire-plan strategies, fire-alarm systems, and, as already mentioned, an application of the code to existing buildings.

It is important to recognize the significance of management in the escape strat-egy. The successful emergency evacuation of a building requires comprehensive management procedures—and this applies whether the occupants of a building are disabled or not. The management procedures for disabled people will, of ne-cessity, include special arrangements for assisting wheelchair-bound people and occupants with walking difficulties, and for supervising the use of an evacuation elevator if provided. Such management is discussed in Section 1 of Part 8 and recommendations are given in Appendix A.

7.3 AUTOMATIC WARNING SYSTEMS AND DISABLED EGRESS

To date there is no code which considers the time savings that are conferred by the new generation of addressable automatic smoke-detection systems. These an-alog systems are equipped with text, voice, or pictorial means of disseminating emergency information throughout a building (Bellamy and Geyer, 1988), for ex-ample, giving the location of a fire and safe exit routes. Early motivation to evac-uate in a fire is a major problem for all building occupants, and studies of human behavior have shown a marked tendency for people to use routes that are familiar to them in preference to those provided for emergency use (Canter, 1985). The time gained in using an automated system would allow the safe use of normal routes for most disabled persons not directly adjacent to the fire origin; and it also motivates able-bodied occupants to exit earlier. Perhaps future codes will permit some relaxation of structural requirements for fire safety in return for the provision of a suitably monitored informative warning system which maximizes the time available for conventional escape. Such a compromise could mitigate, in the foreseeable future, the economic conflicts between code requirements and the wish of the disabled to have free access to most buildings.

7.4 CONDENSED REFERENCES/BIBLIOGRAPHY

Bellamy 1988, *An Evaluation of the Effectiveness of the Components of Informative Fire*
BSI BS 5588 1988, *Fire Precautions in the Design and Construction of Buildings, Part 8*
Canter 1985, *Studies of Human Behavior in Fire: Empirical Results and Their Implications*
Council on Tall Buildings, Committee 56, 1992, *Building Design for Handicapped and Aged*

8

Life-Safety Implementation Through Designed Safe Egress

In building fires, the ability of occupants to egress safely from all threatened spaces is equivalent to a condition of life safety. The life-safety problem can be solved by applying this equivalence principle, strengthened by formulating a program for research and development leading to the implementation of this solution (Council on Tall Buildings, Committee 56, 1992). The solution is embodied in the concept that safe egress can be achieved in buildings designed to have a balance between the available safe egress time (ASET) and the required safe egress time (RSET). Here ASET is defined as the length of the time interval between fire detection or alarm t_{DET} and the time of onset of hazardous conditions t_{HAZ}. RSET is defined as the length of time, subsequent to alarm, which is actually required for safe occupant egress.

The concept of balance between ASET and RSET leads to a quantitative test or criterion for safe building design, the designed safe egress criterion, that is, relative to a potential hazardous fire, a building is of safe design if ASET = $t_{HAZ} - t_{DET} >$ RSET (Cooper, 1983).

This formula can be viewed as an algorithm for evaluating or determining the adequacy of a specific building design vis-à-vis life safety in fires. The application of the algorithm would require assumptions on the nature of likely potential hazardous fires for the type of occupancy in question. A capability for estimating the ASET and RSET for the specific building design under consideration would also be necessary.

ASET and RSET are the key elements of the designed safe egress formulation for life safety. RSET depends on several factors, which include the physical characteristics of egress paths (such as length and width of corridors, aisles, and stairways; sizes, numbers, and construction of doors; and lighting); the occupant density and distribution; and the physiological and psychological characteristics of occupants. To the extent that these physical and human factors impact on the capability to estimate RSET, it is evident that they must be given serious consideration.

An estimate of ASET requires estimates of t_{DET} and t_{HAZ}. t_{DET} depends on the characteristics of available detection and alarm devices, and on the interactions of these with fire-generated environments which develop around them. Be-

113

sides depending on the physiological characteristics of occupants, t_{HAZ} also depends on the fire environment. It is therefore evident that estimates of t_{DET} and t_{HAZ} require predictions of the dynamic environment which develops in buildings during fire conditions. Developing and advancing the capability of making such predictions is consistent with the future goals and ongoing programs of fire-research institutions such as the Building and Fire Research Laboratory of the National Institute of Standards and Technology (NIST).

The concept of designed safe egress involves a wide range of fire phenomena. Successful implementation of the designed safe egress criterion must be based on the formulation of, and solution to, a variety of important fire-research problems. In the past decade, important advances have been made in both the formulation and the solution of some of these problems. There are, however, important problems yet to be addressed. Of interest has been the emergence of a considerable number of fire-growth models and a limited number of people-response models, several of which are listed in the references. Besides leading to the goal of life safety in fire, spinoff products of such research will be helpful in achieving fire-safety objectives in general. In this sense, the designed safe egress criterion can be regarded as a convenient focal point for fire-safety research and development. For example, it would be appropriate to include the following text in a statement of fire-research goals:

One goal of fire research is to develop a capability for evaluating or designing buildings for life safety by a direct application of the designed safe egress criterion which states: Relative to a potential hazardous fire, a building is of safe design if the available safe egress time exceeds the required safe egress time.

8.1 RESEARCH AND DEVELOPMENT STRATEGY

A strategy leading toward implementation of the designed safe egress criterion is conceived as having four major components. These are:

1. Continued proliferation of the designed safe egress criterion within the fire protection and research community.

2. Continued development of analytic tools like those referenced in this chapter to estimate ASET (Cooper, 1982, 1983; Cooper and Davis, 1988 and 1989; Cooper et al., 1990; Evans, 1985; Nelson, 1990; Jones, 1985; Tanaka, 1983; Mitler and Rockett, 1987; Chitty and Cox, 1988).

3. Continued development of analytic tools like those of Mitler and Rockett (1987) and Chitty and Cox (1988) to estimate RSET.

4. Produce more user-oriented products by using the designed safe egress concept to package the above ASET and RESET tools into practical fire-safety evaluation and design systems like those of Nelson (1990), Chitty and Cox (1988), and Bukowski et al. (1989).

Research and development programs and activities which address each of these components have been identified. These are outlined in Table 8.1 and discussed in the following. The framework was laid out in such a manner as to indicate the linkage of all supporting multidisciplinary activities to each other and to the final, pragmatic engineering products that are envisaged.

1 Proliferating the Designed Safe Egress Concept

In general, the tools required to implement the designed safe egress criterion rely heavily on the varied programs carried out in the international fire-research community. By adapting, revising, and supplementing the outputs of these programs, final application-oriented products usable by the fire-safety community would result.

The major measure of success of the overall strategy would be the degree of acceptance, adoption, and implementation of its products by the fire-protection community. For success, the concept of life safety through design of safe egress must be publicized at every opportunity. This can be achieved by preparation and delivery of talks and "popular" publications on the overall concept and on the general goals of the research and development activities. In addition, when the products of research are promulgated by technical talks and papers, they should, at every reasonable opportunity, be placed into the perspective of the designed safe egress criterion. By encouraging familiarity of the fire-safety community with the concept of its criterion, the community would hopefully be ready, indeed, anxious, to use final user-oriented products as they become available.

2 Developing Computational Systems to Estimate ASET

The second major component involves the development of two levels of products. The first includes the major computational components or subsystems, which would be incorporated in total ASET computation schemes. The second level consists of the ASET computation systems themselves. These would be user-oriented products for estimating ASET in different classes of fire scenarios.

Developing Subsystems for ASET Calculations.   Computation subsystems for use in an overall system for estimating ASET would be developed within one of three basic task categories. The first would include the development of mathematical models for simulating actual fire environments. Products developed under this category would include compartment fire model computer codes. Such fire-environment simulators are one of the central building blocks of overall ASET computational schemes. The second task category includes criteria for, and sim-

---

Table 8.1   Implementing life safety through designed safe egress

1. Proliferating the designed safe egress concept
2. Developing computational systems to estimate ASET
   a. Developing subsystems for ASET calculations
      (1) Simulations of enclosure fire environments
         (*a*) Smoke compartment of fire origin
         (*b*) Rest of the building compartment
         (*c*) Simulating smoke leakage of bounding partitions
      (2) Criteria for activation of detector hardware
      (3) Criteria for onset of hazardous conditions
   b. Assembly of ASET subsystems
3. Developing algorithms to estimate RSET
4. Deterministic fire-safety evaluation and design systems

ulation models of, detector hardware activation. In the overall ASET computation scheme, these are used with fire-environment simulation results to estimate t_{DET}. The final task category sets criteria for the onset of hazardous conditions for estimating t_{HAZ}.

Simulations of Enclosure Fire Environments. To develop tractable analyses of ASET, it is convenient to view the fire-generated environment which develops throughout a building in a manner depicted in Fig. 8.1. This figure depicts a building as being made up of two separate compartments with common "leaky" interface partitions. The two types of compartments are characterized according to the dominant physical mechanisms which exist for inter- and intraroom migration of products of combustion. The first of these is the smoke compartment of fire origin. In this compartment, fire-generated buoyancy effects dominate the migration of combustion products. The second of the two compartment types is classified as the rest of the building. Here, combustion products are effectively of uniform concentration in most rooms of the compartment, and the products are driven from room to room by forced ventilation and by pressure differentials generated by stack or wind effects, just as in the case with normal airflow.

Flow phenomena within the compartment of fire origin would be assumed to develop essentially independently of the rest of the building, and they would be analyzed as such. The compartment of origin may be more than one room if there are significant openings between the various rooms in the compartment. The environment throughout the rest of the building is then assumed to be independent of the compartment of fire origin, except for the injection of products of combustion across common partitions. The rate of injection of this smoke would be computed from estimates of cross-partition pressure differentials, and from estimates of the leakage characteristics of the partitions themselves.

In the analysis of a building, a particular space or grouping of spaces would be

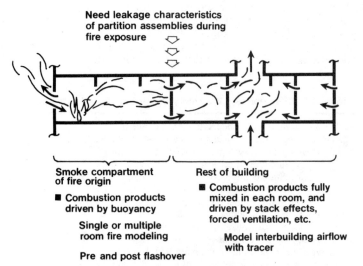

Fig. 8.1 Concept for modeling the development of hazardous conditions in "complex" buildings.

included in the rest of the building compartment if those portions of its bounding partition, being shared with the compartment of fire origin, were appropriately fire-rated, and if these partitions were assured to have relatively low smoke leakage characteristics even under fully developed fire exposure.

When implementing the type of practical computational algorithms envisaged in this component, the fewer the number of building spaces within the compartment of fire origin, the better. Benefits accrue from the use of fire-rated, low-leakage partitions. The reason for this is that it is significantly easier to simulate ventilation/stack effect/wind-driven smoke migration phenomena (in the rest of the building compartment) with confidence than to simulate multiroom, multifloor, fire-driven smoke migration phenomena (in the compartment of fire origin). In addition, when reliable, smoke-impeding partitions and well-designed door-closing hardware are used, safety is likely to be enhanced simply by virtue of increased ASET.

Even for large complex buildings, an analysis of the type suggested in Fig. 8.1 could be carried out with practical ASET calculation schemes developed with limited research, development, and implementation resources. The term *practical calculation* is intended to connote analyses which are carried out by fire-protection practitioners with the use of well documented, user-oriented, cost-effective computation algorithms.

Smoke Compartment of Fire Origin. At this level, reference is to computational models for simulating fire environments in the smoke compartment of fire origin. Activities required involve continued development of single- and multiple-room (including multilevel) compartment fire models, where much progress has been made over the last decade. Also included would be experimental studies which would be used to improve, verify, or identify limitations of such models.

In the most general case, an optimum comprehensive compartment fire model would perhaps involve two stages of fire development. The first would predict environments in all rooms of the compartment up to the time of flashover in the room of fire origin. The second stage would continue the calculation beyond the time of flashover.

Much literature exists on the simulation of postflashover phenomena where ventilation of a single flashed over room is from an outside or ambient environment. The results of these past studies need to be extended to simulate flashed over fire phenomena under multiple-room conditions where the flashed over room of fire origin is (also) ventilated by the nonambient environment of adjacent spaces within the compartment.

Compartment fire models are constructed of separate, elemental algorithms. These building blocks are used to simulate analytically the separate physical processes which take place within the fire environment. Research and development on these elemental algorithms, including fundamental and verification-oriented experimental studies, will constitute major subtasks of the overall strategy. For example, it would be the role of such subtasks to develop and substantiate new and/or improved algorithms that describe plume phenomena, heat transfer, flame spread, and the effects of applying extinguishment or fire-control strategies.

In assembling any particular compartment of fire origin model, judicious choice of the basic building-block algorithms is required. In particular, at this basic assembly level, compromises between accuracy of simulation and ultimate practicality of implementation are of key importance if a model is to have utility in practical design calculations of the type envisaged here. For example, such compromises are required in the development of algorithms to predict flame spread, fire growth, and, in general, the rate of generation of products of combustion.

Research on flame spread and fire growth is clearly of basic, long-term interest. In time, this will lead to the development of a general capability for predicting the dynamics of combustion zones during fires in practical fuel assemblies and under compartment fire conditions. However, for the present it is beyond the state of the art of fire technology to make such predictions accurately. This dilemma is heightened by the fact that combustion zone phenomena drive the basic inter- and intracompartment smoke migration phenomena being simulated. A practical engineering solution lies in the following compromise in simulation accuracy: Prior to flashover, neglect the effect of the enclosure on flame spread, and assume that from the time of ignition to the time of flashover the combustion zone of a particular fuel assembly develops as it would in a free burn situation (Cooper, 1983), [Free burn here is defined as a burn of a fuel assembly in a large (compared to the combustion zone), ventilated space with relatively quiescent atmosphere.] To implement these ideas, one simply uses empirical, free-burn test data to describe the combustion physics of a fire whose hazard is being evaluated.

The implementation of this compartment is, in principle, relatively simple. However, for general use it must be supported by an extensive database acquired from a series of actual free-burn tests. Such a database can take the form of a catalog describing the free-burn characteristics of important real fuel assemblies which are common to general building occupancies of general interest. A limited set of energy release rate data already exists. However, a verified and catalogued database is yet to be assembled. The catalog would be arranged in a manner as to be compatible with input requirements of final user-oriented ASET algorithms.

Rest of the Building Compartment.   At this level, reference is to computational algorithms for simulating fire-generated environments outside of the smoke compartment of fire origin. The basic phenomena to be modeled are the distributions throughout the rest of the building of the time-varying concentration of combustion products. In general, these combustion products are injected from the smoke compartment of fire origin into the rest of the building by virtue of pressure differentials across common partition elements. Once introduced into bounding compartment spaces, continued room-to-room migration, leading to dilution of products of combustion and cooling by heat transfer to partition surfaces, is driven by interspatial pressure differentials. These are generated by forced ventilation, stack effect, or external wind flows.

As with compartments of fire origin, the overall task for the rest of the building compartment consists of three subtasks: continuation of work on computational models for dynamic analyses of intercompartment combustion product migration; reduced-scale and limited full-scale experimental verification of these algorithms; and finally, their modification into user-oriented products.

In the models to be developed and verified under the first two of the subtasks, the rest of the building compartment would be simulated mainly by individual rooms interconnected by moderately leaky partitions. These partitions simulate actual wall and floor construction (including door and window assemblies). The environment of each room would be modeled as being spatially uniform. Certain rooms, with special geometry or flow condition characteristics, would require modeling which accounted for mixed forced-free convection. (For example, the environment within ventilated elevator shafts, with weakly buoyant smoke injected at a relatively high elevation, may not be adequately described by the fully mixed model.) Practical means of analyzing the dynamic environments within these latter spaces is not yet available.

The leakage characteristics of typical partition constructions would be required as inputs to the models. To develop working descriptions of such leakage characteristics, a separate series of subtasks is envisaged. These would include a literature search for existing leakage data, the development of standard leakage test procedures, and the compilation of an applications-oriented catalog of results (Cooper, 1986).

Again, as in the case of compartments of fire origin models, incorporation of appropriate compromises between accuracy of simulation and ultimate practicality of implementation is needed in the rest of the building compartment models.

The simulation problem here is similar to those problems associated with the design of building heating, ventilating, and air-conditioning (HVAC) systems. Indeed, products of that technology have already been used to design and evaluate building smoke-control systems (Klote and Fothergill, 1983). Within the context of the designed safe egress criterion, it is noteworthy that a perfect smoke-control system can be defined as a ventilation system that guarantees safe egress for all occupants outside of a compartment of fire origin. Current smoke-control system designs attempt to accomplish this by always maintaining a favorable pressure differential across building partitions to prevent any significant smoke migration into the rest of the building compartment.

With regard to the last of the three subtasks—the development of user-oriented models—multiple tools are evolving. Reference here is to models for the design of smoke-control systems, which are based on simple steady-state analyses. These models are independent of detailed compartment of fire origin considerations. There may be situations where such analyses would bring buildings into compliance with the designed safe egress criterion. If such is the case, it may be prudent to develop and offer such analytic capabilities, unencumbered by more general dynamic considerations, to the fire-safety community.

Simulating Smoke Leakage of Bounding Partitions. To carry out the general analysis of intrabuilding smoke migration, some knowledge of the leakage properties of wall and floor assemblies during full-fire exposures is required (Cooper, 1986). Reference here is to those partition assemblies which are common to both the compartment of fire origin and the rest of the building smoke migration models discussed in the previous section. The leakage through these assemblies determines the rate at which hazardous combustion products are introduced to the rest of the building. Existing data are sparse.

The goal is to obtain the leakage characteristics of real wall and floor construction elements typically found in building occupancies of interest. Where possible, the work required to achieve this goal would be integrated with the partition leakage tasks discussed earlier in the rest of the building compartment section. However, the present task goes beyond this. Here the development of standard leakage test procedures, carried out under the significantly more difficult conditions of full fire exposures (and different possible cross-partition pressure differential), is needed. Once available, these procedures would have to be applied and, again, an applications-oriented catalog of results compiled. This catalog, developed at relatively high cost per entry, would describe a special category of partitions which, for the purpose of an analysis such as presented in Fig. 8.1, could be regarded as bona fide compartment of fire origin boundary construction elements.

Criteria for Activation of Detector Hardware. A key component for the initiation of occupant egress is successful fire detection and alarm. To predict the time of fire detection in a given fire scenario, it is necessary to develop an overall capability for simulating the response of real fire-detector hardware when immersed in a fire-generated environment. Some advances have been made in this area, but more are needed (Evans, 1985; Cooper and Davis, 1988; Cooper, 1984; Alpert, 1972). Four categories of tasks are involved. The first category consists of experimental investigations on the response of different types of fire detectors when immersed in well-controlled firelike environments. For this task it will be important that the environment descriptors be compatible with the environment descriptors generated by the enclosure fire algorithms discussed in the section on simulations of enclosure fire environments. Second, based on these experiments, continued development and verification of criteria for detector response or detector response simulation algorithms is required. Third, standard test method procedures to measure detector characteristics would be developed and used as inputs to these algorithms. Finally, the validity of the detection criteria would be verified experimentally under real fire-environment conditions.

Criteria for Onset of Hazardous Conditions. The end of the ASET interval is defined by the onset of hazardous conditions. To predict this event, criteria for hazardous fire environments must be established (Purser, 1988). To have utility in the present context, these criteria would have to be compatible with environment descriptors generated by previously discussed compartment fire algorithms. It would then be possible to apply the criteria to predicted fire environments, and to thereby estimate the desired times of onset of hazard in specific fire scenarios.

The wide differences in acceptable levels of hazard, as well as the significant variations in physiological and psychological characteristics expected in any realistic mix of building occupants, complicate any attempt at establishing definitive criteria for hazard. Nonetheless, some definite criteria for a hazardous environment, based on reasonable, global assumptions of occupant characteristics and acceptable risk, must be established if the designed safe egress concept is to have any utility. One model which addresses these issues is presented by Bukowski et al. (1989).

Assembly of ASET Subsystems. The tasks described in the previous section lead to the development of the separate subsystems required for ASET calculations. The present task would be devoted to assembling these subsystems into complete, user-oriented algorithms that would carry out such calculations.

ASET calculation schemes will vary according to the types of individual subsystems used in their assembly. For example, one type of ASET calculation might involve a document which introduces the user to the individual algorithm elements, provides detailed instructions on how to exercise these elements in the analysis of a building, and guides the user toward final ASET calculations. To develop such an ASET calculation scheme, it may be necessary to restructure existing documentation on the individual algorithm elements and, possibly, to modify the elements themselves to make them compatible with one another and user-oriented. Examples of current state of the art are provided by Nelson (1990) and Bukowski et al. (1989).

Another type of tool for calculating ASET is an actual computer code constructed of subroutines taken directly or in part or in modified form from products generated under the various tasks discussed earlier. Such ASET computer codes can vary significantly from one another, depending on the sophistication of their component algorithms. They would require user manuals to make them useful as engineering design and evaluation tools.

3 Developing Algorithms to Estimate RSET

This third major component involves the development of user-oriented algorithms to be used in estimating RSET. Such algorithms would likely be developed within the context of two categories of building space occupancy, with a different basic type of analysis being used to simulate egress of occupants from each of them.

The first category of building space includes those spaces of relatively high occupant density, where the RSET would tend to be constrained by group behavior and capability, by the limiting containment capacity of egress path spaces, and by the limiting flow-rate capacities of egress path elements. Typical of such building spaces would be rooms used for theaters and offices. Purser (1988) provides a current approach to the calculation of egress times.

The second category of building space would be one where the success of evacuation would tend to depend on, and be constrained by, behaviors and capabilities of individual occupants, and by strong one-on-one types of occupant interactions. To develop algorithms for simulating evacuation from this category of building space, occupant capabilities to receive aid from, or provide aid to, fellow occupants must be established. An example of such a behavior model is provided by Levin (1989).

4 Deterministic Fire-Safety Evaluation and Design Systems

Tasks contributing to the fourth and final component of the strategy lead to applications-oriented deterministic fire-safety evaluation and design systems based on the designed safe egress criterion. Using previously developed ASET and RSET algorithms, these systems would be developed as complete user-oriented products. They address specific classes of problems within the general fire-safety community, such as the evaluation and the development of codes and standards, evaluation of the fire safety of specific buildings, design of a fire-safe building, and comparison of fire-safety effectiveness of alternative building components or fire-protection hardware.

Products developed under this component will directly contribute to the basic goal of the overall strategy, namely, to establish a capability for evaluating and designing buildings for fire safety through rational engineering considerations.

8.2   CASE 1: PRISON CELL BLOCKS

Problem: Develop means of assuring that prisoners locked in cells in a multitiered cell block would not be subjected to hazardous conditions if a fire would involve the burnout of a heavily loaded cell (Fig. 8.2).

Solution: Full-scale burn tests and application of compartment of fire origin concepts were used to develop Fig. 8.2. This figure is now included in Cooper and Stroup's (1985) computer program for calculating available safe egress time as a means of determining the design requirements for cell-block design.

8.3 CASE 2: WINDOWLESS MUSEUM

Problem: Determine the available safe egress time due to the potential burn of a fuel assembly likely to be found in a 3700-m² (39,826-ft²) museum having a 4.5-m (14.76-ft)-high ceiling.

Solution: Assume that during the growth stage of the burning fuel assembly the energy release rate would not exceed that from free-burning typical high-stacked warehouse fuel assemblies. Use available data on such free burns as input, and exercise the ASET computer program, thereby generating Fig. 8.3, where the history of the temperature and the descent of the upper smoke layer are plotted (Cooper, 1982; Walton, 1985; Nelson and MacLennan, 1988).

8.4 CASE 3: SLEEPING-ROOM CORRIDOR

Problem: In a jail-type situation where the ceiling height is approximately 3 m (9.8 ft) and sleeping rooms (cells) are open to a common corridor, what is the ASET under sleeping-room-originating fire conditions?

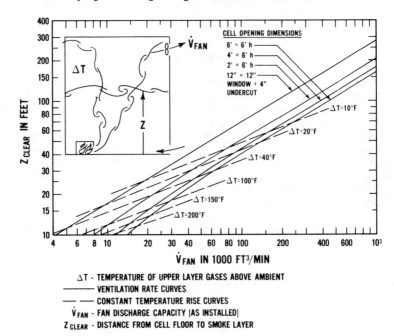

ΔT - TEMPERATURE OF UPPER LAYER GASES ABOVE AMBIENT
——— VENTILATION RATE CURVES
— — CONSTANT TEMPERATURE RISE CURVES
$\dot{V}_{FAN}$ - FAN DISCHARGE CAPACITY (AS INSTALLED)
Z_{CLEAR} - DISTANCE FROM CELL FLOOR TO SMOKE LAYER

Fig. 8.2   Cell-block smoke ventilation curves (*From NFPA, 1988, Fig. A-15-3.1.1(a)*.)

Solution: Assume the threatening fire to be either a subflashover mattress fire (free-burn data available) or a fully involved fire (like those fires considered in the prison cell block of case 1). Assume detection and successful alarm occur at the instant of ignition. Use free-burn data and exercise the ASET computer program (Cooper, 1982; Walton, 1985; Nelson and MacLennan, 1988). Figure 8.4 is a plot of ASET as a function of the total area, assuming onset of hazard to occur at the instant the smoke layer drops to within 1.5 m (4.92 ft) of the floor. Figure 8.5 shows the ASET results when the condition of the smoke layer as well as its elevation are the basis of the hazard criterion.

8.5 CONDENSED REFERENCES/BIBLIOGRAPHY

Alpert 1972, *Calculation of Response Time of Ceiling-Mounted Fire Detectors*
Bukowski 1989, *HAZARD I Technical Reference Guide*
Chitty 1988, *ASKFRS*
Cooper 1982, *A Mathematical Model for Estimating Available Safe Egress Time in Fires*
Cooper 1983, *A Concept for Estimating Safe Available Egress Time in Fires*
Cooper 1984, *A Buoyant Source in the Lower of Two, Homogeneous Stably Stratified Lay-*
Cooper 1985, *ASET—A Computer Program for Calculating Available Safe Egress Time*
Cooper 1986, *The Need and Availability of Test Methods for Measuring the Smoke Leak-*

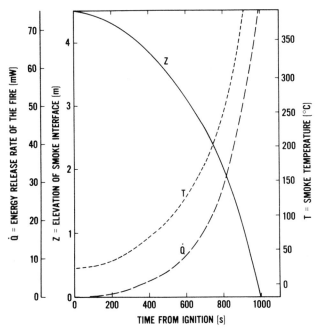

Fig. 8.3 Conditions in 3700- by 4.5-m (12,000- by 15-ft) museum after ignition of a "semi-universal" fire.

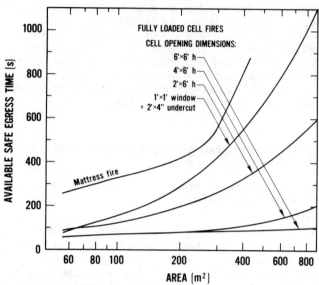

Fig. 8.4 Available safe egress time (ASET) in single-level prison 3 m (9 ft) in height; immediate detection. Criterion for onset of hazard: either $T_{upper} = 183°C$ or $Z_{interface} = 1.5$ m (5.0 ft).

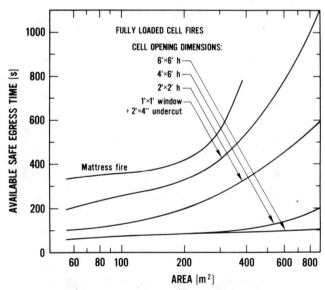

Fig. 8.5 Available safe egress time (ASET) in single-level prison 3 m (9 ft) in height; immediate detection. Criterion for onset of hazard: either $Z_{interface} > 1.5$ m (5.0 ft) *and* $T_{upper} = 183°C$ or $Z_{interface} \leq 1.5$ m (5.0 ft) *and* $T_{upper} = 93°C$ or CO concentration = 2000 ppm.

Cooper 1988, 1989, *Estimating the Environment and the Response of Sprinkler Links in*
Cooper 1990, *The Consolidated Compartment Fire Model (CCFM) Computer Code Appli-*
Council on Tall Buildings, Committee 56, *Building Design for Handicapped and Aged Per-*
Evans 1985, *Calculating Sprinkler Actuation Times in Compartments*
Harwell Laboratory 1990, *HARWELL-FLOW3D Release 2.3: User Manual*
Jones 1985, *A Multicompartment Model for the Spread of Fire, Smoke, and Toxic Gases*
Kisko 1985, *EVACNET+: A Computer Program to Determine Optimal Building Evacua-*
Klote 1983, *Design of Smoke Control Systems for Buildings*
Levin 1989, *EXIT—A Simulation Model of Occupant Decisions and Actions in Residential*
Mitler 1987, *User's Guide to FIRST, A Comprehensive Single-Room Fire Model*
National Fire Protection Association 1988, *Figure A-15-3.1.1(a) Cell Block Smoke Control*
Nelson 1988, *Emergency Movement*
Nelson 1990, *FPETOOL: Fire Protection Engineering Tools for Hazard Estimation*
Purser 1988, *Toxicity Assessment of Combustion Products*
Tanaka 1983, *A Model of Multiroom Fire Spread*
Walton 1985, *ASET-B: A Room Fire Program for Personal Computers*

9

A Design Model
for Planning Pedestrian
Evacuation Routes

Given the available safe-egress-time concept, escape routes for emergency evacuation of tall buildings deserve special attention as part of the design process. As described earlier, the feasibility of the total evacuation of a high-rise building has been discussed frequently. In technical literature on egress and alternative evacuation, concepts such as partial or selective evacuation, or evacuation into refuge areas in the building, are described. These considerations are also reflected in the national code requirements on means of escape in some countries. If an emergency is considered in a broader sense, including life-threatening aspects other than fire, there are situations that support the necessity of total evacuation. These include the threat of earthquake, a toxic or natural gas leak, a power blackout or elevator failure, a bomb threat or a civil defense emergency, and, finally, the natural reaction of people to escape danger. It is not always possible to convince the occupants of a building to wait to evacuate or simply remain in a "safe area," once they are aware of the existence of a fire emergency [for example, the World Trade Center fire, 1975; see Paulsen (1981)].

During the last 20 years a number of design methods for calculating the movement of people in buildings have been developed. This study briefly reviews some of those modeling efforts. In addition, it discusses some aspects of the pedestrian movement and addresses a design method based on the data provided by Predtechenskii and Milinski (1978). A model is presented which is applicable to the evacuation of multistory buildings via staircases and predicts the flow movement, including the building's physical structure and the interdependencies between adjacent egress-way elements.

9.1 INVESTIGATIONS ON HUMAN SPATIAL REQUIREMENTS

The individual space is an important aspect of pedestrian activity and a satisfactory means of measuring quality of design. A pedestrian occupies a block of

127

space and, during ambulation, systematically occupies and relinquishes additional space. The extra space is utilized, first for the activity of moving, and second as a buffer against physical contact with other parts of the environment and against interactive contact with other people (Templer, 1975).

The space required for pedestrian movement depends on the physical dimensions of the human frame, standing at rest, which is described in this discussion as the perpendicular projected area of a person (or the projected area factor). Moreover, it also depends on the volumetric requirements of the body during locomotion (Fruin 1970; Templer, 1975).

The human ecology addresses these zones in the concept of *individualraum* (individual space), defined as the stretch between an individual and a specific event, and the concept attributes to every environmental relation both an "informatory aspect" as well as a "material-energetic aspect" (Knötig, 1980).

Fruin recommends, for the design of some types of spaces, a body ellipse 450 by 610 mm (18 by 24 in.), equivalent to an area of 0.21 m² (2.3 ft²). This figure has been used to develop the capacity of New York City subway cars but, as discussed later, it would be too high for building-evacuation purposes. Fruin states that the human body can be crowded into an area of about 0.14 m² (1.5 ft²) for the average man and 0.09 m² (1 ft²) for the average woman, representing the limit of body dimensions. His recommendation for the dimensions of the average man is fairly similar to the author's investigations.

Based on the analysis of body dimensions and observation of waiting lines, Fruin proposed standards for these types of spaces, as summarized in Table 9.1. Table 9.2 shows the values for the perpendicular projected areas of different age groups given by Knötig (1980). In Table 9.3 the mean values for the perpendicular projected areas of persons by means of anthropometric measurements of a randomly selected Austrian group of people of different ages are represented (Kendik, 1984).

There, by using an artificial sun which provided parallel rays of light, and a mirror arrangement set up with an angle of 45°, the body frames of test persons were projected to the floor, drawn, and measured. Thus approximately 600 drawings of different test persons in standing position, with and without coats, and taking a step have been evaluated. The results from this Austrian statistical analysis do not comply with the Russian data given in Table 9.2. This might indicate that the projected area per person (projected area factor) f varies in terms of population.

The spatial requirements of pedestrians also change in terms of velocity. Re-

Table 9.1   Pedestrian waiting spaces—level of service standards

Level of service category	Average area per person		Interperson spacing		Circulation through queue
	m²	ft²	m	ft	
A	< 1.2	< 13	1.2	4	Unrestricted
B	0.9–1.2	10–13	1.1–1.2	3.5–4	Slightly restricted
C	0.7–0.9	7–10	0.9–1.1	3–3.5	Restricted but possible by disturbing others
D	0.3–0.7	3–7	0.6–0.9	2–3	Severely restricted
E	0.2–0.3	2–3	0.6	2	Not possible
F	> 0.2	> 2	—	—	Not possible

Source: After Fruin (1970).

searchers such as Predtechenskii and Milinski (1978), Fruin (1970), Templer (1975), Pauls (1980), and Seeger and John (1978) conducted speed surveys outside or inside buildings on horizontal walkways as well as on stairs. Predtechenskii and Milinski measured the flow density and velocity in different types of buildings nearly 3600 times under normal environmental conditions. Their observations indicated that the flow velocity shows wide variations, especially in the range of lower densities. Hence they assumed the values of the flow velocity and capacity above the mean walking speeds and capacities under normal environmental conditions to be analogous to the pedestrian parameters in case of emergency. Therefore three movement levels have been defined:

1. Normal flow conditions
2. Comfortable flow conditions
3. Emergency flow conditions

Table 9.2   Projected area per person

	Projected area per person	
Person	m^2	ft^2
Children	0.04–0.06	0.4–0.6
Teenagers	0.06–0.09	0.6–1.0
Adults in:		
Summer clothes	0.10	1.1
Spring clothes	0.11	1.2
Winter clothes	0.125	1.35
Adults in spring clothes and carrying:		
A briefcase	0.18	1.9
A suitcase	0.24	2.6
Two suitcases	0.39	4.2

Source: After Knötig (1980).

Table 9.3   Anthropometric measurements of an Austrian group of people, in square meters

	5 years, all	10–15 years			15–30 years			> 30 years, all
		Women	Men	All	Women	Men	All	
A(Du); x	0.705	1.300	1.290	1.291	1.683	1.894	1.825	1.872
Standard deviation	0.171	0.175	0.203	0.208	0.115	0.379	0.334	0.252
$f(N)$; x	0.0696	0.1092	0.1126	0.1113	0.1383	0.1484	0.1458	0.1740
Standard deviation	0.0078	0.0202	0.0174	0.0187	0.0172	0.0171	0.0172	0.0315
$f(M)$; x	—	0.1453	0.1326	0.1386	0.1809	0.1892	0.1862	—
Standard deviation	—	0.0178	0.0191	0.0186	0.0213	0.0296	0.0272	—
$f(S)$; x	—	0.1262	0.1221	0.1238	0.1508	0.1645	0.1600	0.1918
Standard deviation	—	0.0198	0.0170	0.0180	0.0163	0.0191	0.0193 .	0.0356

A(Du); x DuBois-area, mean value
$f(N)$; x Mean projected area per person, standing and without coats
$f(M)$; x Mean projected area per person, standing and wearing coats
$f(S)$; x Mean projected area per person, walking
Source: After Pauls (1984).

The mean values of velocity under comfortable flow conditions have been estimated from the lower range of the measured walking speeds, and for emergency
flow conditions, from the upper range of the measured values. Figure 9.1 compares the evacuees' speed down the stairs in terms of the density given by Fruin
(1970), Pauls (1980), and Predtechenskii and Milinski (1978).

9.2 MODELING PEDESTRIAN MOVEMENT

The current models evolving from people movement may be classified as follows
(Kendik, 1985a):

1. Flow models based on the carrying capacity of independent egress-way components, or the unit exit-width concept
2. Flow models based on empirical studies of crowd movement
3. Computer simulation models
4. Network optimization models

The more important of these models are discussed in this section.

1 Flow Models Based on the Carrying Capacity of Independent Egress-Way Components: Unit Exit-Width Concept

The historical development of carrying-capacity investigations has already been
broadly reviewed in several publications by Stahl and Archea (1977, 1982) and
Pauls (1980, 1984).

An early NFPA document recommended as a guideline for stair design an average flow rate of 45 persons per minute per 0.56-m (22-in.) width unit (Stahl and
Archea, 1977). In 1935, in a publication of the U.S. National Bureau of Standards

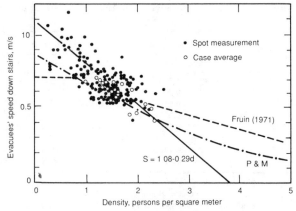

Fig. 9.1   Relation between speed and density on stairs.

[now National Institute of Standards and Technology (NIST)], test results about measurements of flow rates through doors, corridors, and on stairs under non-emergency conditions were presented. For different types of occupancies the measured maximum flow rates varied between 23 and 60 persons per minute per unit stair width, and 21 and 58 persons per minute per unit door or ramp width (National Bureau of Standards, 1935). To date, the NFPA Life Safety Code 101 (NFPA, 1982) utilizes the unit exit-width concept together with travel distances and occupant-load criteria. However, for some reason, the time component is absent in the present code.

In the United Kingdom, the first national egress guidance for places of public entertainment was produced in 1934 (Home Office, 1935). The recommendations were "based not only on experience gained in the U.K., but on a study of disasters which have happened abroad and of the steps taken by the authorities of foreign countries" (Read, to be published). In this document, the following formulas were provided for the determination of the total width of exits required from each portion of a building, reflecting the concept of the unit exit width:

$$A = \frac{Z(\text{floor area, ft}^2)}{EBCD} \qquad (9.1)$$

where A = number of units of exit width required
   B = constant, referring to type of construction of building
   C = constant for arrangement and protection of stairs
   D = constant for exposure hazard
   E = factor dependent on height of floor above or below ground level
   Z = class of building use (closely seated audience, for example)

Then

$$N = \frac{A}{4} + 1 \qquad (9.2)$$

where N is the number of exits required.

In this document it was also stated that about 40 persons per minute per unit exit width down the stairs or through exits is an appropriate figure in connection with these formulas.

Forty persons per minute per unit exit width is also recommended in Post-War Building Studies No. 29 (HMSO, 1952). In this report, another calculation method is suggested in Appendix II. The width of staircases in the current GLC Code of Practice (Greater London Council, 1974) as well as in BS 5588 Part 3 (BSI, 1983) is computed by this method (Tidey, 1983). Calculation of the total population that a staircase can accommodate is based on the following assumptions:

1. Rate of flow through an exit is 40 persons per unit width per minute.
2. Each story of the building is evacuated onto the stairs in not more than 2.5 min. This average clearance time was proposed after an evacuation experience during a fire in the Empire Palace Theatre in Edinburgh in 1911 (HMS, 1952).
3. There is the same number of people on each story.
4. Evacuation occurs simultaneously and uniformly from each floor.
5. In moving at a rate of 40 persons per unit width per minute, a staircase can

accommodate 1 person per unit width on alternate stair treads and 1 person per each 0.3 m^2 (3 ft^2) of landing space.

6. The story height is 3 m (10 ft).

7. The exits from the floors onto the stairs are the same width as the stairs.

8. People leaving the upper floors are not obstructed at the ground floor exit by persons leaving the ground floor.

Then

$$P = \text{staircase capacity} \times \text{number of upper stories} + (t_e - t_s)rw \qquad (9.3)$$

where P = total population that a staircase can accommodate

t_e = maximum permissible exit time from any one floor onto staircase, taken as 2.5 min

t_s = time taken for a person to traverse a story height of stairs at standard rate of flow, predicted as 0.4 min

r = standard rate of flow, taken as 40 persons per unit width per minute

w = width of staircase, units

The staircase capacity is predicted after point 5 of the preceding assumptions.

This method of calculation predicts that with an increasing number of stories, the building has fewer persons per floor.

In 1955, K. Togawa in Japan was apparently the first researcher who attempted to model mathematically the people movement through doorways and on passageways, ramps, and stairs (Pauls, 1984; Stahl and Archea, 1977; and Kobayashi, 1981). He provided the following equation:

$$v = V_0 D^{-0.8} \qquad (9.4)$$

where v = crowd walking velocity or flow velocity

V_0 = constant velocity, = 1.3 m/s, which is apparently the velocity under free-flow conditions

D = density in persons per square meter

Hence the flow rate N is given by

$$N = V_0 D^{0.2} \qquad (9.5)$$

(Note: N here is the same as the specific flow q referred to later.)

Based on data from the investigations of Togawa and the London Transport Board (1958), Melinek and Booth (1975) analyzed the flow movement in buildings and provided an estimation of the total evacuation time from buildings using the following formulas.

The maximum population M, which can be evacuated to a staircase, assuming a permitted evacuation time of 2.5 min, is given approximately by

$$M = 200b + (18b + 14b^2)(n - 1) \qquad (9.6)$$

where b = staircase width, m

n = number of stories served by staircase

This equation predicts a higher number of persons than the method presented in Post-War Building Studies No. 29 (HMSO, 1952).

The minimum evacuation time T_e for a multistory building is given by

$$T_e = \frac{\sum\limits_{i=r}^{n} Q_i}{N'b_{r-1}} + rt_s \qquad (9.7)$$

where r = floor number (1 to n), which gives the maximum value of T_e
Q_i = population of floor i
b = width of staircase between floors $r - 1$ and r
N' = flow rate of people per unit width down the stairs
t_s = time for a member of an unimpeded crowd to descend one story

If the population Q and the staircase width b are the same for each floor, then T_e is the larger of T_1 and T_n, where

$$T_1 = \frac{nQ}{N'b} + t_s \qquad (9.8)$$

$$T_n = \frac{Q}{N'b} + nt_s \qquad (9.9)$$

T_1 corresponds to congestion on all floors and T_n to no congestion. Melinek and Booth suggested as typical values on N' and t_s 1.1 persons per second per minute and 16s. Compared with evacuation tests in multistory buildings, the method predicted in most cases evacuation times that were too low.

A further application of the unit-width concept has been the mathematical model of Müller (1966, 1968, 1969). Assuming a flow rate of 30 persons per minute per 0.6-m (2.0-ft) stair width, Müller provided the following equations for the assessment of total evacuation time in multistory buildings:

$$t = t_1 + \frac{P}{bf_0/0.6} \qquad (9.10)$$

$$t = \frac{3h_G}{v} + \frac{P}{bf_0/0.6} \qquad (9.11)$$

where t_1 = travel time on stairs to descend one story
h_G = floor height
P = number of persons in building
b = stair width, m
v = flow velocity down the stairs, = 0.3 m/s (1.0 ft/s)
f_0 = flow rate per unit stair width of 0.6 m (2.0 ft)

Müller presumed that the flow velocity would decrease to 0.2 m/s (0.7 ft/s) due to interaction of flows from each upper floor, which would cause an increase in density on the stairs. Hence the travel time between two upper stories is given by

$$t_2 = \frac{3h_G}{0.2} = 15h_G \qquad (9.12)$$

and the minimum evacuation time via the staircase is

$$t = 10h_G + 15h_G n \qquad (9.13)$$

The maximum required stair width is given by

$$b_{max} = \frac{0.08P}{nh_G} \qquad (9.14)$$

Müller suggested limiting the building height rather than widening the staircases.

2 Flow Models Based on Empirical Studies of Crowd Movement

During the last decade Pauls developed the "effective width" model. This model is based on his extensive empirical studies of crowd movement on stairs as well as on data about the mean egress flow as a function of stair width. In this context he conducted several evacuation drills in high-rise office buildings and observed normal crowd movement in large public-assembly buildings. The model describes the following phenomena (Pauls, 1980a,b, 1982, 1984).

1. The usable portion of a stair width, or the effective width of a stair, begins approximately at a distance of 150 mm (0.5 ft) from a boundary wall or 88 mm (0.29 ft) from the centerline of a graspable handrail (edge effect).
2. The relation between mean evacuation flow and stair width is a linear function and not a step function, as assumed in traditional models based on lanes of movement and units of exit width. The evacuation flow is directly proportional to the effective width of a stair.
3. The mean evacuation flow is influenced in a nonlinear fashion by the total population per effective width of a stair.

Pauls provides the following equation for the evacuation flow in persons per meter of effective stair width:

$$f = 0.206p^{0.27} \qquad (9.15)$$

where p is the evacuation population per meter of effective stair width. The total evacuation time is given by

$$t = 0.68 + 0.081p^{0.73} \qquad (9.16)$$

This calculation method has recently been accepted as an appendix to the National Fire Protection Association's Life Safety Code (1982, 5th edition).

The calculation method of Predtechenskii and Milinski is a deterministic flow model which predicts the movement of an egress population on a horizontal or a sloping escape route, instantaneously, in terms of its density and velocity.

In general, the mean concentration, or the density of flow, is often defined as the number of persons per unit area, and sometimes the reciprocal has been used (Fruin, 1970). These definitions are based on the implicit assumption that the physical dimensions of the human frame are identical for all people, or the differences might be negligible.

The following equation relates the ratio between the sum of the persons' perpendicular projected areas and the available floor area for the flow, and it estimates the flow density homogeneously over the area of an escape route:

$$D = \frac{Pf}{bl} \qquad (9.17)$$

where p = number of persons in flow
  f = perpendicular projected area of a person
  b = flow width, identical to width of escape route
  l = flow length

D is dimensionless.
  The egress population passing a definite cross section on an escape route of width b is referred to as flow capacity,

$$Q = Dvb \quad \text{m}^2 \text{ per min} \qquad (9.18)$$

where v is the flow velocity.
  Predtechenskii and Milinski define the flow velocity on horizontal escape routes, through doors, and on the stairs in terms of the flow density, given in Eq. 9.17:

$$v = f(D) \qquad (9.19)$$

Being a steady-state expression, Eq. 9.17 implies that the flow density remains constant provided neither the escape route configurations nor the flow outlines change. Theoretically the density distribution of an infinitive flow may change during the lateral displacement of the crowd, that is, the density increase escalates in the direction of flow as persons running into part of the flow with increased density are moving faster. Inversely the density diminution in the course of flow leads to decomposition of the flow into separate parts of distinct densities. Hence a correction factor, which is proportional to the density increase, should be added to Eq. 9.19,

$$v = f(D) - c\left(\frac{dD}{dx}\right), \qquad c > 0 \qquad (9.20)$$

This approach has been applied to a computer simulation model for emergency evacuation of buildings on the basis of data of Predtechenskii and Milinski (1978). However, a sensitivity analysis of this model, which changed the projected area factor, did not produce the expected variation in evacuation time (Kendik, 1984). The model predicted higher evacuation times for the tested building as the value of the projected area factor (inherently, the flow density) was decreased.

On the other hand, observations of crowd movement with limited flow lengths, as in the case of egress from buildings, do not corroborate the previous considerations. This might indicate that Eq. 9.17 provides sufficient proximity for the determination of flow density on escape routes in buildings. However, further research is needed on this subject.

Another important flow parameter is the flow capacity per meter of escape route width, which is defined as the specific flow,

$$q = Dv \quad \text{m per min} \qquad (9.21)$$

The specific flow is a function of density. It increases over an interval, and after passing an absolute maximum q_{max} it decreases again. The value of q_{max} is different for distinct kinds of escape routes. Figure 9.2 illustrates the variation of the specific flow in terms of density.

The efficiency of an evacuation depends on the continuity of flow between three restrictions, namely, horizontal passages, doors, and stairs. Hence the main condition for free flow is the equivalence of flow capacities on successive parts of the escape route:

$$Q(i) = Q(i + 1) \tag{9.22}$$

or from Eqs. 9.18 and 9.21,

$$q(i)b(i) = q(i + 1)b(i + 1) \tag{9.23}$$

Figure 9.3 illustrates a scheme for the merging of three partial flows coming from different directions. In this case the condition of the free flow can be described as follows:

$$Q(i; 1) + Q(i; 2) + Q(i; 3) = Q(i + 1)$$

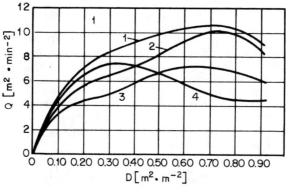

Fig. 9.2 Specific flow versus flow density for different kinds of escape routes under normal environmental conditions. 1—doorways; 2—horizontal routes; 3—stairs (upward); 4—stairs (downward). (After Kendik, 1984.)

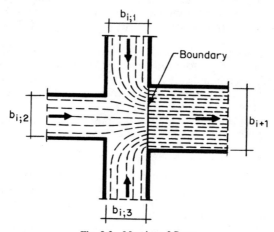

Fig. 9.3 Merging of flows.

or

$$q(i + 1) = \frac{Q(I)}{b(i + 1)} \qquad (9.24)$$

where $Q(I)$ is the sum of the capacities of all partial flows.

If the value of the specific flow $q(i + 1)$ exceeds the maximum, that is, $q(i + 1) > q_{max}$, the flow density increases spontaneously to its maximum value ($D_{max} = 0.92$), which leads to a line at the boundary to the main route $i + 1$.

Due to this congestion, not all persons may attempt to participate in the merging process simultaneously. It is presumed that the contribution of the partial flows to the main flow is proportional to their capacity Q. The percentage of the contribution of each flow to the main flow can be obtained from the ratio between the width of each partial flow and the sum of the widths of all partial flows,

$$p(1) = \frac{b(1; i)}{B(i)}$$

$$p(2) = \frac{b(2; i)}{B(i)} \qquad (9.25)$$

$$p(n) = \frac{b(n; i)}{B(i)}$$

where $B(i)$ is the sum of the widths of all partial flows.

If during the merging process of the partial flows the specific flow $q(i + 1)$ does not exceed the maximum value, that is, $q(i + 1) < q_{max}$, no congestion occurs on escape routes.

Observations of crowd movement under normal environmental conditions show that during the merging of two flows with distinct density and velocity, the movement parameters of the incoming flow will be changed by adjusting its density and speed to the parameters of the uptaking flow. According to this context, a boundary will be formed between the flows with the parameters $D(i)$, $q(i)$, $D(i + 1)$, and $q(i + 1)$, and its location changes at the following speed.

If $v(i) < v(i + 1)$, then

$$v'' = q(i) - \frac{q(i + 1)}{D(i)} - D(i + 1) \qquad (9.26)$$

If $v(i) > v(i + 1)$, then

$$v'' = q(i + 1) - \frac{q(i)}{D(i + 1)} - D(i) \qquad (9.27)$$

A graphic representation of this process is shown in Fig. 9.4. The merging of flows terminates at point C on the graph.

One may consider this phenomenon to be also appropriate for emergency evacuations, for example, where groups of persons initially have different flow densities and velocities, but a common purpose such as leaving the buildings, trying to reach people moving ahead, or searching for contact or information.

The formation of a congested flow (queuing) is an analogous process. If $Q(i) > Q(i + 1)$, lines begin at the boundary between the passages of distinct flow capacities. At the beginning of congestion, the flow consists of two parts, namely, a group of persons with the maximum flow concentration ($D_{max} = 0.92$), which has already arrived at the critical section of the escape route, and the rest of the evacuees approaching with a higher velocity and a density less than D_{max}. In this case the rate of congestion is given by the following equation:

$$v''(\text{STAU}) = \frac{q(D_{max})b(i + 1)/b(i) - q(i)}{D_{max} - D(i)} \qquad (9.28)$$

where $q(D_{max})$ = specific flow at maximum density
$b(i + 1)$ = width of congested flow
$b(i)$ = initial width of flow
$q(i)$ = initial value of specific flow
$D(i)$ = initial flow density

After the last person moving at the higher velocity has reached the end of the line, the congestion diminishes at the following rate:

$$v(\text{STAU}) = \frac{v(D_{max})b(i + 1)}{b(i)} \qquad (9.29)$$

where $v(D_{max})$ is the flow velocity at the maximum density.

This calculation method, of which the basic considerations have been summarized, has been applied by Predtechenskii and Milinski mainly to the evacuation

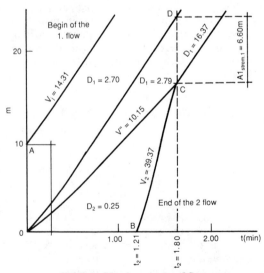

Fig. 9.4   Merging process of flows.

of auditoriums and halls. Section 9.3 deals with the calibration of the method and its application to high-rise buildings.

9.3 EVALUATION OF UNIT EXIT WIDTH IN HIGH-RISE BUILDINGS

For the prediction of flow movement in multistory buildings, an egress model was developed based on the work presented in Section 9.2 (Kendik, 1982). This has been calibrated against the data from the evacuation tests carried out by the Forschungsstelle für Brandschutztechnik at the University of Karlsruhe, Germany, in three high-rise administration buildings (Seeger and John, 1978).

The egress model presented here addresses the time sequence (within the available safe egress time) from when people start to evacuate floors until they finally reach the exterior or an approved refuge area in a high-rise building. Hence it does not consider the time prior to their awareness of the fire nor their decision-making processes. However, it can cope with the problem of potential congestion on stairs and through exits (including the interdependencies of adjacent egress-way elements, which appears to be a major problem), especially in cases of high population densities.

The method differs from other egress models mainly in its flexibility in predicting variations of the physical flow parameters during the course of movement. In this it neither assigns fixed values to the flow density or velocity nor assumes fixed flow rates. Hence it provides a performance-oriented predictive tool for the evaluation of a building's means of egress in terms of pedestrian movement.

The model presumes that during the evacuation of a multistory building, the occupants build up a partial flow on the corridor of each floor, which is defined between the first and the last persons of the flow.

If the following simplifications are introduced into the general model, the flow movement via staircases shows some regularities:

1. The length of the partial flow on each floor is assumed to be equivalent to the greatest travel distance along the corridor.
2. The number of persons as well as the escape route configurations are identical on each story.
3. Each partial flow attempts to evacuate simultaneously and enters the staircase at the same instant.

These simplifications result in the following regularities.

1. If the evacuation time on the corridor of each floor $t(F)$ is less than the evacuation time on the stairs per floor $t(TR)$, then the partial flows from each floor can leave the building without interaction, as shown in Fig. 9.5. In this case, the total evacuation time is given by the following equation:

$$t(\text{Ges}) = t(F) + nt(\text{TR}) \qquad (9.30)$$

where $t(F)$ = evacuation time on corridor of each floor
n = number of floors
$t(\text{TR})$ = evacuation time on stairs per floor

2. If the evacuation time on the corridor of each floor $t(F)$ exceeds the evacuation time on the stairs per floor $t(\text{TR})$, then the partial flows from each floor encounter the rest of the evacuees entering the staircase on the landing of the story below. Even though this event causes an increase in density on the stairs, the capacity of the main flow remains below the maximum value Q_{max}. This indicates that the stair width is still appropriate to take up the merged flow. In other words, if $t(F) > t(\text{TR})$ and

$$q(\text{TR}; n-1) = \frac{Q(T; n-1) + Q(\text{TR})}{b(\text{TR})} < q(\text{TR}; \text{max}) \qquad (9.31)$$

where $q(\text{TR}; n-1)$ = value of specific flow on stairs after merging process
$Q(T; n-1)$ = flow capacity through door to staircase on each floor
$Q(\text{TR})$ = initial flow capacity on stairs
$q(\text{TR}; \text{max})$ = maximum flow capacity on stairs

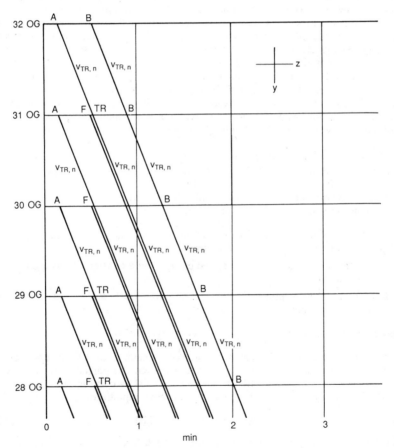

Fig. 9.5 Partial flows move down the stairs without interaction. Due to brief delay in protected lobby on each floor level, a merger of partial flows does not occur.

then the total evacuation time is given by

$$t(\text{Ges}) = t(F) + nt(\text{TR}) + m \, dt \qquad (9.32)$$

where the last term relates the delay time of the last person from the top floor. The factor m is the number of patterns of higher density, which is reduced during the course of the evacuation process. (These are the areas between the dashed lines on Fig. 9.6. The dashed lines are representing the boundaries be-

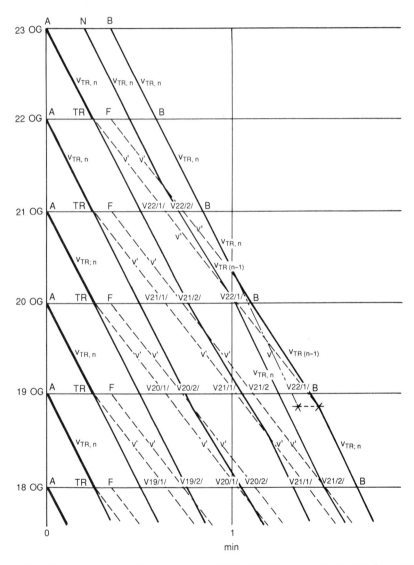

Fig. 9.6   Flow movement is delayed due to periodical increase in density on each floor level.
$t(\text{Ges}) = 6.67$ min.

tween the populations of distinct flow parameters.) m can be assessed by an iteration.

The value of the delay time dt is given by the following equation:

$$dt = v''x[t(F) - t(\text{TR})]\,\frac{v(\text{TR}; n) - v(\text{TR}; n - 1)}{v(\text{TR}; n - 1) - v''}\,v(\text{TR}; n) \qquad (9.33)$$

where $t(F)$ = travel time of last evacuee along corridor
  $t(\text{TR})$ = travel time of person going from the top floor of the stairs to the adjoining story
  v'' = velocity by which the boundary between the initial flow on the stairs with the parameters $D(\text{TR}; n)$ and $q(\text{TR}; n)$ and the merged flow with the parameters $D(\text{TR}; n - 1)$ and $q(\text{TR}; n - 1)$ changes its location
  $v(\text{TR}; n)$ = velocity of flow at density $D(\text{TR}; n)$
  $v(\text{TR}; n - 1)$ = velocity of flow on stairs at density $D(\text{TR}; n - 1)$

3. If the value of the specific flow on the stairs exceeds the maximum during the merging of the partial flows at story $n - 1$, congestion occurs on the stairs as well as at the entry to the staircase. In this case

$$q(\text{TR}; n - 1) = \frac{Q(T; n - 1) + Q(\text{TR}; n)}{b(\text{TR})} > q(\text{TR}; \text{max}) \qquad (9.34)$$

where $q(\text{TR}; n - 1)$ = value of specific flow on stairs after merging process
  $Q(T; n - 1)$ = flow capacity through door to staircase on each floor
  $Q(\text{TR}; n)$ = initial flow capacity on stairs
  $b(\text{TR})$ = stair width
  $q(\text{TR}; \text{max})$ = maximum flow capacity on stairs

From Eq. 9.7 the percentage of contribution of each partial flow to the main flow can be obtained as follows:

$$p(T) = \frac{b(T)}{b(T) + b(\text{TR})} \qquad (9.35)$$

$$p(\text{TR}) = \frac{b(\text{TR})}{b(T) + b(\text{TR})} \qquad (9.36)$$

where $b(T)$ = door width to staircase
  $b(\text{TR})$ = stair width

To determine the new widths of the partial flows on the stairs, the main width of the flow $b(\text{TR})$ is multiplied by the following fractions:

$$b(T_1) = p(T)b(\text{TR}) \qquad (9.37)$$

$$b(\text{TR}_1) = p(\text{TR})b(\text{TR}) \qquad (9.38)$$

Due to the congestion on the stairs, the evacuees from other floors cannot enter the staircase immediately. A backup occurs at the floor exit and the partial flow on the corridor extends backward at a speed of:

$$v''(T; \text{STAU}) = \frac{q(T; D_{max})b(T_1)}{b(T)} - \frac{q(T)}{D_{max}} - D(T) \qquad (9.39)$$

where v'' $(T; \text{STAU})$ = speed of congestion on corridor
$q(T; D_{max})$ = specific flow at maximum density through doorways
$b(T_1)$ = width of partial flow from each floor in main flow on stairs
$b(T)$ = door width to staircase, or width of flow from each floor under free-flow conditions
$q(T)$ = specific flow through door to staircase under free-flow conditions on stairs
D_{max} = maximum flow density
$D(T)$ = density through door to staircase without congestion on stairs

After the last person of the flow on the corridor reaches the queue at the entry to the staircase, the congestion diminishes at the following speed:

$$v(T; \text{STAU}) = \frac{v(T; D_{max})b(T_1)}{b(\text{TR})} \qquad (9.40)$$

where $v(T; D_{max})$ is the velocity of a flow through a doorway at the maximum density.

The flow movement on the corridor ends at the instant $t(F; \text{STAU})$ indicated as F; STAU in Fig. 9.7; $t(F; \text{STAU})$ is the egress time from a story.

From the beginning of the merging process of the partial flows until the end of queuing at the exit door to the stairway, congestion also occurs on the stairs. Due to this, the flow on the stairs extends backward at a rate of:

$$v'(\text{TR}; \text{STAU}) = \frac{q(\text{TR}; D_{max})b(\text{TR}_1)/b(\text{TR}) - q(\text{TR}; n)}{D_{max}} - D(\text{TR}; n - 1) \qquad (9.41)$$

where $v'(\text{TR}; \text{STAU})$ = speed of congestion on stairs
$q(\text{TR}; D_{max})$ = specific flow on stairs at maximum density
$b(\text{TR}_1)$ = width of partial flow from top floor n in main flow on stairs
$b(\text{TR})$ = stair width, or width of flow on stairs under free-flow conditions
$q(\text{TR}; n)$ = specific flow on stairs without congestion
D_{max} = maximum density
$D(\text{TR}; n)$ = flow density on stairs without congestion, or density of partial flow from top floor

Here two different situations may arise:

1. $t(F) + t(\text{TR}) > t(F; \text{STAU})$. For example, if the last person from the floor under the top story enters the staircase before the last person from the top floor reaches the line on the stairs, then the partial flow from the top floor can use the total stair width after $t(F; \text{STAU})$ (Fig. 9.7). In this case, the speed of congestion on the stairs changes as follows:

$$v''(\text{TR}; \text{STAU}) = \frac{q(\text{TR}; D_{max}) - q(\text{TR}; n)}{D_{max}} - D(\text{TR}; n) \qquad (9.42)$$

After the last person from the top floor reaches the line on the stairs, the congestion diminishes at the speed $v(\text{TR}; D_{max})$, which corresponds to the flow velocity on the stairs at the maximum density.

2. If $t(F) + t(\text{TR}) > t(F; \text{STAU})$, then the last person from the top floor reaches the line on the stairs before the last person from the story below enters the staircase. In this case the congestion on the stairs diminishes at the following speed until $t(F; \text{STAU})$:

$$v(\text{TR}; \text{STAU}) = \frac{v(\text{TR}; D_{max})b(\text{TR}_1)}{b(\text{TR})} \qquad (9.43)$$

where $v(\text{TR}; \text{STAU})$ = flow velocity on stairs at maximum density
$b(\text{TR}_1)$ = width of partial flow from top floor in main flow

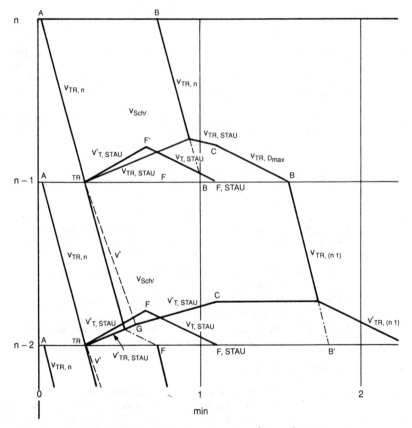

Fig. 9.7   Flow motion process calculated using $f = 0.14 \text{ m}^2$ (1.5 ft²). Delay time due to congestion is $\Delta\tau = 0.43$ min, repeated on each floor level. $t(\text{Ges}) = 15.34$ min.

$b(\text{TR})$ = stair width, or width of flow from top floor under uncongested-flow conditions

After the end of congestion at the entry to the staircase, the partial flow from the top floor can use the total width of the stair again. The movement process at floor $(n - 1)$ is complete at the point indicated as TR; STAU in Fig. 9.7. The flow moving down the stairs from floor $(n - 1)$ consists of two different groups of people. The movement parameters of the part ahead are q (TR; n) and $v(\text{TR}; n)$, which are the initial flow parameters. This group is followed by the evacuees emanating from the overcrowded area at the level $(n - 1)$ and moving by the specific flow $q(\text{TR}; D_{\text{max}})$, but at a lower density than the maximum. [After Fig. 9.1 there are two different density values, corresponding to the specific flow at the maximum density $q(\text{TR}; D_{\text{max}})$.]

During the merging process of both groups, the boundary between them changes its location at a speed of v'', which can be determined by the following equation:

$$v'' = \frac{q(\text{TR}; D_{\text{max}}) - q(\text{TR}; n)}{D(\text{TR}; n - 1) - D(\text{TR}; n)} \tag{9.44}$$

where $q(\text{TR}; D_{\text{max}})$ = specific flow on stairs at maximum density
   $q(\text{TR}; n)$ = specific flow on stairs at beginning of flow movement
   $D(\text{TR}; n - 1)$ = density of group of people emanating from overcrowded area at level $(n - 1)$
   $D(\text{TR}; n)$ = initial flow density on stairs

Simultaneously, on the corridor of story $(n - 2)$ and on the stairs between floors $(n - 1)$ and $(n - 2)$, the flow motion forms in a manner similar to the flow movement on the upper flows. There the velocity of the flow extending backward on the stairs will be different according to whether it reaches the boundary mentioned before or after $t(F; \text{STAU})$. [$t(F; \text{STAU})$ is the egress time from a floor.]

If the end of the line arrives at the boundary before $t(F; \text{STAU})$, then the rate of congestion is determined by Eq. 9.41. Otherwise the rate of congestion is predicted by the following equation:

$$v''(\text{TR}; \text{STAU}) = \frac{q(\text{TR}, D_{\text{max}})b(\text{TR}_1)}{b(\text{TR})} - \frac{q(\text{TR}; D_{\text{max}})}{D_{\text{max}} - D(\text{TR}; n - 1)} \tag{9.45}$$

It should be noted that, in this case, the values of the specific flows for both groups, the incoming flow as well as the uptaking flow, are the same. After the last evacuee from floor $(n - 2)$ enters the staircase, the length of the congested flow on the stairs remains constant until the last person from story $(n - 1)$ reaches the waiting population on the stairs. Then the congestion diminishes at the rate $v(\text{TR}; D_{\text{max}})$, which is the flow velocity at the maximum density. The flow movement at story $(n - 2)$ ends at the point $t(\text{TR}; \text{STAU})$, indicated as B in Fig. 9.7.

If the last person from floor $(n - 1)$ had reached the adjoining story without any delay due to congestion, he or she would arrive there after the time t_1. The delay time due to congestion on escape routes dt, repeated at each floor level, is predicted by

$$dt = t''(\text{TR}; \text{STAU}) - t_1 \tag{9.46}$$

where $t''(\text{TR; STAU})$ is the length of time required for the flow to leave the floor level $(n - 2)$.

In case of congestion on escape routes, the total evacuation time of a multi-story building is determined by the following equation:

$$t(\text{Ges}) = t(\text{TR; STAU}) + \frac{(n - 1)l(\text{TR})}{v(\text{TR; } n - 1)} + (n - 2)\, dt \qquad (9.47)$$

where $t(\text{TR; STAU})$ = length of time required for flow to leave floor level $(n - 1)$

$l(\text{TR})$ = travel distance on stairs between adjoining stories

$v(\text{TR; } n - 1)$ = velocity of flow emanating from congested area at floor level $(n - 1)$

dt = delay time due to congestion

n = number of upper floors in building

The total evacuation time $t(\text{Ges})$ is influenced in a nonlinear fashion by the projected area factor (or the density increase). Figure 9.8 illustrates the change in evacuation time in three high-rise administration buildings, plotted against the projected area factor f and the number of persons per floor $P(G)$.

Within the range of experimental data from real evacuation tests and by using the average value of $f = 0.12$ to 0.14 m^2 (1.29 to 1.51 ft^2) per person, the predictions of the presented egress model are likely to provide an adequate basis for the assessment of flow movement on escape routes. Improvement of the model is certainly possible, but it would require additional specific data in terms of flow density.

Figure 9.9 shows the diagram of an office occupancy floor of one of the high-rise administration buildings where a real evacuation test was conducted. This building consists of a ground floor, one mezzanine, 21 upper floors, and two tower stories. The height between two floors was measured as 3.60 m (11.8 ft). The building is arranged around a triangular core with a staircase at each corner of the core layout. Each staircase is approached by a protected lobby with a length of 0.90 m (2.6 ft). The continuous corridor leading to a staircase has a width of $b(F) = 1.87$ m (6.13 ft). The greatest travel distance along the corridor measures $l(F) = 29.40$ m (96.4 ft). The doorway opening between the corridor and the protected lobby, as well as the exit door to the staircase, has a width of $b(T) = 0.82$ m (2.68 ft). Due to the triangular form of the ground plan, the staircases are also arranged around a triangular pillar, with three flights between two floors. The width of the stairs is $b(\text{TR}) = 1.25$ m (4.1 ft). During the evacuation test, 567 persons were evacuated via the observed staircase. The average number of persons per floor was $p(G) = 26$ [approximately 20 m^2 (215 ft^2) per person]. Although the building was apparently underoccupied, the flow downstairs was delayed for the first 3 to 4 min because of the door with fairly inappropriate dimensions swinging into the protected lobby.

In this case the model predicts the flow from the floors through the protected lobby into a staircase and downstairs to the final exit. Furthermore, it presumes that the evacuation has already been initiated, and at the time 0, indicated as point A in Fig. 9.7, the first person of the partial flow on each floor passes through the doorway into the protected lobby.

By changing the number of persons per floor per staircase, or changing the corridor width leading to the staircase or the width of the floor exits, the egress model predicts the results listed in Table 9.4 (Kendik, 1985a).

Given the data corresponding to the real evacuation test (Table 9.4, column I), the model predicts the total evacuation time to be 10.29 min, which is fairly close to the measured time of 10.47 min. The floor egress time from an upper floor (except the top and ground floors) is estimated to be 0.79 min under congested-flow conditions. The length of the congestion would be 0.71 m (2.3 ft) at the floor exit and 1.11 m (3.6 ft) on the stairs.

By increasing the number of persons per floor per staircase from 26 to 40, the total evacuation time increases significantly (15.34 min from Table 9.4, column II). By decreasing the corridor width from 1.87 to 1.25 m (6.13 to 4.1 ft), the total evacuation time of 40 persons per floor per staircase is predicted to be 14.50 min, which is less than the time estimated in the previous example, since due to higher

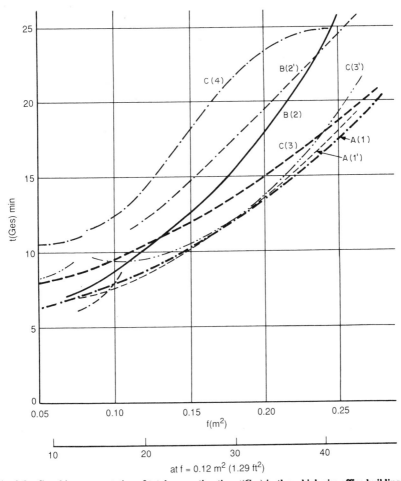

Fig. 9.8   Graphic representation of total evacuation time t(Ges) in three high-rise office buildings in terms of projected area factor f and number of persons per floor $P(G)$. When determining the total evacuation time in terms of the number of persons per floor $P(G)$, the projected area factor is assumed to stay constant at $f = 0.12$ m^2 (1.29 ft^2).

density in the corridor, less persons can enter the staircase in the same time period. Hence the initial stair density and the speed of congestion decrease.

Widening the floor exit from 0.82 to 1.25 m (2.7 to 10.8 ft) (which corresponds to the stair width) does not change the evacuation pattern significantly. In this case the floor egress time decreases, which leads to a greater congestion on the stairs (Table 9.4, column IV).

For the building the total length of the corridor on each floor is approximately 1.10 m (3.6 ft). Decreasing the corridor width would not threaten the flow movement. (The predictions show that the total evacuation time slightly decreases.)

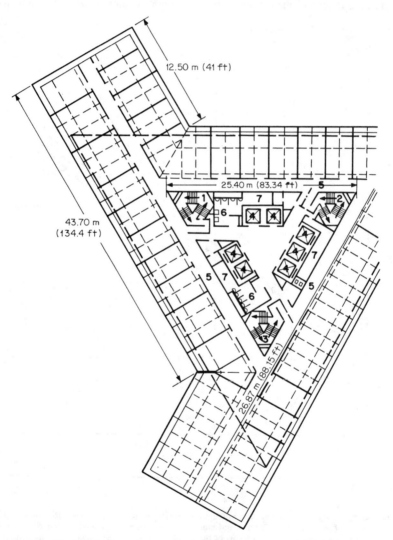

Fig. 9.9 Office occupancy floor of high-rise administration building.

Less corridor width would mean approximately 70 m² additional space on each story and approximately 1540 m² (16,575 ft²) more rental area for this entire building (corresponding to the equivalent area of one floor).

9.4 COMPUTER PROGRAM FOR THE DESIGN AND EVALUATION OF ESCAPE ROUTES

Recently Kendik wrote a computer program in BASIC language for an HP 150 personal computer based on the egress model described. It is written as a dialogue between the user and the computer, where the escape route configurations (the width and length of each section) as well as the number of occupants are entered during the course of the computation. The program easily enables the user to change the dimensions of the building's means of egress and the occupant load, and it works out the influence of these variations on the complete circulation system.

9.5 CONDENSED REFERENCES/BIBLIOGRAPHY

BSI BS 5588 1983, *Fire Precautions in the Design and Construction of Buildings: Part 3*
Fruin 1970, *Designing for Pedestrians: A Level of Service Concept*
Garkisch unpublished, *Simulation der Räumung eines Gebäudes im Gefahrenfall*
Greater London Council 1974, *Code of Practice*
HMSO 1952, *Post-War Building Studies No. 29*

Table 9.4   Results predicted by egress model

	I	II	III	IV
Number of persons	26	40	40	40
Corridor length, m	29.4	29.4	29.4	29.4
Corridor width, m	1.87	1.87	1.87	1.87
Door width, m	0.82	0.82	0.82	0.82
Flow density on corridor	0.07	0.10	0.15	0.15
Flow velocity on corridor, m/min	44.43	39.00	32.73	32.73
Specific flow on corridor, m/min	2.94	3.97	4.99	4.99
Egress time from top floor, min	0.69	0.80	0.93	0.93
Initial flow density on stairs	0.10	0.16	0.12	0.12
Speed of congestion at floor exit of any upper floor, m/min	−1.75	−6.76	−3.17	−0.63
Maximum length of congestion at floor exit, m	0.71	2.48	1.87	0.41
Egress time from any upper floor, m/min	0.79	1.10	1.18	1.00
Speed of congestion on stairs, m/min	−2.11	−4.31	−2.90	−3.48
Maximum length of congestion on stairs, m	1.11	2.57	2.23	2.59
Total evacuation time via this staircase, min	10.29	15.34	14.50	14.55
Real evacuation time, min (Seeger, 1978)	10.47			

Source: After Kendik (1985a).

Home Office 1935, *Manual of Safety Requirements on Theatres and Other Places of Public*
Kendik 1982, *Die Berechnung der Räumungszeit in Abhängigkeit der Projektionsfläche bei*
Kendik 1983, *Determination of the Evacuation Time Pertinent to the Projected Area Factor*
Kendik 1984, *Die Berechnung der Personenströme als Grundlage für die Bemessung von*
Kendik 1985a, *Assessment of Escape Routes in Buildings and a Design Method for Calcu-*
Kendik 1985b, *Methods of Design for Means of Egress: Towards a Quantitative Compar-*
Knötig 1980, *Generelles Inter-Aktions-Schema*
Kobayashi 1981, *Design Standards of Means of Egress in Japan*
London Transport Board 1958, *Second Report of the Operational Research Team on the*
Melinek 1975, *An Analysis of Evacuation Times and the Movement of Crowds in Buildings*
Müller 1966, *Die Beurteilung von Treppen als Rückzugsweg in mehrgeschossigen Gebäuden*
Müller 1968, *Die Überschneidung der Verkehrsströme bei dem Berechnen der Räumungszeit*
Müller 1969, *Die Darstellung des zeitlichen Ablaufs bei dem Räumen eines Gabäudes*
National Bureau of Standards 1935, *Design and Construction of Building Exits*
National Fire Protection Association 1982, *Code for Safety to Life from Fire in Buildings*
National Fire Protection Association 1985, *Code for Safety to Life from Fire in Buildings*
Pauls 1980a, *Building Evacuation: Research Findings and Recommendations*
Pauls 1980b, *Building Evacuation: Research Methods and Case Studies*
Pauls 1982, *Effective-Width Model for Crowd Evacuation*
Pauls 1984, *Development of Knowledge about Means of Egress*
Paulsen 1981, *Human Behaviour and Fire Emergencies: An Annotated Bibliography*
Predtechenskii 1978, *Planning of Foot Traffic Flow in Buildings*
Read, to be published, *Means of Escape in Case of Fire: The Development of Legislation*
Seeger 1978, *Untersuchung der Räumungsabläufe in Gebäuden als Grundlage für die*
Stahl 1977, *An Assessment of the Technical Literature on Emergency Egress from Build-*
Stahl 1982, *Time Based Capabilities of Occupants to Escape Fires in Public Buildings: A*
Templer 1975, *Stair Shape and Human Movement*
Tidey 1983, *Private communication*

Part 3 – Design and Construction of Tall Buildings—Contemporary Fire Safety Engineering Topics

Numerous contemporary design tools have been developed or refined in concert with advanced numerical techniques, which have evolved in the past 20 years. The purpose of these is to better describe the fire performance of building components and structures. They are available in both generalized and highly specialized forms. The more general analytical methods relate to general building design, separation, and such, while the more specific ones may, for example, describe the performance of particular building elements, sections, or assemblies. Finally, and most specific, are those analytical tools which elaborate on the anticipated behavior of specific materials or composite structures of unique design, which are critical components within a larger overall design.

The chapters in Part 3 consider these new tools as well as the experimental techniques which are used to complement them. In addition, specific materials and test results are presented as they relate to tall building design methods.

10

Analytical Tools to Evaluate Building Spacing, Compartment Sizing, and Exterior Walls in Tall Buildings

A primary way to reduce fire loss in neighboring buildings is to provide adequate fire barriers if the buildings are close together, or to provide adequate separation with sized and controlled openings where the buildings are spaced apart (Barnett, 1988). Once the buildings are far enough apart, no measures are required to protect neighboring property. Conversely, the only ways to reduce fire loss *within* an owner's building are to install sprinklers or to reduce the areas at risk, that is, to reduce the fire-compartment size. However, few, if any, fire codes state how fire-compartment sizes are to be determined—they merely prescribe maximum sizes.

Two fire-engineering design methods are discussed in this chapter. They have considerable flexibility and determine fire separation in conjunction with the fire-compartment size. Both use the common factor of fire-resistance rating (FRR), which, with suitable factors, can be linked to a nominal fire duration (NFD). Both design methods encourage the use of smaller fire compartments, which in turn will reduce the cost of property lost by fire. Both separation distance and compartment size can be related to fire duration, which in turn can be linked to FRR. Hereafter, the fire duration is referred to as the nominal fire duration, or NFD.

10.1 FIRE-RESISTANCE RATING

FRR is determined in a fire test laboratory by subjecting a loaded or unloaded construction to a fire environment controlled to a standard time-temperature, as in BS 476 (BSI, 1986), ISO 834 (1975), or ASTM E-119 (1988). For the fire design of a particular building, four types of FRR need to be considered. First, there are what may be termed *collapse* and *noncollapse* FRRs. Attention must be paid to their purpose, the aim of the first being to protect life, that of the second, to protect property. Second, there can be minimum and maximum FRRs.

A collapse FRR refers to one in which the building element may collapse after the FRR time has been reached. This is generally prescribed by fire authorities for life-safety construction, protecting egress ways such as in stairs and corridors. Once the life-safety function has been completed, collapse may not concern the authorities. For example, the egress fire barriers and their supports may be specified as requiring a ½-hr FRR in a fire compartment likely to have a fire of 2-hr duration.

A noncollapse FRR refers to one in which the building element is not intended to collapse during fire. This is generally prescribed by fire authorities for property-safety construction. For example, an external wall within critical distance of a neighboring building should remain stable for the duration of the fire, NFD. If its openings increase in size or the wall collapses, the neighboring building can be subjected to increased radiation at that point in time. Similarly, a wall acting as a fire separation for property protection between two fire compartments should remain stable for the NFD to prevent horizontal fire spread through the building. Also, a floor installed as a fire separation for property protection should remain stable for the NFD to prevent vertical fire spread through the building.

A minimum FRR may be required for any building element supporting a life-safety fire barrier. For example, a stairway in a multistory building may require a 1-hr FRR and pass down through various fire compartments requiring a range of FRRs for property protection, such as ½, 1, or 2 hr. Any structural member inside these fire compartments supporting the 1-hr separations of the stairway needs to have a minimum of 1-hr FRR.

A maximum FRR related to the available water supply may need to be considered if reliance is placed on the fire service to control the fire with water. The control of fire intensity can be related to the available flow of water. The NFD and, hence, the maximum permissible FRR can be linked to the available storage of water. The two examples in Table 10.1 may make this clear. The background behind the coefficients k_{11} and k_{13} is given in Barnett (1989).

The first example is able to control and extinguish a 133-MW fire, but only up to a ½-hr FRR. The second example, with three times the stored water but only half the water flow, can control a smaller 67-MW fire with up to a 4-hr FRR.

10.2　FIRE-COMPARTMENT-SIZING METHOD

Of necessity, fire-code writers have to oversimplify the results of fire research in order to produce design methods for use in their codes. To extinguish a fire, the

Table 10.1　Examples of control of fire intensity related to available flow of water

	Example 1	Example 2
Available water storage S	500,000 l	1,500,000 l
Available water flow F	160 l/s	80 l/s
Water-flow duration time $t_w = S/60F$	52 min	312 min
Nearest permissible FRR	30 min	240 min
Time safety factor $k_{11}, = FRR/t_w$	1.74	1.30
Theoretical cooling $Q_w = 2.605F$	417 MW	209 MW
Cooling efficiency k_{13}	0.32	0.32
Effective cooling capacity $Q_\sigma = Q_w k_{13}$	133 MW	67 MW

primary factors to consider are air and water flow. During a fire in a building, the rate of pyrolysis—or the heating power of the fire—depends on airflow, which in turn is controlled by the height and width of the openings. Therefore to extinguish the fire with water, it follows that the rate of cooling—or the cooling power and, hence, water flow—may also be related to airflow. The fire duration also depends on airflow as the flow controls the rate of pyrolysis of the total fire load. The total fire load in turn depends on the fire area and its corresponding fire-load density. Simple algebraic formulas link all these factors.

Total fire loads can be quoted in three forms: mass G, mass in "wood equivalent" values B, or stored-energy values E based on the net calorific value at ambient moisture content. In reality, fire intensities are determined not by stored energy but by energy-release rates from a given specific surface. While the energy-release rate system can be used for the more sophisticated fire-engineering designs, it is likely to be too cumbersome for everyday use by designers and building officials. Therefore, this discussion concentrates on the stored-energy basis using the term e_F as the fire-load energy density:

$$e_F = \frac{E}{A_F} \quad \text{MJ/m}^2 \tag{10.1}$$

where E = total net energy at ambient moisture content
A_F = nominal fire compartment area

1 Energy Release

The common approach is to regard fire as a burning process whereby fuel appears to be consumed. The process may be referred to in mass-release or energy-release terms.

Mass release may be termed *burning rate*. Because not all of the released fuel actually burns, a better term is *pyrolysis rate*, which can be defined as the mass rate of evaporation of material from a liquid pool or as the mass-loss rate of a solid or gas which accompanies combustion during a fire. In mathematical terms, and at any given point in time, the rate of pyrolysis is $R = dB/dt$ kg/s. Figure 10.1a is a typical mass-loss graph, while curve B in Fig. 10.1b is the corresponding pyrolysis rate. It is important to distinguish between different rates of pyrolysis. R_{peak} is the maximum value reached during a fire, and it generally exceeds $R_{80/30}$ by a small amount. When R is quoted in normal fire-research literature, it generally refers to $R_{80/30}$, which is regarded as the mass-loss rate; it occurs in the time interval between 80 and 30% of the fuel's original weight. This is shown as curve A in Fig. 10.1b and termed the $R_{80/30}$ curve. Curve A also includes the loss rates before and after the 80/30 period, which are shown as $R_{100/80}$ and $R_{30/0}$. Fire engineers may, however, find it more convenient to work with average rates and apply safety factors to relate maximum to average values. This is illustrated as curve C, termed the $R_{100/0}$ curve. An important point is that, at the end of the fire, the areas under curves A, B, and C are the same and equal the total weight in kilograms of fuel consumed. This fact can be used to cross-check the mathematics used to derive curves A, B, and C. In the simplest and basic mass form,

$$B = R_{100/0}t_d \quad \text{kg} \tag{10.2}$$

In the strictest sense, fire is an energy-release or energy-loss process. The *energy-loss rate* may be defined as the rate of loss of energy from a pool, gas, or solid fire, including any unburnt combustible material. In mathematical terms, the energy loss rate Q is equal to dE/dt watts. Figure 10.2 illustrates the same concepts as Fig. 10.1. The areas under curves D, E, and F are all the same and equal the total energy E in joules released from the fuel.

Using the average energy flow $Q_{100/0}$ as an example and applying various adjustment factors, the algebra will look like the following:

1. basic energy form,

$$E = A_z e_F = Q_{100/0} t_d \quad \text{J} \tag{10.3}$$

2. modified form,

$$A_z e_F = (k_{10} Q_{80/30})(k_{11} t_f) \tag{10.4}$$

Rearranging Eq. 10.4, the design fire size A_z can be determined as

$$A_z = (k_{10} k_{11}) \frac{Q_{80/30}}{e_F} (t_f) \quad \text{m}^2 \tag{10.5}$$

It can be seen from Eq. 10.5 that the fire size A_z can be related directly to the FRR t_f and also to the $Q_{80/30}/e_F$ ratio. If acceptable maximum values can be es-

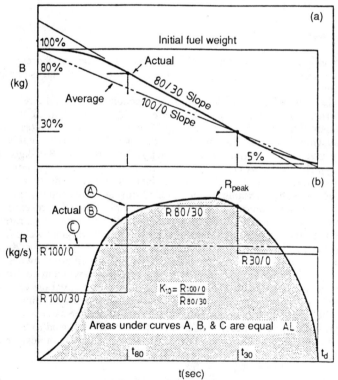

Fig. 10.1 Differences between rates of pyrolysis R. Area under all three curves A, B, and C in (b) should equal total weight loss B in (a).

tablished for $Q_{80/30}$ and e_F, each in a series of stepped categories, parameters can be set up for each fire zone. However, to develop Eq. 10.5 for fire-design sizes, the conversion factors k_{10}, k_{11}, k_{12}, and k_{13}, used for fire-zoning purposes, need to be discussed.

Factor k_{10}. A factor k_{10} is required to link the 80/30 and 100/0 rates of pyrolysis because $R_{80/30}$ will be used to determine the required minimum water flow F, whereas $R_{100/0}$ will be used to determine the NFD t_d. Thus,

$$k_{10} = \frac{R_{100/80}}{R_{80/30}} = \frac{Q_{100/80}}{Q_{80/30}} \tag{10.6}$$

Typical conservative values of k_{10} would be 0.67 for wood and 0.91 for gasoline. They can be readily established from existing test data in which full burnout has occurred.

Factor k_{11}. FRR times and fire-temperature duration times are not necessarily the same thing. Fire duration also has different meanings, depending on different fire researchers and authors. In a standard fire test furnace, the heat is supplied by fuel, and once the intended FRR test time is reached, say after 2 hr, the fuel is

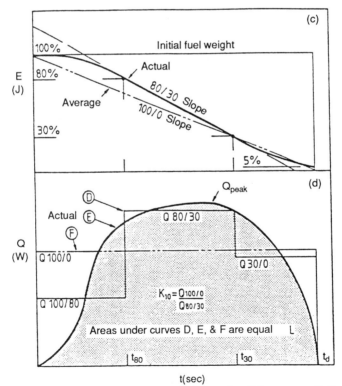

Fig. 10.2 Differences between energy loss rates Q. Area under all three curves D, E, and F in b should equal total energy loss E in a.

shut off almost immediately. While the time-temperature values are recorded during the 2-hr heat-up period, they are not normally recorded during cool-down. On the other hand, once the fuel is consumed in a building fire (or shut off so to speak), the fire duration is not considered complete until the temperature in the cool-down or decay period falls below a critical temperature. This critical temperature can be a percentage of the maximum value, or it can be an absolute limit of, say, 400 to 600°C (752 to 1100°F) depending on whether the material is steel, concrete, or a reinforcing medium. Therefore a factor k_{11} can be used to link fire-temperature duration and FRRs. Typical values for k_{11} can be derived from cooling condition curves such as those set out in ISO 834 (1975) (see Fig. 10.3).

Values for k_{11} can also be derived by using semilog graph paper on which heat-up and cool-down curves become straight lines. Plotting the standard time-temperature curve as heat-up gives an upward straight line. From each temperature at the 1-, 2-, 3-, and 4-hr positions, plot a cool-down line at an appropriate slope taken from actual experiments. Some authorities use 7°C (44.6°F) per minute for 1 hr or less, and 10°C (50°F) for above 1 hr. Each cool-down line can be finished at 80 or 50% of the 1-, 2-, 3-, or 4-hr temperature or at any other critical temperature as desired.

The suggested method illustrated in Fig. 10.4 uses the standard time-temperature relationship in ISO 834 (1975). The combined heat-up and cool-down times give t_d. Thus,

$$k_{11} = \frac{t_d}{t_f} \qquad (10.7)$$

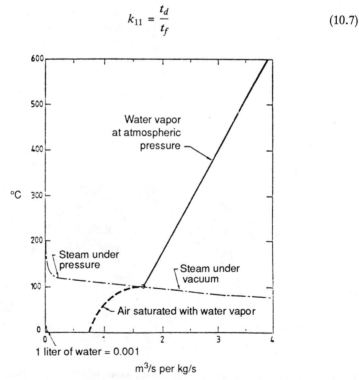

Fig. 10.3 Volume of water after being applied to a fire at the rate of 1 l/s and heated to fire temperature.

Barnett suggests that the 10°C/min (50°F/min) cool-down rate finishing at the 400°C (752°F) value would be appropriate for a fire code. Hence typical k_{11} values would be 1.92, 1.54, 1.39, and 1.31, respectively (Table 10.2).

Factor k_{12}. The maximum *theoretical,* or gross-calorific, energy output Q_F of a fire burning oven-dried wood can be summed up as follows: 1 kg/s = 20

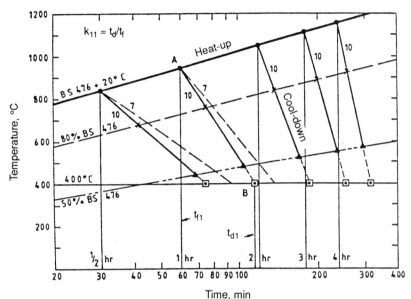

Fig. 10.4 Method of determining k_{11}.

Table 10.2 Maximum building areas based on water flow, openings, and fire-load densities

Fire zone	Flow factor	Maximum fire-load-energy densities e_F, MJ/m² Q_H/e_F									
		0.008	0.017	0.025	0.033	0.050	0.067	0.100	0.133	0.200	0.267
1	40	4000	2000		1000		500				
2	80		4000		2000		1000		500		
3	120			4000		2000		1000		500	
4	240				4000		2000		1000		500

FRR, hr	k_{11}	Design fire areas, A_z, m²									
½	2.47	49	99	148	198	296	395	593	790	1186	1581
¾	2.11	63	127	190	253	380	506	760	1013	1519	2000
1	1.92	77	154	230	307	461	614	922	1229	1843	2000
1½	1.68	101	202	302	403	605	806	1210	1613	2000	2000
2	1.54	123	246	370	493	739	986	1478	1971	2000	2000
3	1.39	167	334	500	667	1001	1334	2000	2000	2000	2000
4	1.31	210	419	629	838	1258	1677	2000	2000	2000	2000

MJ/s = 20 MW. The *effective* heat output Q_H of a fuel is less than the theoretical value Q_F by a thermal efficiency factor k_{12}. Thus,

$$k_{12} = \frac{Q_H}{Q_F} \qquad (10.8)$$

Oven-dry wood burning at 1 kg/s has a theoretical *gross* heating capacity of 20 MW. Wood at ambient moisture content has a theoretical *net* heating capacity of 16.7 MW/kg. After taking into account moisture in the wood and losses due to air preheat, ash, unburnt but pyrolyzed fuel, and surplus air passing out via the openings, less than half the total available heat is left within the fire compartment to weaken the structure. A k_{12} factor of 0.50 applied to the net normal-moisture calorific value represents a calorific contribution of 8.4 MJ/kg. Thus normal wood burning in a building of average ventilation characteristics leaves behind an *effective* energy flow of only 8.4 MJ/kg to heat the structure and fabric (Fig. 10.5).

Factor k_{13}. The maximum *theoretical* cooling energy Q_w of water flow can be summed up as follows: 1 l/s = 1 kg/s = 2.6 MJ/s = 2.6 MW. The actual, or *effective*, cooling input Q_c of the water which arrives directly on the fire is less than the theoretical value Q_w by an efficiency factor k_{13}. Thus,

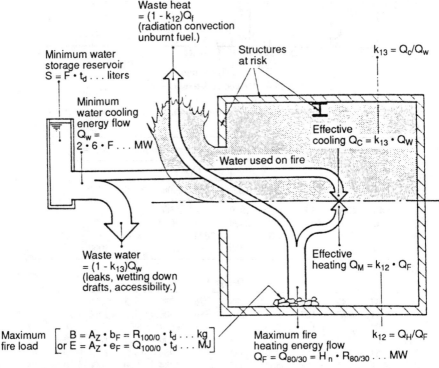

Fig. 10.5 Thermal balancing between effective cooling Q_c and effective heating Q_H for fire-zoning purposes.

$$k_{13} = \frac{Q_c}{Q_w} \qquad (10.9)$$

Most textbooks on fire fighting seem to indicate that to overwhelm a fire, the efficiency of water as a cooling medium is about one-third, or 0.32. Thus the effective cooling capacity of a flow of 1 l/s is 0.84 MW, or a *standard* 10-l/s fire hose is 8.4 MW.

Water as a quenching agent has two principal advantages. First, in the form of heated water vapor, it can displace oxygen, and second, it can cool the fire. The first advantage is illustrated in Fig. 10.3, which shows that at atmospheric pressure, 1 liter of water converts to water vapor by expanding more than 1600 times at 100°C (212°F), but it expands more than 4000 times at typical fire temperatures. The second advantage is illustrated in Fig. 10.6, which shows that the cooling power of each liter of water per second applied to a fire increases with temperature. Therefore the selection of an effective cooling power of only 0.84 MW for design purposes can be seen as conservative.

Figure 10.5 illustrates the balance between effective cooling Q_c and effective heating Q_H, each being only a portion of the theoretical cooling Q_w and theoretical heating Q_F, respectively.

10.3 FIRE-ZONING SYSTEM

One of the main difficulties in fire-engineering calculations is determining the total energy-loss rate Q. When one varies Q, it frequently affects temperature and

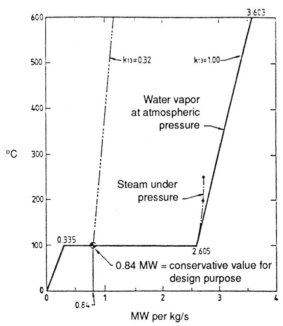

Fig. 10.6 Cooling power of water applied to a fire at the rate of 1 l/s and thermal efficiency factor k_{13}.

vice versa. By setting standard power zone levels, and hence target rates of energy release, the work of fire engineers, researchers, fire fighters, and building designers can be simplified as they will all be producing calculations, research results, extinguishing systems, and building designs within the same frame of reference and using the same units.

It would seem logical to draft fire regulations on a fire-zone classification system, which puts buildings into classes of potential heat output, or effective power. By law, limits could be established for building openings so that the effective heat generated during a fire will not be greater than the opening limit set for that zone. In this way, the fire would be unlikely to overwhelm the effective fire-extinguishing capacity provided for that zone. On the same basis, fire brigades and water-supply systems could be designed to have the necessary effective cooling capacity for that fire zone. Taken together, these policies would enable the fire-service and water-supply agencies to provide each zone with sufficient water-cooling resources, that is, not less than Q_c minimum for that zone. In turn, building regulations would limit the heat output of fires (for example, by controlling openings for fuel surface area) to be not greater than Q_H maximum for that zone.

1 Fire-Zone Water Requirements

If an engineer was asked to design a water-extinguishing system using four standard fire hoses of 10 l/s each, this would result in a properly sized system all the way from reservoir to hose nozzles, including water mains and pumps. Assuming that the minimum number of fire engines in the first fire zone would be one *standard* 40-l/s fire engine equipped with four *standard* 10-l/s fire hoses and matching water-mains facilities, a convenient power design level of 33.3 MW arises. For four fire zones, the water-extinguishing systems result in Q_c values in multiples of 33.3 MW, as shown in Table 10.3.

2 Fire-Zone Building Requirements

The energy-loss rate of a building fire depends on air supply to the perimeter of the fuel. Openings play a significant part in this airflow and, hence, in the rate-of-pyrolysis formula from which energy-loss rates are derived. Openings may be expressed in geometrical terms, such as w, h_v, A_v, or in ventilation terms, such as $A_v(h_v)^{0.5}$ or F_v:

Table 10.3 Water-extinguishing systems for four fire zones

Fire zone	Engines	Hydrants	Hoses	Water flow, F, l/s	Q_w, MW	k_{13}	Q_c, MW
1	1	2	4	40	104	0.32	33
2	2	4	8	80	208	0.32	67
3	3	6	12	120	312	0.32	100
4	4	8	16	160	416	0.32	133

$$F_v = A_v(h_v)^{0.5} = w(h_v)^{1.5} \quad m^{2.5} \tag{10.10}$$

Most buildings have opening heights that vary. The *effective* vertical opening height h_v is a single-height value that can be substituted for a number of varied opening heights in a building to give the same results in a fire. Vertical openings are considered and horizontal openings are usually ignored,

$$h_v = \frac{A_1 h_1 + A_2 h_2 + \cdots}{A_1 + A_2 + \cdots} \quad m \tag{10.11}$$

For R, any rate-of-pyrolysis formula can be selected by fire engineers to determine Q_F. Note that the calculated Q_F will be in terms of either peak or average values that are derived from $R_{80/30}$ or $R_{100/0}$. A few formulas are discussed in the following.

A simple, well-known, and easily applied formula is

$$R_{80/30} = 0.0917 A_v (h_v)^{1/2} \quad kg/s \tag{10.12}$$

It would matter very little for fire-code-writing purposes if the 0.0917 factor was increased by 9% to a value of 0.1. In view of other inexactitudes in fire-code writing, 9% is not significant. Revising Eq. 10.12 gives

$$F_v = 10R \quad l/s \tag{10.13}$$

The second well-known, but less easy to apply formula can be derived from an equation derived by Law (Ove Arup, 1977):

$$t_m = \frac{B}{(A_v A_{T3})^{1/2}} \quad min \tag{10.14}$$

which can be rewritten via Eqs. 10.2 and 10.6 as

$$R_{80/30} = \frac{(A_v A_{T3})^{1/2}}{60 k_{10}} \tag{10.15}$$

A more accurate formula, although difficult to simplify for fire-code purposes (it is readily handled by a programmable calculator or a PC), is

$$R_{80/30} = k_1 \left(\frac{W}{D}\right)^{1/2} F_v \quad kg/s \tag{10.16}$$

where k_1 can be derived from

$$k_1 = \frac{A_{T3}}{F_v(105 + 3.6 A_{T3}/F_v)} \tag{10.17}$$

The *effective* energy-loss rate Q_H is then derived from $k_{12}Q_F$, where Q_F is in turn derived from R as

$$Q_F = H_n^1 R_{80/30} \quad MW \tag{10.18}$$

Fire codes should allow designers to choose any suitable rate-of-pyrolysis formula—and the corresponding F_v value—to determine the power level Q_H for each fire zone. As an illustration, Table 10.4 has been prepared for four fire zones using the simplest rate-of-pyrolysis method, Eq. 10.13. It uses R in suggested multiples of 4 kg/s, which results in Q_H values of multiples of 33.3 MW.

3 Combined Fire-Zone Results

When Tables 10.3 and 10.4 are combined, the reasons for selecting the design power levels in multiples of 33.3 MW become obvious. As stated, oven-dry wood burning at 1 kg/s has an effective net heat transfer of about 8.4 MW. Therefore to "overwhelm" 8.4 MW of effective fire heat would require about 10 l/s of water having an effective cooling capacity of 8.4 MW. If F is the cooling flow in liters per second and R the fuel flow in kilograms per second, then F, like F_v, needs to be about 10 times the value of R. With R in selected multiples of 4 kg/s for the fire zones, the combined results become useful in the relatively simple form of Table 10.5.

For fuels other than cellulosic, that is, petrols or plastics, Tables 10.4 and 10.5 would need revision using the appropriate H_n^1 in Eq. 10.18.

4 Application to Fire Code

Older fire codes were based on fire-load densities in wood equivalents with imperial values of 12½, 25, and 50 lb/ft². Using a net calorific value of 16.7 MJ/kg to convert from a mass system to an energy system, the results for e_F become 1000, 2000, and 4000 MJ/m². As discussed previously, the suggested design values for the maximum permissible effective heat flux Q_H for the fire zones should be in multiples of 33.3 MW. Dividing Q_H by e_F produces the 10 ratios shown across the top of Table 10.2. These, in turn, can be applied to Eq. 10.5 using FRR values of

Table 10.4   Rate-of-pyrolysis method for four fire zones

Fire zone	Airflow F_v, m$^{2.5}$	Fuel flow R, kg/s	Q_F, MW	k_{12}	Q_H, MW
1	40	4	67	0.50	33
2	80	8	133	0.50	67
3	120	12	200	0.50	100
4	160	16	267	0.50	133

Table 10.5   Combined results for four fire zones

Fire zone	Design power levels Q_c and Q_H, MW	Typical fuel flow R_{max}, kg/s	Typical water flow F_{min}, l/s
1	33	4	40
2	67	8	80
3	100	12	120
4	133	16	160

½, ¾, 1, 1½, 2, 3, and 4 hr to produce the maximum permissible building square footage.

As an example, using a k_{10} value of 0.67, a k_{12} value of 0.50, $Q_{80/30}$ = $Q_F = Q_H/k_{12}$, and the FRR value t_f in seconds, Eq. 10.5 simplifies to

$$A_z = 1.333 k_{11} \frac{Q_H}{e_F} t_f \quad \text{m}^2 \tag{10.19}$$

Table 10.2 has been produced from Eq. 10.19. To use Table 10.2, enter at the appropriate fire-zone level and move right to select the appropriate fire-load density value e_F for the fire compartment. Check that the ventilation factor F_v for the fire compartment does not exceed the flow factor value shown in column 2. Below the appropriate e_F value, any of the seven design fire areas A_z can be chosen to suit the desired FRR. A small FRR calls for small floor areas, while a large FRR allows larger areas. Smaller floor areas also create smaller radiation surfaces on external walls, which in turn affect fire-separation barriers between buildings. The same FRR is then used in conjunction with the design method for fire separation between external walls described in Section 10.4. If separation proves to be unsafe, the FRR must be revised, and a new design fire area must be selected from Table 10.2; otherwise opening sizes must be reduced.

5 Tradeoff Coefficients

As it is customary in most fire codes to allow tradeoffs for fire-extinguishing facilities and fire-service accessibility, coefficients such as C_1, C_2, and C_3 can be considered.

C_1 can be regarded as an external fire-extinguishing-facilities coefficient based on how large an average fire is when put out by water applied by a fire service. The floor area of an extinguished fire A_z is not always as large as the entire fire compartment A_F. Thus,

$$C_1 = \frac{A_F}{A_z} \tag{10.20}$$

A typical fire-service annual report might show, for example, that in 80% of all property fires, only 50% of the property is damaged.

C_2 can be regarded as an internal fire-extinguishing-facilities coefficient. For example, use C_2 = 1.0 for no facilities, 1.1 for 12-mm (0.5-in.) hose reels, 1.2 for 18-mm (0.7-in.) hose reels, 2.0 for automatic gas-extinguishing systems, 5.0 for single-supply sprinkler systems, and 10.0 for double-supply systems.

C_3 can be regarded as a fire-compartment-accessibility coefficient. For example, C_3 would be 1.0, 1.1, 1.2, and 1.3 for fire-fighter access on foot via different remote entrances; or C_3 would be 1.0, 1.2, 1.4, and 1.6 for approved fire-engine access around 1, 2, 3, or 4 sides of a building, respectively; but a single coefficient cannot represent both fire-fighter and fire-engine access.

6 Fire-Compartment Size

If desired, a nominal fire-compartment area A_F can be derived from Table 10.2 as follows:

$$A_F = C_1 A_z \quad \text{m}^2 \tag{10.21}$$

Alternatively, using all three coefficients, the maximum permissible fire-compartment area A_{FM} can be derived as

$$A_{FM} = C_1 C_2 C_3 A_z \quad \text{m}^2 \tag{10.22}$$

10.4 FIRE-SEPARATION METHOD

If they use the concept at all, most fire codes use only one or two radiation-flux values when considering the external walls of buildings. If it is satisfactory to assume that fire-compartment temperatures are proportional to time, then it should be equally satisfactory to assume that radiation-flux levels are proportional to time. On this premise, and using the standard time-temperature curve, a wider than usual range of design radiation values can be derived, which are in step with standard FRRs. The method allows building designers more flexibility in the choice of ordinary or fire windows, both in mirror-image and in non-mirror-image situations (see discussion on distance in Subsection 1).

Subdivided into five categories, the hazards likely to cause a neighbor's property to ignite are external wall collapse, flying brands, flame contact, emitted heat radiation, and received heat radiation.

In fire-engineering terms, an external wall is a special kind of wall that is different from ordinary internal walls, and may be different from fire walls and fire partitions. Within flame-contact range, the external wall needs to function like a fire wall and cope with fire from both sides. Beyond flame-contact range, but within radiation danger range, the external wall needs to cope with fire from inside and radiation on the outside. Once the distance between buildings is large enough, no measures need to be taken for the safety of the neighbor's property.

1 Meaning of "Distance"

Because the positions of owner and neighbor can be reversed in a fire, that is, either building can ignite, all wall criteria apply equally to both parties. In traditional fire codes, each building situation is regarded as being the mirror image of the other (see Fig. 10.7). Where the owner has no control over the neighbor's construction, mirror-imaging, or equal distancing, may not occur, but safety is not necessarily impaired (see Fig. 10.8). Furthermore, mirror imaging does not necessarily err on the side of safety for unequal limiting distances, as illustrated in Fig. 10.9. The meaning of distance is important in fire-code applications and needs to be clearly understood. For the purpose of this discussion, the following definitions shall be used:

1. Radiation distance R shall be the distance between the flame front of a burning building and the face of any exposed building (owner's or neighbor's).
2. Separation distance S shall be the distance between the wall face of the burning building and the wall face of any exposed building (owner's or neighbor's).
3. Limiting distance L shall be the distance between the face of a burning building and an adjacent building. L_0 shall refer to the owner's limiting dis-

tance and L_N to the neighbor's limiting distance. (Note that limiting distance is not necessarily half of separation distance or radiation distance; refer to Fig. 10.7.)

4. Protected limiting distance L_x shall be the value of the limiting distance spec-

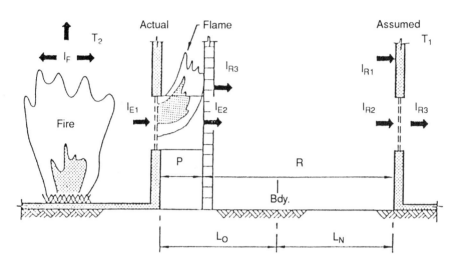

(a) Ordinary windows - mirror imaged

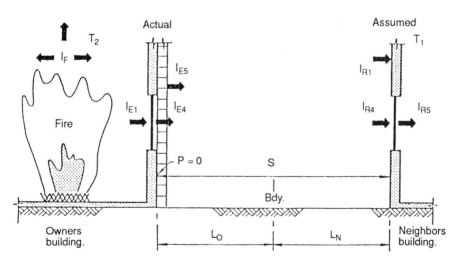

(b) Fire windows - mirror imaged

Fig. 10.7   Diagrams illustrating separation distance S, flame-projection distance P, and limiting distance L. Radiation distance is R. (a) Ordinary windows, mirror-imaged. (b) Fire windows, mirror-imaged.

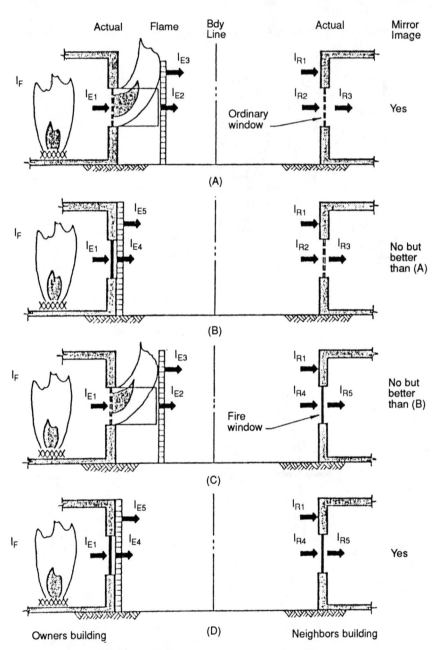

Fig. 10.8 Diagrams illustrating differences between ordinary and fire windows at equal limiting distance when based on mirror imaging.

ified by the writers of the fire code in which fire-resistance closures shall be mandatory.

5. Flame-projection distance P shall be either 2 m (6 ft) or less if calculated in an approved manner under full wind conditions, or zero if suitable fire windows are installed.

Based on the foregoing definitions and as illustrated in Fig. 10.7,

$$S = R + P \tag{10.23}$$

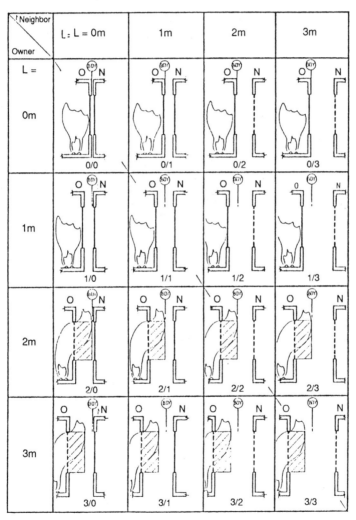

Fig. 10.9   Flame contact for $P = 2$ m and $L_x = 1$ m. Flame contact occurs in 1 case (2/0) out of 16. Two cases (0/1 and 2/3) are unsafe from radiation hazard when owner and neighbor are reversed.

$$L = \frac{1}{2} S(R + P) \qquad (10.24)$$

2 External Wall Collapse

In designing an owner's building, the designer should ensure that the exterior walls remain standing for the duration of a fire because they act as heat barriers to prevent spread of fire to a neighbor's building. In addition, ignition of the neighbor's property by the owner's wall can be controlled by ensuring that any external wall of the owner's building within danger range (assumed as the height of the owner's wall) should also remain standing. If any parts of the external wall collapse too soon, projecting flame sizes may be increased and emitted radiation areas will certainly be increased. Both effects may cause ignition of the neighbor's property, which otherwise might not occur.

3 Flying Brands

Combustible materials on or inside neighboring buildings can be ignited from a fire by any of three forms of ignition, namely, spontaneous, pilot, and contact ignition. The most common combustible material found on the exterior of buildings is wood, and its ignition behavior when preheated by radiation is representative of a large variety of combustible building materials. This is shown as I_{R1} in Figs. 10.7 and 10.8. Typical values of radiation intensities for pilot and spontaneous ignition of wood are 12.5 and 25 kW/m^2, respectively. Where a neighboring building does not have any combustible material on the exterior and has fire windows installed, neither contact ignition nor pilot ignition will apply, and flying brands can be ignored. Where an adjacent building has ordinary glazing in its windows, which is liable to crack under radiant heat, broken openings may admit flying brands. These flying brands may cause pilot ignition of combustibles inside the window, such as curtaining, wood paneling, or stored goods. This is indicated by I_{R3} in Figs. 10.7 and 10.8.

4 Flame Contact

The danger range for flame contact depends on the dimensions of the projecting flame. In general it can be said that under no wind conditions, flames will project a distance varying from half the window height for long windows to 1½ times the window height for square windows. Ignition control between buildings depends on the selected design values of the flame-projection distance P and the specified protected limiting distance L_x. The values of P and L_x vary in different codes and are not always easy to determine from the codes themselves. For example, the British system seems to work from building face to building face, that is, $P = 0$ m and $L_x = 0$ m; the Canadian system seems to work from flame front to building face, using $P = 1.5$ m and $L_x = 1.2$ m; for New Zealand's proposed new fire code, values of $P = 2$ m and $L_x = 1.5$ m have been recommended.

5 Emissivity

The intensity of radiant heat energy given off by a heated object depends on its emissivity ϵ. The perfect emitter is a blackbody and has an emissivity of unity. A fire compartment, when heated, acts as a cavity radiator with holes in it and thus approximates a blackbody with an emissivity of 1.00.

6 Configuration Factors

The intensity of radiant energy I_R falling on a surface from an emitter can be found by using an appropriate configuration factor ρ_n, which takes into account the geometrical relationship between emitter and receiver. For exterior facades, an enclosing rectangle, or rectangular radiating facade, F is assumed, having height H and width W, which in turn is divided into four equal rectangles A, B, C, and D. The configuration factor ρ_A for rectangle A is taken perpendicular to the central corner of rectangle A (and similarly for B, C, and D). This method of presentation takes advantage of the fact that configuration factors are additive. The value of ρ_A can vary from 0 to 0.25 and the value of ρ_n from 0 to 1.0. The total configuration factor ρ_n for the whole rectangle is made up of the individual configuration factors for rectangles A, B, C, and D. Thus,

$$\rho_n = \rho_A + \rho_B + \rho_C + \rho_D = 4\rho_A \qquad (10.25)$$

The following formula for a receiver parallel to the radiator is applicable:

$$\phi_A = \frac{1}{360}\left[\frac{x}{(1 + x^2)^{1/2}}\tan^{-1}\frac{y}{(1 + x^2)^{1/2}} + \frac{y}{(1 + y^2)^{1/2}}\tan^{-1}\frac{x}{(1 + y^2)^{1/2}}\right] \qquad (10.26)$$

where $x = H_A/R = H/2R$; $y = W_A/R = W/2R$
   H_A = height of quarter rectangle A, = $H/2$
   W_A = width of quarter rectangle A, = $W/2$
   R = radiation distance between emitter and receiver
   $\tan^{-1}$ = degree mode

Equations 10.25 and 10.26 do not depend on the use of water and, hence, can be applied in nonwater regions such as rural areas, small islands, and frozen climates.

7 Emitted Heat Radiation

Increasing the separation distance between buildings reduces the radiation hazard, and for a given separation distance, the intensity of received radiation I_R on the nonburning building depends on the intensity of emitted radiation I_E from the burning building. The maximum fire-compartment temperature T_2 in degrees centigrade can be determined in a number of ways from a standard time-temperature formula such as BS 476 or ISO 834 as follows:

$$T_2 = 345 \log_{10}(8t_m + 1) + T_1 \quad °C \qquad (10.27)$$

The radiant energy flux I_F present inside a fire compartment can be considered proportional to the difference in temperature of the fire compartment T_2 and its surrounding T_1 in kelvins. Thus,

$$I_F = \epsilon\sigma(T_2^4 - T_1^4) \quad \text{kW/m}^2 \qquad (10.28)$$

I_F can be considered as radiating in all directions inside the fire compartment at once. Therefore it can be expected to be the same value as I_{E1} inside a compartment's opening, as illustrated in Fig. 10.7. Thus $I_{E1} = I_F$. By using Eqs. 10.27 and 10.28, values of I_{E1} corresponding to standard FRR times can be determined, as shown in Fig. 10.10.

Because of the emitted radiation just inside the opening I_{E1}, the question arises as to what critical value of emitted radiation I_{EC} should be used for outside

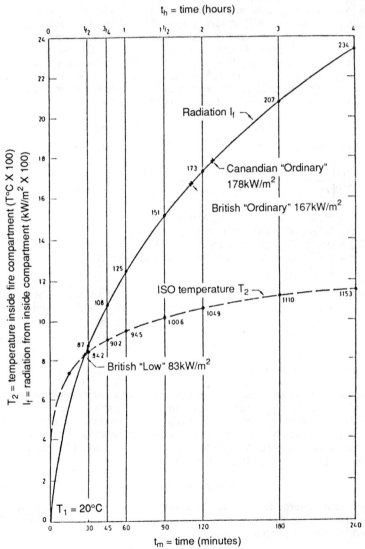

Fig. 10.10   ISO temperature T_2 and corresponding radiation I_F versus time t_m for a fire compartment.

a burning building. This depends on whether or not the glazing remains in position. Figure 10.7 shows the outside critical value from a burning building as I_{E2} for ordinary windows and I_{E4} for fire windows. When the glass remains in position, the radiation on the outside of the glass reduces to a value of between 10 and 50% of the impressed radiant heat. Thus a radiation-reduction factor k_1 of 0.50 through wired glass would appear to be a reasonable design assumption. Provided the glass remains in position, that is, fire-resistant glazing is used, this reduced radiation effect can be utilized in building-separation calculations.

When ordinary glazing cracks and falls out of an opening, flames begin projecting outside the building. The area of flame greater than the openings is ignored, and these radiating openings are assumed to be acting at a distance out from the building face (not at the face of the building wall) equal to the flame-projecting distance P. Radiation from projecting flames is usually neglected in fire-separation calculations. This is considered to be a conservative adjustment, which greatly simplifies the design procedure for radiation-distance calculations.

For ordinary glazing $k_1 = 1.0$, and thus, $I_{EC} = I_{E2} = I_{E1} = I_P$. For fire-resistant glazing $k_1 = 0.50$, and thus, $I_{EC} = I_{E4} = 0.5I_{E1} = 0.5I_P$. These values for I_{EC} refer to the unit radiation from one opening. Most external walls have a number of openings, but the problem can be reduced to that of a single opening or radiator, termed the *enclosing rectangle,* which is considered as emitting radiation over its whole area at a reduced intensity I_{E3} (or I_{E5}) (refer to Figs. 10.7 and 10.8). The enclosing rectangle should be calculated for a number of selected large and small cases, as shown on Fig. 10.11, and the worst case should apply. The reduction factor k_v is taken as the ratio of the sum of the areas of all the vertical openings in the wall A_v to the total area of the enclosing rectangle A_E, which equals $H \times W$. k_v is also equal to the ratio between I_{E3} (or I_{E5}) and I_{EC}. Thus,

$$k_v = \frac{A_v}{A_E} = \frac{I_{E3}}{I_{EC}} \quad \text{or} \quad \frac{I_{E5}}{I_{EC}} \tag{10.29}$$

8 Received Heat Radiation

When a neighbor's building receives heat radiation I_R, there is not only the hazard to any combustible material on the outside of the building I_{R1}, but there is also a hazard to the combustible contents in any room likely to receive radiation

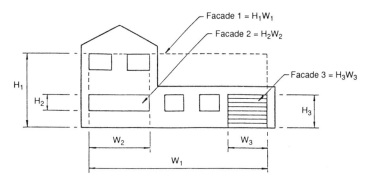

Fig. 10.11   Various types of facades from which separation distances are calculated.

Table 10.6 Table showing how, for a given FRR, the radiation factors for individual cases A, B, C, and D are in the ratio of 1, 2, 4, and 8*

CASE	A			B			C			D		
FRR hrs	I_{RC} kW/m²	I_{EC} kW/m²	k_R	I_{RC} kW/m²	I_{EC} kW/m²	k_R	I_{RC} kW/m²	I_{EC} kW/m²	k_R	I_{RC} kW/m²	I_{EC} kW/m²	k_R
1/2	12.5	87	0.144	12.5	43.5	0.287	50	87	0.575	50	43.5	1.149
3/4	12.5	108	0.116	12.5	54	0.231	50	108	0.463	50	66.7	0.926
1	12.5	125	0.100	12.5	62.5	0.200	50	125	0.400	50	62.5	0.800
1-1/2	12.5	151	0.083	12.5	75.5	0.166	50	151	0.331	50	75.5	0.662
2	12.5	173	0.072	12.5	86.5	0.145	50	173	0.289	50	86.5	0.578
3	12.5	207	0.060	12.5	103.5	0.121	50	207	0.242	50	103.5	0.483
4	12.5	234	0.053	12.5	117	0.107	50	234	0.214	50	117	0.427
k_1	1.0			0.5			1.0			0.5		
k_2	1.0			1.0			0.5			0.5		

*For cases refer to Fig. 10.8.

Table 10.7 Case factors k_c as per Fig. 10.8

Case	Case factor k_c
A	1
B	2
C	4
D	8

Table 10.8 Fire facade factors k_F

Facade size			Radiation distance R, m								
H, m	W, m	A_E, m²	0.01	0.25	0.50	0.75	1.00	1.25	1.50	2	3
1	1	1	1.00	1.20	1.81	2.80	4.18	5.95	8.11	13.6	29.3
	2	2	2.00	2.26	2.99	4.09	5.54	7.37	9.57	15.1	30.9
	3	3	3.00	3.37	4.32	5.68	7.35	9.34	11.7	17.4	33.3
	4	4	4.00	4.48	5.71	7.38	9.36	11.6	14.2	20.2	35.5
	5	5	5.00	5.60	7.10	9.13	11.5	14.1	16.9	23.5	40.3
	6	6	6.00	6.71	8.51	10.9	13.6	16.6	19.8	27.0	44.7
	8	8	8.00	8.95	11.3	14.5	18.0	21.8	25.8	34.4	54.4
	10	10	10.0	11.2	14.1	18.1	22.4	27.1	32.0	42.2	65.0
	15	15	15.0	16.8	21.2	27.1	33.6	40.5	47.6	62.3	93.4
	20	20	20.0	22.4	26.3	36.1	44.7	53.9	63.3	82.7	123
1.5	1	1.5	1.50	1.72	2.36	3.38	4.78	6.57	8.74	14.2	30.0
	2	3	3.00	3.22	3.85	4.87	6.29	8.08	10.3	15.8	31.5
	3	4.5	4.50	4.77	5.54	6.72	8.28	10.2	12.5	18.1	34.0
	4	6	6.00	6.34	7.29	8.71	10.5	12.6	15.1	21.1	37.2
	5	7.5	7.50	7.92	9.06	10.8	12.9	15.3	18.0	24.4	41.1
	6	9	9.00	9.49	10.9	12.8	15.3	18.0	21.1	28.0	45.5
	8	12	12.0	12.7	14.4	17.0	20.2	23.6	27.4	35.7	55.4
	10	15	15.0	15.8	18.0	21.3	25.1	29.4	34.0	43.8	66.1
	15	22.5	22.5	23.7	27.0	31.8	37.5	43.8	50.5	64.6	95.0
	20	30	30.0	31.6	36.1	42.4	50.0	58.4	67.2	85.7	125
2	1	2	2.00	2.27	2.99	4.09	5.54	7.37	9.57	15.1	30.9
	2	4	4.00	4.20	4.81	5.82	7.22	9.00	11.2	16.7	32.4
	3	6	6.00	6.23	6.89	7.98	9.45	11.3	13.5	19.1	34.9
	4	8	8.00	8.27	9.06	10.3	11.9	14.0	16.3	22.2	38.3
	5	10	10.0	10.3	11.3	12.7	14.6	16.8	19.4	25.6	42.2
	6	12	12.0	12.4	13.5	15.2	17.3	19.8	22.7	29.4	46.7
	8	16	16.0	16.5	17.9	20.1	22.8	26.0	29.5	37.4	56.7
	10	20	20.0	20.6	22.4	25.1	28.4	32.3	36.5	45.9	67.7
	15	30	30.0	30.9	33.6	37.5	42.5	48.1	54.3	67.6	97.2
	20	40	40.0	41.2	44.7	50.0	56.6	64.1	72.2	89.8	128

through glazed or unglazed openings (I_{R3}) or fire windows (I_{R5}). Reference to Figs. 10.7 and 10.8 will show the radiation received just outside the openings as I_{R2} for ordinary windows and I_{R4} for fire windows. For design purposes, the critical design value of received radiation I_{RC} will be the least value of I_{R1}, I_{R2}, and I_{R4}. To avoid ignition by received radiation, limits have to be set on the values of I_{RC}, depending on whether the situation is likely to be exposed to pilot ignition or to spontaneous ignition. The critical design value I_{RC} can be compared with the radiant energy I_R received by a surface remote from a heated emitter as follows:

$$I_R = \phi_n \cdot I_{E3} \leq I_{RC} \tag{10.30}$$

There are three possible values to consider. First, if the neighbor's building has a combustible material on the exterior walls, pilot ignition of 12.5 kW/m² is the likely criterion. Thus for design purposes, $I_{RC} = I_{R1} = 12.5$ kW/m². Second, ordinary glazing is likely to crack and fall out. This means pilot ignition of combustible materials just inside the openings must be used. The recommended radiant-flux value based on pilot ignition should be 12.5 kW/m². Thus for design purposes, $I_{RC} = I_{R2} = I_{R3} = 12.5$ kW/m². Third, for fire-resistant glazing likely to remain in position for the design period or duration of the building fire, a spontaneous ignition value such as 25 kW/m²—or other selected value—will apply for I_{R5} *inside* the window. If $k_2 = 0.50$, resulting in 50% radiation reduction *through* the fire window, the design value for I_{R4} *outside* the window can be twice I_{R5}, that is, 50 kW/m². Thus for design purposes,

					Radiation distance R, m									
4	5	6	7	8	9	10	12	14	16	18	20	22	25	30
52.9														
55.4	83.7													
58.8	87.2	122												
62.9	91.5	126												
67.7	96.6	132	173											
78.9	101	144	186	233										
91.6	123	160	202	250										
127	164	205	251	302	358									
165	210	257	308	364	424									
53.5	81.8													
56.1	84.4	119												
59.5	87.9	122												
63.7	92.2	127	168											
68.5	97.3	132	173	221										
79.8	110	145	187	234	288									
92.6	124	160	203	251	305									
128	165	206	252	303	359	421	562							
167	211	258	309	365	425	490	637							
52.9														
54.4	82.7													
57.0	85.3	120												
60.4	88.8	123	164											
64.7	93.2	128	169	216										
69.6	98.3	133	174	222	275									
80.9	111	146	188	235	289	349								
93.9	125	161	204	252	306	366	505							
130	167	207	253	304	360	422	563	729						
169	213	260	311	366	426	492	638	807						

$$I_{RC} = I_{R4} = \frac{I_{R5}}{k_2} = 50 \text{ kW/m}^2 \qquad (10.31)$$

The reason for selecting different reduction coefficients k_1 and k_2 is that while they may be the same for cases A and D illustrated in Fig. 10.8, they will be different when cases B and C are applied.

9 Combination of Radiation and Glazing

Figure 10.8 demonstrates not only mirror and nonmirror situations, but it also illustrates the differences between four cases A, B, C, and D. If critical design values for I_{RC} and I_{EC} are set out for the four cases, as in Table 10.6, it will be seen that the radiation factors k_R fall into the ratios of 1, 2, 4, and 8. Hence the four situations can conveniently be compared against case A under the one term k_O called the case factor.

Table 10.8 Fire facade factors k_F (Continued)

H, m	W, m	A_E, m²	0.01	0.25	0.50	0.75	1.00	1.25	1.50	2	3
2.5	1	2.5	2.50	2.81	3.64	4.86	6.41	8.31	10.6	16.2	32.0
	2	5	5.00	5.21	5.83	6.87	8.29	10.1	12.3	17.8	33.6
	3	7.5	7.50	7.71	8.33	9.36	10.8	12.6	14.8	20.4	36.2
	4	10	10.0	10.2	10.9	12.1	13.6	15.5	17.8	23.6	39.5
	5	12.5	12.5	12.8	13.6	14.9	16.6	18.7	21.1	27.2	43.6
	6	15	15.0	15.3	16.2	17.7	19.6	22.0	24.6	31.1	48.2
	8	20	20.0	20.4	21.6	23.4	25.9	28.7	32.0	39.6	58.4
	10	25	25.0	25.5	27.0	29.2	32.2	35.7	39.6	48.4	69.7
	15	37.5	37.5	38.2	40.4	43.8	48.1	53.2	58.8	71.4	99.9
	20	50	50.0	51.0	53.9	58.3	64.1	70.8	78.2	94.7	131
3	1	3	3.00	3.37	4.32	5.68	7.35	9.34	11.7	17.4	33.3
	2	6	6.00	6.23	6.89	7.98	9.45	11.3	13.5	19.1	35.0
	3	9	9.00	9.20	9.82	10.8	12.2	14.0	16.2	21.8	37.6
	4	12	12.0	12.2	12.9	13.9	15.4	17.2	19.5	25.1	41.1
	5	15	15.0	15.2	16.0	17.1	18.7	20.7	23.0	29.0	45.2
	6	18	18.0	18.3	19.1	20.4	22.1	24.3	26.9	33.1	49.9
	8	24	24.0	24.3	25.4	27.0	29.2	31.8	34.8	42.0	60.5
	10	30	30.0	30.4	31.7	33.6	36.3	39.4	43.0	51.4	72.0
	15	45	45.0	45.6	47.5	50.4	54.2	58.7	63.9	75.7	103
	20	60	60.0	60.8	63.3	67.1	72.2	78.2	85.0	100	136
3.5	1	3.5	3.50	3.92	5.01	6.52	8.34	10.5	12.9	18.8	34.8
	2	7	7.00	7.25	7.97	9.13	10.7	12.6	14.9	20.6	36.5
	3	10.5	10.5	10.7	11.3	12.4	13.8	15.6	17.8	23.4	39.2
	4	14	14.0	14.2	14.8	15.8	17.3	19.1	21.3	26.9	42.8
	5	17.5	17.5	17.7	18.4	19.5	21.0	22.9	25.2	31.0	47.1
	6	21	21.0	21.2	22.0	23.2	24.8	26.9	29.3	35.4	51.9
	8	28	28.0	28.3	29.2	30.7	32.6	35.1	37.9	44.8	62.8
	10	35	35.0	35.4	36.4	38.2	40.6	43.4	46.8	54.7	74.7
	15	52.5	52.5	53.0	54.6	57.2	60.6	64.7	69.5	80.5	107
	20	70	70.0	70.7	72.8	76.2	80.7	86.1	92.4	107	141

Factor k_c. Values for case factors using case A as a base value of 1 are shown in Table 10.7. It can be seen that case factor k_c is a geometry factor unaffected by distance or the enclosing rectangle of the facade. For design purposes the combustibility of the neighbor's facade and the type of assembly in the owner's and neighbor's openings are the only factors considered. By varying these factors, the area of the openings can be increased to 2, 4, or 8 times that permitted for case A.

Factor k_F. Combining Eqs. 10.29 and 10.30 results in the maximum permitted area of openings A_v being

$$A_v = \left(\frac{A_E}{\phi_n}\right)\left(\frac{I_{RC}}{I_{EC}}\right) \ \text{m}^2 \tag{10.32}$$

The terms in parentheses can be replaced by two separate factors. Thus,

$$A_v = k_F \cdot k_R \ \text{m}^2 \tag{10.33}$$

where k_F = fire-facade factor, = A_E/ϕ_n
k_R = radiation-ratio factor, = I_{RC}/I_{EC}

						Radiation distance R, m								
4	5	6	7	8	9	10	12	14	16	18	20	22	25	30
54.0														
55.6	83.9	118												
58.2	86.5	121												
61.7	90.0	125	165	213										
65.9	94.4	129	170	217	271									
70.9	99.6	134	175	223	276	336								
82.4	112	147	189	236	290	350	488							
95.5	126	163	205	253	307	367	506							
132	168	209	255	305	361	423	565	730	920					
172	215	262	319	368	428	493	639	808	1001					
53.4	83.7													
57.0	85.3	120												
59.6	87.9	122	163	210										
65.2	91.5	121	167	214	268									
67.5	96.0	131	172	219	272	332								
72.5	101	136	177	224	278	337	476							
84.2	114	149	190	238	292	351	490							
97.6	129	165	207	255	309	369	508	672						
135	171	211	257	307	363	425	566	732	922					
175	218	264	315	370	430	495	641	810	1003	1219				
57.0	85.3													
58.6	86.9	122	162											
61.3	89.6	124	165	212										
64.9	93.2	128	169	216	269	329								
69.3	97.7	132	173	220	274	334	472							
74.4	103	138	179	226	279	339	477							
86.3	116	151	192	240	293	353	492	655						
99.9	130	166	209	256	310	371	510	673	862					
138	173	214	259	309	365	427	568	734	923	1138				
179	221	267	318	373	432	497	643	812	1005	1221	1462			

k_F is another geometric factor, which depends only on the dimensions of the facade height H and width W and the radiation distance R. The results can be reduced to tabular form, as shown in Table 10.8.

Factor k_R. As shown in Table 10.6, the ratio of the initial design values for radiation I_{RC}/I_{EC} depends on FRR and the case factor. Using the k_R values of case A as the base results in Table 10.9.

If building materials are used that have I_{RC} values higher or lower than 12.5 kW/m², then revised versions of Tables 10.6 and 10.9 would be required. For example, plastic wall claddings may have an I_{RC} of only 10 kW/m².

10 Permissible Area of Openings

Equation 10.33 can now be modified to become a new equation for the permissible area of openings A_v as follows:

$$A_v = k_C \cdot k_F \cdot k_R \quad \text{m}^2 \tag{10.34}$$

Table 10.8 Fire facade factors k_F (Continued)

| Facade size | | | Radiation distance R, m | | | | | | | | |
H, m	W, m	A_E, m²	0.01	0.25	0.50	0.75	1.00	1.25	1.50	2	3
4	1	4	4.00	4.48	5.71	7.38	9.36	11.6	14.8	20.2	36.5
	2	8	8.00	8.27	9.06	10.3	11.9	14.0	16.3	22.2	28.3
	3	12	12.0	12.2	12.9	13.9	15.4	17.2	19.5	25.1	41.0
	4	16	16.0	16.2	16.8	17.8	19.2	21.1	23.3	28.9	44.7
	5	20	20.0	20.2	20.8	21.9	23.3	25.2	27.5	33.1	49.1
	6	24	24.0	24.2	24.9	26.0	27.6	29.5	31.9	37.8	54.1
	8	32	32.0	32.3	33.1	34.4	36.2	38.5	41.2	47.8	65.4
	10	40	40.0	40.3	41.3	42.9	45.0	47.7	50.8	58.3	77.8
	15	60	60.0	60.5	61.9	64.1	67.2	71.0	75.4	85.7	111
	20	80	80.0	80.6	82.5	85.5	89.5	94.5	100	114	146
5	1	5	5.00	5.60	7.10	9.13	11.5	14.1	16.9	23.5	40.4
	2	10	10.0	10.3	11.3	12.7	14.6	16.8	19.4	25.6	42.2
	3	15	15.0	15.2	16.0	17.1	18.7	20.7	23.1	29.0	45.2
	4	20	20.0	20.2	20.8	21.9	23.3	25.2	27.5	33.1	49.1
	5	25	25.0	25.2	25.8	26.8	28.3	30.1	32.3	37.9	53.9
	6	30	30.0	30.2	30.8	31.9	33.3	35.2	37.4	43.1	59.2
	8	40	40.0	40.2	40.9	42.1	43.7	45.8	48.2	54.4	71.3
	10	50	50.0	50.3	51.1	52.4	54.3	56.6	59.4	66.3	84.6
	15	75	75.0	75.4	76.5	78.4	81.0	84.2	88.0	97.1	121
	20	100	100	101	102	105	108	112	117	129	158
6	1	6	6.00	6.71	8.51	10.9	13.6	16.6	19.8	27.0	44.7
	2	12	12.0	12.4	13.5	15.2	17.3	19.8	22.7	29.4	46.7
	3	18	18.0	18.3	19.1	20.4	22.1	24.3	26.9	33.1	49.9
	4	24	24.0	24.2	24.9	26.0	27.6	29.5	31.9	37.8	54.1
	5	30	30.0	30.2	30.8	31.9	33.3	35.2	37.4	43.1	59.2
	6	36	36.0	36.2	36.8	37.8	39.3	41.1	43.3	49.0	65.0
	8	48	48.0	48.2	48.9	49.9	51.4	53.3	55.7	61.5	78.0
	10	60	60.0	60.2	61.0	62.2	63.8	65.9	68.5	74.9	92.3
	15	90	90.0	90.3	91.3	92.9	95.1	98.0	101	110	131
	20	120	120	120	122	124	127	130	135	145	172

where k_C = case factor (from Table 10.7)
k_F = fire-facade factor (from Table 10.8)
k_R = radiation-ratio factor (from Table 10.9)

If further desired, a wetting-down factor, for example, k_W = 2, can be introduced into Eq. 10.34. This factor is appropriate if the fire service is able to effectively wet down the walls of the neighboring building.

10.5 APPLICATION OF DESIGN METHODS

The separation design method described in this chapter offers more choices than usual in the fire-engineering design of external walls. It is also linked directly to the standard fire-resistive construction ratings ranging from ½ to 4 hr and to fire-compartment-area design methods. Furthermore, the design method is readily adaptable to tables, graphs, and hand-held programmable calculators. It can be used not only to design new buildings, but also to check critical situations between existing buildings for both mirror-image and non-mirror-image situations.

						Radiation distance R, m								
4	5	6	7	8	9	10	12	14	16	18	20	22	25	30
58.8	87.2													
60.4	88.8	123	164											
63.1	91.5	126	167	214	267									
66.8	95.2	130	171	218	271	331								
71.3	99.7	134	175	222	276	336	474							
76.5	105	140	181	228	281	341	479							
88.7	118	153	194	242	295	355	494	657						
103	133	169	211	259	313	373	512	675	864					
142	177	216	261	312	368	429	570	736	925	1140	1380			
183	225	271	321	375	435	500	646	814	1007	1223	1464	1730		
62.9	91.5	126												
64.7	93.2	128	169	216										
67.5	95.9	131	171	219	272	332								
71.3	99.7	134	175	222	276	335	474							
76.0	104	139	180	227	280	340	478	642						
81.5	110	145	186	233	286	346	484	648						
94.2	123	158	199	247	300	360	498	662	851					
109	139	174	216	264	318	378	516	680	869	1083				
150	184	223	267	317	373	435	576	741	951	1145	1384	1649		
194	234	279	328	382	441	506	651	820	1012	1228	1469	1735	2170	
67.7	96.6	132	173	220	273	333	472	635	823	1037	1276			
69.6	98.3	133	174	222	275	335	473	636	825	1039	1278	1541	1984	2848
72.5	101	136	177	224	278	337	476	639	828	1041	1280	1544	1987	2851
76.5	105	140	181	228	281	341	479	643	831	1045	1284	1548	1991	2855
81.5	110	145	186	233	286	346	484	648	836	1050	1289	1552	1995	2860
87.2	116	150	191	238	292	352	490	653	842	1056	1294	1558	2001	2865
101	129	164	205	253	306	366	504	668	856	1070	1309	1573	2016	2880
116	145	181	222	270	324	384	522	686	875	1088	1327	1591	2034	2898
159	192	231	275	325	380	441	582	747	937	1151	1390	1655	2098	2963
206	244	288	337	390	449	513	658	826	1018	1235	1475	1741	2185	3051

The methods do not rely on the use of water and are suitable for areas having little or no stored or running water, such as rural areas, remote islands, or frozen climates. Where running water is readily available for extinguishing or wetting-down purposes, concessions to design values would be reasonable, such as using a wetting-down factor k_W.

By making fire windows mandatory within a protected limiting distance I_x from the boundary, flame contact through openings can be virtually eliminated. Mirror imaging does not always produce safe solutions for radiation hazards, but it does allow critical situations to be readily checked. Non-mirror-image situations, including mixtures of fire windows on one building and ordinary windows on the other buildings, can also be readily checked. Where fire windows are used and external cladding is noncombustible, spontaneous ignition values can be used in the design process. At least 50% reduction in radiation can occur through fire windows. Combined, these two factors can lead to considerable increases in traditional, allowable opening areas in external walls, not only where fire windows are compulsory, but also where they are voluntarily used beyond the protected limiting distance I_x.

The separation design method depends only on the radiation distance R and not on the distance to the boundary L. Any existing opening configuration or new window design can be checked, regardless of the boundary location, such as two buildings on the same or different sites, or the walls of atriums and large vertical shafts. The separation design method can also be used to check the radiation-distance hazard for individual windows, in which case A_v is made to equal A_E.

Equation 10.34 and Tables 10.7 to 10.9 provide a facade-factor design method which is simple and accurate and can be readily incorporated into fire codes. The combined use of a case factor k_C, a radiation-ratio factor k_R, and a fire-facade factor k_F greatly reduces the amount of tabular data needed for fire-design purposes. The use of the case factor k_C based on case A, graphically illustrates how openings can be increased in size for cases B, C, and D when k_R and k_F remain constant.

The fire-compartment design method aids understanding in both the provision of fire-extinguishing facilities and the fire design of buildings. For example, the

Table 10.9 Fire resistance ratings and radiation ratio factors

FRR hrs	Radiation Ratio Factor k_R
1/2	0.144
3/4	0.116
1	0.100
1-1/2	0.083
2	0.072
3	0.060
4	0.053

maximum available flow rate of a water supply determines its cooling power. If the peak heating power of a fire is not to exceed this cooling power, the openings in a fire compartment, or the surface area of the fuel, must be controlled in size in order to keep the rate of pyrolysis under control. The rate of pyrolysis not only determines the heating power at which a fire will burn, it also determines the fire's duration for a given total fire load. This, in turn, enables the maximum fire-compartment area to be calculated for a given fire rating and fire-load density (see Table 10.2). However, while the methods described are based on a simplistic approach, it is recommended that code writers should not enshrine the various equations (for example, Eq. 10.19) in "concrete."

More importantly, the fire-compartment-sizing design method can be linked directly to the design method for fire separation between external walls of buildings, thereby providing considerable flexibility to building designers. Both methods depend on the common factor of fire duration. Used together, the two design methods can lead to a greater understanding of the fire-safety design process for buildings by both building designers and building officials. This would result in a more rational approach to the design and building-permit process. For example, allowances could be made in new fire codes to encourage more sophisticated solutions to fire-compartment sizing if a building designer can demonstrate the appropriate technology.

10.6 CONDENSED REFERENCES/BIBLIOGRAPHY

ASTM E-119 188, *Standard Test methods for Fire Tests of Building Construction and Ma-*
Barnett 1988, *Fire Separation between External Walls of Buildings*
Barnett 1989, *Fire Compartment Sizing Design Method*
BSI 1986, *Fire Tests on Building Materials and Structures*
ISO 1975, *Fire Resistance Tests—Elements of Building Construction*
Ove Arup 1977, *Design Guide for Fire Safety of Bare Structural Steel*

11

Deterministic Assessment of the Fire Exposure of Exterior Walls

An alternative method to describe fire-exposure levels for exterior walls involves relating these to the source or compartment fire causing the exposure. Conventionally the fire resistance of exterior walls, or any enclosing element for that matter, is determined by utilizing an exposure based on a standard fire-endurance test. The accepted standard fire severity of the "typical" compartment fire is represented by the time-temperature relationship given in ASTM E-119 (1988) or in ISO 834 (1975). These methods are almost universally accepted for both design and building-regulation purposes for those structural elements and fire-rated enclosures directly exposed to a fully developed postflashover room fire. This standard fire, however, defined as it is in terms of time-temperature, cannot be translated into any specific room fire, nor can the time-temperature curve be used directly as a scalar for the fire severity generated by any specific room fire.

This definition does not pose any intractable problems if all that is desired is the acceptability of a given structural element or enclosure exposed to the "standard" fire, because the specimen in question can be tested. However, one can easily visualize a typical room fire in a building: after the fire has developed to the postflashover condition, the window openings in the exterior walls would be venting flames out the top and feeding combustion air to the fire through the bottom of the window. These venting flames would expose the exterior face of the wall above the window to both radiant and convective thermal flux. In addition, the venting flames and window openings would be "radiators," exposing any nearby buildings to heat. Furthermore, if the adjacent building is quite close to the building on fire, the venting flames could wash over its surface, subjecting it to convective thermal flux. While the fire may be exposing the assemblies in the room to something akin to a standard fire, the venting flames and window radiators are exposing the external assemblies to something significantly less than the standard fire.

This scenario is pertinent to the design of exterior walls and to the siting of buildings on a lot. If two buildings are close enough to each other that one could expose the other to enough thermal flux as to cause significant damage to its exterior wall, there is an even greater danger of fire transfer to the combustible contents of the exposed building through its windows. Good fire-protection practice calls for building setbacks on a lot and limiting window areas or other transparent openings in exterior facing walls. Such practices minimize the potential for fire spread from building to building.

The exterior wall above a venting window will be subjected to some level of thermal flux, and this situation is immutable. As long as room fires develop to the postflashover condition and ordinary glazing is used for windows of exterior walls, fire exposure to these walls must be considered. For fire escaping from a window, a quantitative technique is desired which assesses the level of fire severity incident on an exterior surface above a venting window. A drawback to developing such a technique is that the venting flame exposure occurs when a "real" room burns in the postflashover mode. The standard interior exposure, however, comes from a standard fire identified only by a time-temperature relationship. Two problems arise: first, the standard fire is not associated with a standard room and, second, the standard time-temperature does not provide a proper scalar for transposing exposed components of a real venting flame to similar components of a standard fire.

One step toward determining a more equitable exposure would be to determine the total incident thermal flux inflicted on a test assembly in a standard fire-endurance test. This would create a time–thermal-flux relationship equivalent to the standard time-temperature relationship and would partly alleviate the second problem noted. This, at least, would overcome the inconsistency that the convective energy component of incident flux is a linear function of the gas and flame temperature while the radiant energy component of incident flux is a function of the fourth power of the radiator temperature. Thus the radiant flux and the convective flux, in consistent units of energy per unit area per unit time, could be calculated and summed to obtain the total incident flux. It is generally accepted that the proportionate level of the radiant and convective components in the total incident flux is not defined in the standard fire, and this mix of components varies from standard furnace to standard furnace. Furthermore, the impact inflicted on an exposed assembly by these components is material-specific: the effect of the incident radiant flux on an exposed assembly is a function of the emissivity of the surface of that assembly, while the effect of the incident convective flux on an exposed assembly is a function of the convective transfer coefficient of the surface of the assembly.

The fact that no "standard" room is currently defined when assessing fire exposure poses another problem. Even a cursory study of the progress of a fully developed fire in a room will reveal a myriad of parameters which affect the burning process and the generation of thermal energy on a time-dependent base. The geometry of the room; the geometry, number, and location of ventilating openings; and the mass distribution, specific surface, and composition of the fuel are only some major variables that affect the time release of thermal energy. Similar parameters affect the venting flame plume, which is a major source of exterior fire exposure.

11.1 ROOM-FIRE MODEL

In spite of all the deficiencies mentioned, it is desirable to quantify the severity of exposure on an exterior wall above a venting window. One way to examine this is to consider a postflashover room-fire model which includes a reasonable number of significant parameters affecting fire development. The model selected for this discussion was developed by Margaret Law at Ove Arup & Partners, resulting from a study (1977) performed for the American Iron and Steel Institute (AISI). That study produced a design guide (AISI, 1979), which is useful for organizing calculations of specified assessments. A spread-sheet program is used on a microcomputer to perform the calculations.

The program's required input is the geometry and location of the ventilation openings; the depth, width, and height of the fire room; and the specific density and specific surface of the fuel in the room. With this information the program calculates the rate of burning, the height, width, and thickness of the venting flame, and an estimate of the fire duration. The program then calculates the geometry factor and the temperature of the venting flame. From this, the emissivity and the average incident radiant flux from the venting flame are calculated. Likewise, the convective transfer coefficient and the average incident convective flux from the venting flame are also calculated. Adding these two components of incident flux results in the total average incident flux on the exterior wall area above the venting window.

Calculations can be made using this technique for a range of room geometries, window geometries, and fuel conditions. In this study, the rooms tested varied from 3.05 by 3.05 m (10 by 10 ft) to 24.38 by 24.38 m (80 by 80 ft). The room height was set at 2.44 m (8.0 ft), but in specific rooms the height was varied up to 8.53 m (28.0 ft). The windows varied from 0.91 m wide by 0.91 m high (3.0 by 3.0 ft) to full wall width by full room height. The windows were placed in one wall defined by the room width and also in one perpendicular wall defined by the room depth. A total of 82 separate combinations of room and window geometries were calculated for a fuel density of 48.8 kg/m^2 (10 lb/ft^2) and for a fuel density of 97.6 kg/m^2 (20 lb/ft^2). The specific surface of the fuel was taken so that, for the fuel-surface-controlled fire, the entire fuel load is consumed in 20 min.

The total average incident flux on the exterior wall area above the venting window can be compared to the total incident flux imposed on a test assembly when exposed to the standard fire. The graph in Fig. 11.1 is taken from a thesis research project from Ohio State University (Oberg, 1977). Oberg measured and recorded the incident radiant flux and the total incident flux imposed on a 3.05- by 3.05-m (10- by 10-ft) concrete-block wall assembly exposed to the standard ASTM E-119 fire-endurance test. The measurements were taken at the center of the exposed face of concrete block and at the center of each of the four quarter sections of the specimen. The measurements were taken at intervals throughout the exposure test period. The average of the results of four separate tests is shown in Fig. 11.1. Inspection of Fig. 11.1 shows that the total incident flux follows the general characteristics of the standard time-temperature curve. It can also be noted that the incident convective flux is essentially constant after the early portion of the test.

The equivalent severity of the fire exposure incident on a wall above a venting window is determined by comparing the area under the ASTM E-119 time-flux curve obtained by Oberg to the area under the time-flux curve calculated from the computer model. In both cases the area below a thermal flux of 1.25 W/cm^2 (66 Btu/ft^2 · min) is excluded. This flux is the threshold flux for pilot ignition of ordinary combustible materials. The assumption is that flux levels below this would have no significant damaging effect on the wall assembly. Figure 11.2 illustrates this relationship for one typical room-fire situation. The area under the ASTM E-119 time-flux curve is calculated for the actual fire duration by dividing the total fuel by the rate of burning or by the duration, based on a specific fuel

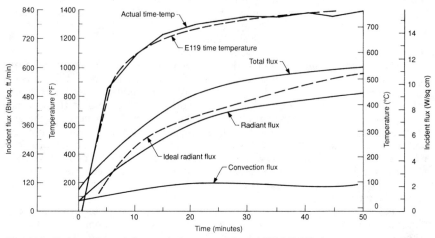

Fig. 11.1   Incident thermal flux equivalent to standard ASTM E-119 time-temperature. (*From Oberg, 1977.*)

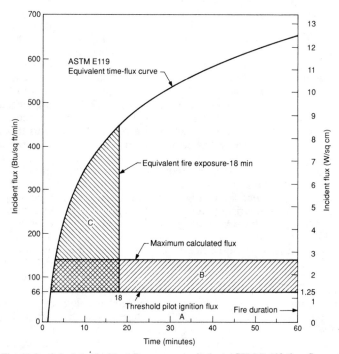

Fig. 11.2   Calculated incident flux versus equivalent ASTM E-119 time-flux.

density first suggested by Ingberg, whichever is less (1928). The fire duration based on Ingberg is 60 min for a 48.8-kg/m^2 (10-lb/ft^2) fuel density and 120 min for a 97.6-kg/m^2 (20-lb/ft^2) fuel density.

The area under the maximum calculated flux curve is indicated in Fig. 11.2 by the crosshatched area B. This area is then reconstructed under the ASTM E-119 time-flux curve to equal the crosshatched area C. The time required under the ASTM curve to obtain an equal area is the equivalent fire exposure.

11.2 ROOM-MODEL TEST RESULTS

Tables 11.1 through 11.6 show the results of room-fire calculations. Table 11.1 gives the fire duration, the average total incident flux on the wall above the venting window, and the equivalent fire exposure on that wall for 82 distinct room fires for a specific fuel load of 48.8 kg/m^2 (10 lb/ft^2). Table 11.2 shows comparable data for a fuel load of 97.6 kg/m^2 (20 lb/ft^2).

Tables 11.3 and 11.4 contain the same information as Table 11.1 [room-fire results for a fuel density of 48.8 kg/m^2 (10 lb/ft^2)], while Tables 11.5 and 11.6 contain the same information as Table 11.2 [room-fire results for a fuel density of 97.6 kg/m^2 (20 lb/ft^2)]. In Tables 11.3 and 11.5 the room-fire cases are listed by fire duration, in Tables 11.4 and 11.6 they are listed by equivalent fire exposure.

In Table 11.3 there are four room cases (15, 16, 17, and 73) which had fire durations of 20 min. These four cases were provided with sufficient ventilation air so that the surface area of the fuel controlled the mass burning rate. In all the other cases shown in Tables 11.3 and 11.5 [at both 48.8- and 97.6-kg/m^2 (10- and 20-lb/ft^2) fuel density], the mass burning rate was controlled by the ventilation air available. Those cases where the mass-burning rate was so slow that it resulted in unusually long fire durations [those in excess of 480 min for the 48.8-kg/m^2 (10-lb/ft^2) fuel density and 960 min for the 97.6-kg/m^2 (20-lb/ft^2) fuel density] were unrealistic and perhaps should be excluded. They are included in the results, however, because they are within the range of room and window geometries being examined.

Table 11.4 shows that the most severe exterior fire exposure was 19 min, and it occurred from the equivalent of a 1-hr standard fire. All of the more severe cases occurred in rooms ventilated by a single narrow window. This was to be expected since an almost unventilated fire would result in a large proportion of released volatiles in the room, and these would be starved for combustion air. Consequently the volatiles would burn after venting out the window, where combustion air is available.

Table 11.6 shows that the most severe cases of exterior fire exposure was 30 min for what is defined as a 2-hr standard room fire. Comparing Tables 11.4 and 11.6 shows essentially the same order of room cases for the more severe exposures. However, doubling the fuel density does not result in twice the equivalent exposure.

The fact that some measurable exterior fire exposure can be expected when a room fire vents out the windows strongly suggests that some minimum level of fire resistance be required of all exterior walls. This is particularly true in multistory buildings where requirements already exist which prevent fire spread from

Table 11.1 Effect of room geometry and window size and location on the exposure of exterior walls above venting window openings from 10-lb/ft² fuel-load compartment fires

Room case	Room size, ft			Window size, ft				Fire duration, min	Incident flux, Btu/ft² · min	Equivalent exposure, min
	D	W	H	w_1	h_1	w_2	h_2			
1	10.0	10.0	8.0	3.0	3.0			60	103	12
2	10.0	10.0	8.0	3.0	5.0			39	88	8
3	10.0	10.0	8.0	5.0	5.0			34	80	6
4	10.0	10.0	8.0	10.0	5.0			31	59	0
5	10.0	10.0	8.0	3.0	3.0	3.0	3.0	29	82	6
6	10.0	10.0	8.0	5.0	5.0	5.0	5.0	22	72	4
7	10.0	10.0	8.0	10.0	5.0	10.0	5.0	23	61	0
8	10.0	10.0	8.0	10.0	8.0	10.0	8.0	25	36	0
9	10.0	20.0	8.0	3.0	3.0			76	108	13
10	10.0	20.0	8.0	3.0	5.0			43	104	10
11	10.0	20.0	8.0	5.0	5.0			32	97	8
12	10.0	20.0	8.0	10.0	5.0			27	85	6
13	10.0	20.0	8.0	10.0	8.0			25	70	3
14	10.0	20.0	8.0	3.0	3.0	3.0	3.0	32	97	8
15	10.0	20.0	8.0	5.0	5.0	5.0	5.0	20	82	5
16	10.0	20.0	8.0	10.0	5.0	10.0	5.0	20	73	4
17	10.0	20.0	8.0	10.0	8.0	10.0	8.0	20	65	0
18	20.0	10.0	8.0	3.0	3.0			153	104	12
19	20.0	10.0	8.0	10.0	8.0			50	42	0
20	20.0	10.0	8.0	3.0	3.0	3.0	3.0	63	80	8
21	20.0	10.0	8.0	10.0	8.0	10.0	8.0	37	48	0
22	20.0	20.0	8.0	3.0	3.0			211	104	12
23	20.0	20.0	8.0	10.0	5.0			53	86	8
24	20.0	20.0	8.0	20.0	5.0			45	71	4
25	20.0	20.0	8.0	20.0	8.0			43	48	0
26	20.0	20.0	8.0	3.0	3.0	3.0	3.0	78	89	10
27	20.0	20.0	8.0	10.0	5.0	10.0	5.0	32	74	5
28	20.0	20.0	8.0	10.0	8.0	10.0	8.0	31	70	4
29	20.0	40.0	8.0	3.0	3.0			297	109	13
30	20.0	40.0	8.0	3.0	8.0			75	122	15
31	20.0	40.0	8.0	10.0	5.0			54	103	11
32	20.0	40.0	8.0	20.0	5.0			41	92	8
33	20.0	40.0	8.0	40.0	5.0			36	79	6
34	20.0	40.0	8.0	40.0	8.0			34	58	0
35	40.0	20.0	8.0	3.0	3.0			595	104	12
36	40.0	20.0	8.0	10.0	5.0			108	85	9
37	40.0	20.0	8.0	20.0	5.0			81	75	6
38	40.0	20.0	8.0	3.0	3.0	3.0	3.0	211	83	8
39	40.0	20.0	8.0	20.0	8.0	20.0	8.0	49	61	0
40	40.0	40.0	8.0	3.0	3.0			841	104	12
41	40.0	40.0	8.0	10.0	5.0			127	98	11

Table 11.1 Effect of room geometry and window size and location on the exposure of exterior walls above venting window openings from 10-lb/ft² fuel-load compartment fires (Continued)

Room case	Room size, ft			Window size, ft				Fire duration, min	Incident flux, Btu/ft² · min	Equivalent exposure, min
	D	W	H	w_1	h_1	w_2	h_2			
42	40.0	40.0	8.0	20.0	5.0			81	92	10
43	40.0	40.0	8.0	40.0	5.0			63	82	8
44	40.0	40.0	8.0	40.0	8.0			56	66	0
45	40.0	40.0	8.0	3.0	3.0	3.0	3.0	297	91	10
46	40.0	40.0	8.0	10.0	5.0	10.0	5.0	58	90	10
47	40.0	40.0	8.0	20.0	5.0	20.0	5.0	45	80	6
48	40.0	40.0	8.0	20.0	8.0	20.0	8.0	40	77	6
49	40.0	40.0	8.0	40.0	5.0	20.0	5.0	48	77	6
50	40.0	80.0	8.0	3.0	3.0			1190	109	13
51	40.0	80.0	8.0	3.0	8.0			273	127	16
52	40.0	80.0	8.0	80.0	8.0			43	76	5
53	40.0	80.0	8.0	3.0	3.0	3.0	3.0	421	104	12
54	40.0	80.0	8.0	20.0	5.0	20.0	5.0	42	101	10
55	40.0	80.0	8.0	40.0	5.0	20.0	5.0	42	92	8
56	40.0	80.0	8.0	80.0	5.0	40.0	5.0	36	81	6
57	80.0	40.0	8.0	3.0	3.0			2380	104	12
58	80.0	40.0	8.0	5.0	5.0			664	91	10
59	80.0	40.0	8.0	10.0	5.0			335	91	10
60	80.0	40.0	8.0	20.0	5.0			183	89	10
61	80.0	40.0	8.0	40.0	8.0			94	72	5
62	80.0	40.0	8.0	3.0	3.0	3.0	3.0	841	83	8
63	80.0	40.0	8.0	40.0	8.0	40.0	8.0	60	71	5
64	80.0	80.0	8.0	3.0	3.0			3365	104	12
65	80.0	80.0	8.0	10.0	5.0			469	100	12
66	80.0	80.0	8.0	80.0	8.0			71	81	8
67	80.0	80.0	8.0	3.0	3.0	3.0	3.0	1190	91	10
68	80.0	80.0	8.0	20.0	5.0	20.0	5.0	93	97	11
69	80.0	80.0	8.0	40.0	5.0	40.0	5.0	63	88	9
70	80.0	80.0	8.0	80.0	5.0	40.0	5.0	62	83	8
71	80.0	80.0	8.0	80.0	8.0	80.0	8.0	46	75	5
72	10.0	80.0	8.0	3.0	3.0			149	137	18
73	10.0	80.0	8.0	80.0	8.0			20	52	0
74	80.0	10.0	8.0	3.0	3.0			1190	101	12
75	80.0	10.0	8.0	10.0	8.0			143	56	0
76	80.0	10.0	8.0	10.0	8.0	10.0	8.0	86	64	0
77	20.0	80.0	8.0	3.0	3.0			421	119	15
78	20.0	80.0	8.0	3.0	8.0			98	148	19
79	20.0	80.0	8.0	80.0	8.0			26	68	3
80	80.0	20.0	8.0	3.0	3.0			1683	106	13
81	80.0	20.0	8.0	10.0	8.0			156	78	7
82	80.0	20.0	8.0	20.0	8.0	20.0	8.0	74	67	2

Table 11.2 Effect of room geometry and window size and location on the exposure of exterior walls above venting window openings from 20-lb/ft^2 fuel-load compartment fires

Room case	Room size, ft			Window size, ft				Fire duration, min	Incident flux, Btu/ft^2 · min	Equivalent exposure, min
	D	W	H	w_1	h_1	w_2	h_2			
501	10.0	10.0	8.0	3.0	3.0			119	103	18
502	10.0	10.0	8.0	3.0	5.0			78	88	10
503	10.0	10.0	8.0	5.0	5.0			67	80	8
504	10.0	10.0	8.0	10.0	5.0			61	59	0
505	10.0	10.0	8.0	3.0	3.0	3.0	3.0	57	82	6
506	10.0	10.0	8.0	5.0	5.0	5.0	5.0	43	72	4
507	10.0	10.0	8.0	10.0	5.0	10.0	5.0	45	61	0
508	10.0	10.0	8.0	10.0	8.0	10.0	8.0	51	36	0
509	10.0	20.0	8.0	3.0	3.0			153	108	18
510	10.0	20.0	8.0	3.0	5.0			85	104	14
511	10.0	20.0	8.0	5.0	5.0			66	97	10
512	10.0	20.0	8.0	10.0	5.0			54	85	8
513	10.0	20.0	8.0	10.0	8.0			50	70	4
514	10.0	20.0	8.0	3.0	3.0	3.0	3.0	63	97	10
515	10.0	20.0	8.0	5.0	5.0	5.0	5.0	38	83	6
516	10.0	20.0	8.0	10.0	5.0	10.0	5.0	36	73	4
517	10.0	20.0	8.0	10.0	8.0	10.0	8.0	37	67	2
518	20.0	10.0	8.0	3.0	3.0			306	104	18
519	20.0	10.0	8.0	10.0	8.0			100	42	0
520	20.0	10.0	8.0	3.0	3.0	3.0	3.0	126	80	10
521	20.0	10.0	8.0	10.0	8.0	10.0	8.0	74	48	0
522	20.0	20.0	8.0	3.0	3.0			422	104	18
523	20.0	20.0	8.0	10.0	5.0			106	86	12
524	20.0	20.0	8.0	20.0	5.0			91	71	4
525	20.0	20.0	8.0	20.0	8.0			86	48	0
526	20.0	20.0	8.0	3.0	3.0	3.0	3.0	157	89	14
527	20.0	20.0	8.0	10.0	5.0	10.0	5.0	64	74	6
528	20.0	20.0	8.0	10.0	8.0	10.0	8.0	61	70	4
529	20.0	40.0	8.0	3.0	3.0			595	109	20
530	20.0	40.0	8.0	3.0	8.0			150	122	22
531	20.0	40.0	8.0	10.0	5.0			108	103	16
532	20.0	40.0	8.0	20.0	5.0			81	92	12
533	20.0	40.0	8.0	40.0	5.0			71	79	6
534	20.0	40.0	8.0	40.0	8.0			69	58	0
535	40.0	20.0	8.0	3.0	3.0			1190	104	18
536	40.0	20.0	8.0	10.0	5.0			216	85	12
537	40.0	20.0	8.0	20.0	5.0			162	75	8
538	40.0	20.0	8.0	3.0	3.0	3.0	3.0	423	83	10
539	40.0	20.0	8.0	20.0	8.0	20.0	8.0	98	61	0
540	40.0	40.0	8.0	3.0	3.0			1683	104	18
541	40.0	40.0	8.0	10.0	5.0			254	98	16

Table 11.2 Effect of room geometry and window size and location on the
exposure of exterior walls above venting window openings from 20-lb/ft² fuel-load
compartment fires (*Continued*)

Room case	Room size, ft			Window size, ft				Fire duration, min	Incident flux, Btu/ft² · min	Equivalent exposure, min
	D	W	H	w_1	h_1	w_2	h_2			
542	40.0	40.0	8.0	20.0	5.0			163	92	14
543	40.0	40.0	8.0	40.0	5.0			127	82	10
544	40.0	40.0	8.0	40.0	8.0			112	66	0
545	40.0	40.0	8.0	3.0	3.0	3.0	3.0	595	91	14
546	40.0	40.0	8.0	10.0	5.0	10.0	5.0	115	90	14
547	40.0	40.0	8.0	20.0	5.0	20.0	5.0	90	80	8
548	40.0	40.0	8.0	20.0	8.0	20.0	8.0	79	77	6
549	40.0	40.0	8.0	40.0	5.0	20.0	5.0	96	77	8
550	40.0	80.0	8.0	3.0	3.0			2380	109	20
551	40.0	80.0	8.0	3.0	8.0			547	127	24
552	40.0	80.0	8.0	80.0	8.0			85	76	6
553	40.0	80.0	8.0	3.0	3.0	3.0	3.0	841	104	18
554	40.0	80.0	8.0	20.0	5.0	20.0	5.0	85	101	14
555	40.0	80.0	8.0	40.0	5.0	20.0	5.0	84	92	12
556	40.0	80.0	8.0	80.0	5.0	40.0	5.0	72	81	8
557	80.0	40.0	8.0	3.0	3.0			4759	104	18
558	80.0	40.0	8.0	5.0	5.0			1327	91	14
559	80.0	40.0	8.0	10.0	5.0			669	91	14
560	80.0	40.0	8.0	20.0	5.0			365	89	14
561	80.0	40.0	8.0	40.0	8.0			188	72	6
562	80.0	40.0	8.0	3.0	3.0	3.0	3.0	1683	83	10
563	80.0	40.0	8.0	40.0	8.0	40.0	8.0	120	71	6
564	80.0	80.0	8.0	3.0	3.0			6730	104	18
565	80.0	80.0	8.0	10.0	5.0			939	100	16
566	80.0	80.0	8.0	80.0	8.0			141	81	10
567	80.0	80.0	8.0	3.0	3.0	3.0	3.0	2380	91	14
568	80.0	80.0	8.0	20.0	5.0	20.0	5.0	186	97	16
569	80.0	80.0	8.0	40.0	5.0	40.0	5.0	125	88	12
570	80.0	80.0	8.0	80.0	5.0	40.0	5.0	125	83	10
571	80.0	80.0	8.0	80.0	8.0	80.0	8.0	92	75	6
572	10.0	80.0	8.0	3.0	3.0			297	137	26
573	10.0	80.0	8.0	80.0	8.0			31	62	0
574	80.0	10.0	8.0	3.0	3.0			2380	101	16
575	80.0	10.0	8.0	10.0	8.0			286	56	0
576	80.0	10.0	8.0	10.0	8.0	10.0	8.0	171	64	0
577	20.0	80.0	8.0	3.0	3.0			841	119	22
578	20.0	80.0	8.0	3.0	8.0			195	148	30
579	20.0	80.0	8.0	80.0	8.0			52	68	2
580	80.0	20.0	8.0	3.0	3.0			3365	106	18
581	80.0	20.0	8.0	10.0	8.0			312	78	8
582	80.0	20.0	8.0	20.0	8.0	20.0	8.0	149	67	2

Table 11.3 Effect of room geometry and window size and location on the exposure of exterior walls above venting window openings from 10-lb/ft² fuel-load compartment fires (sorted on fire duration)

Room case	Room size, ft			Window size, ft				Fire duration, min	Incident flux, Btu/ft² · min	Equivalent exposure, min
	D	W	H	w_1	h_1	w_2	h_2			
16	10.0	20.0	8.0	10.0	5.0	10.0	5.0	20	73	4
73	10.0	80.0	8.0	80.0	8.0			20	52	0
15	10.0	20.0	8.0	5.0	5.0	5.0	5.0	20	82	5
17	10.0	20.0	8.0	10.0	8.0	10.0	8.0	20	65	0
6	10.0	10.0	8.0	5.0	5.0	5.0	5.0	22	72	4
7	10.0	10.0	8.0	10.0	5.0	10.0	5.0	23	61	0
13	10.0	20.0	8.0	10.0	8.0			25	70	3
8	10.0	10.0	8.0	10.0	8.0	10.0	8.0	25	36	0
79	20.0	80.0	8.0	80.0	8.0			26	68	3
12	10.0	20.0	8.0	10.0	5.0			27	85	6
5	10.0	10.0	8.0	3.0	3.0	3.0	3.0	29	82	6
4	10.0	10.0	8.0	10.0	5.0			31	59	0
28	20.0	20.0	8.0	10.0	8.0	10.0	8.0	31	70	4
14	10.0	20.0	8.0	3.0	3.0	3.0	3.0	32	97	8
11	10.0	20.0	8.0	5.0	5.0			32	97	8
27	20.0	20.0	8.0	10.0	5.0	10.0	5.0	32	74	5
3	10.0	10.0	8.0	5.0	5.0			34	80	6
34	20.0	40.0	8.0	40.0	8.0			34	58	0
56	40.0	80.0	8.0	80.0	5.0	40.0	5.0	36	81	6
33	20.0	40.0	8.0	40.0	5.0			36	79	6
21	20.0	10.0	8.0	10.0	8.0	10.0	8.0	37	48	0
2	10.0	10.0	8.0	3.0	5.0			39	88	8
48	40.0	40.0	8.0	20.0	8.0	20.0	8.0	40	77	6
32	20.0	40.0	8.0	20.0	5.0			41	92	8
55	40.0	80.0	8.0	40.0	5.0	20.0	5.0	42	92	8
54	40.0	80.0	8.0	20.0	5.0	20.0	5.0	42	101	10
10	10.0	20.0	8.0	3.0	5.0			43	104	10
52	40.0	80.0	8.0	80.0	8.0			43	76	5
25	20.0	20.0	8.0	20.0	8.0			43	48	0
24	20.0	20.0	8.0	20.0	5.0			45	71	4
47	40.0	40.0	8.0	20.0	5.0	20.0	5.0	45	80	6
71	80.0	80.0	8.0	80.0	8.0	80.0	8.0	46	75	5
49	40.0	40.0	8.0	40.0	5.0	20.0	5.0	48	77	6
39	40.0	20.0	8.0	20.0	8.0	20.0	8.0	49	61	0
19	20.0	10.0	8.0	10.0	8.0			50	42	0
23	20.0	20.0	8.0	10.0	5.0			53	86	8
31	20.0	40.0	8.0	10.0	5.0			54	103	11
44	40.0	40.0	8.0	40.0	8.0			56	66	0
46	40.0	40.0	8.0	10.0	5.0	10.0	5.0	58	90	10
1	10.0	10.0	8.0	3.0	3.0			60	103	12
63	80.0	40.0	8.0	40.0	8.0	40.0	8.0	60	71	5

Table 11.3   Effect of room geometry and window size and location on the exposure of exterior walls above venting window openings from 10-lb/ft² fuel-load compartment fires (sorted on fire duration) (Continued)

Room case	Room size, ft			Window size, ft				Fire duration, min	Incident flux, Btu/ft² · min	Equivalent exposure, min
	D	W	H	w_1	h_1	w_2	h_2			
70	80.0	80.0	8.0	80.0	5.0	40.0	5.0	62	83	8
20	20.0	10.0	8.0	3.0	3.0	3.0	3.0	63	80	8
43	40.0	40.0	8.0	40.0	5.0			63	82	8
69	80.0	80.0	8.0	40.0	5.0	40.0	5.0	63	88	9
66	80.0	80.0	8.0	80.0	8.0			71	81	8
82	80.0	20.0	8.0	20.0	8.0	20.0	8.0	74	67	2
30	20.0	40.0	8.0	3.0	8.0			75	122	15
9	10.0	20.0	8.0	3.0	3.0			76	108	13
26	20.0	20.0	8.0	3.0	3.0	3.0	3.0	78	89	10
37	40.0	20.0	8.0	20.0	5.0			81	75	6
42	40.0	40.0	8.0	20.0	5.0			81	92	10
76	80.0	10.0	8.0	10.0	8.0	10.0	8.0	86	64	0
68	80.0	80.0	8.0	20.0	5.0	20.0	5.0	93	97	11
61	80.0	40.0	8.0	40.0	8.0			94	72	5
78	20.0	80.0	8.0	3.0	8.0			98	148	19
36	40.0	20.0	8.0	10.0	5.0			108	85	9
41	40.0	40.0	8.0	10.0	5.0			127	98	11
75	80.0	10.0	8.0	10.0	8.0			143	56	0
72	10.0	80.0	8.0	3.0	3.0			149	137	18
18	20.0	10.0	8.0	3.0	3.0			153	104	12
81	80.0	20.0	8.0	10.0	8.0			156	78	7
60	80.0	40.0	8.0	20.0	5.0			183	89	10
38	40.0	20.0	8.0	3.0	3.0	3.0	3.0	211	83	8
22	20.0	20.0	8.0	3.0	3.0			211	104	12
51	40.0	80.0	8.0	3.0	8.0			273	127	16
45	40.0	40.0	8.0	3.0	3.0	3.0	3.0	297	91	10
29	20.0	40.0	8.0	3.0	3.0			297	109	13
59	80.0	40.0	8.0	10.0	5.0			335	91	10
77	20.0	80.0	8.0	3.0	3.0			421	119	15
53	40.0	80.0	8.0	3.0	3.0	3.0	3.0	421	104	12
65	80.0	80.0	8.0	10.0	5.0			469	100	12
35	40.0	20.0	8.0	3.0	3.0			595	104	12
58	80.0	40.0	8.0	5.0	5.0			664	91	10
40	40.0	40.0	8.0	3.0	3.0			841	104	12
62	80.0	40.0	8.0	3.0	3.0	3.0	3.0	841	83	8
50	40.0	80.0	8.0	3.0	3.0			1190	109	13
74	80.0	10.0	8.0	3.0	3.0			1190	101	12
67	80.0	80.0	8.0	3.0	3.0	3.0	3.0	1190	91	10
80	80.0	20.0	8.0	3.0	3.0			1683	106	13
57	80.0	40.0	8.0	3.0	3.0			2380	104	12
64	80.0	80.0	8.0	3.0	3.0			3365	104	12

Table 11.4 Effect of room geometry and window size and location on the exposure of exterior walls above venting window openings from 10-lb/ft^2 fuel-load compartment fires (sorted on equivalent exposure)

Room case	Room size, ft			Window size, ft				Fire duration, min	Incident flux, Btu/ft^2 · min	Equivalent exposure, min
	D	W	H	w_1	h_1	w_2	h_2			
21	20.0	10.0	8.0	10.0	8.0	10.0	8.0	37	48	0
73	10.0	80.0	8.0	80.0	8.0			20	52	0
44	40.0	40.0	8.0	40.0	8.0			56	66	0
4	10.0	10.0	8.0	10.0	5.0			31	59	0
25	20.0	20.0	8.0	20.0	8.0			43	48	0
8	10.0	10.0	8.0	10.0	8.0	10.0	8.0	25	36	0
7	10.0	10.0	8.0	10.0	5.0	10.0	5.0	23	61	0
76	80.0	10.0	8.0	10.0	8.0	10.0	8.0	86	64	0
19	20.0	10.0	8.0	10.0	8.0			50	42	0
17	10.0	20.0	8.0	10.0	8.0	10.0	8.0	20	65	0
75	80.0	10.0	8.0	10.0	8.0			143	56	0
34	20.0	40.0	8.0	40.0	8.0			34	58	0
39	40.0	20.0	8.0	20.0	8.0	20.0	8.0	49	61	0
82	80.0	20.0	8.0	20.0	8.0	20.0	8.0	74	67	2
13	10.0	20.0	8.0	10.0	8.0			25	70	3
79	20.0	80.0	8.0	80.0	8.0			26	68	3
6	10.0	10.0	8.0	5.0	5.0	5.0	5.0	22	72	4
16	10.0	20.0	8.0	10.0	5.0	10.0	5.0	20	73	4
28	20.0	20.0	8.0	10.0	8.0	10.0	8.0	31	70	4
24	20.0	20.0	8.0	20.0	5.0			45	71	4
63	80.0	40.0	8.0	40.0	8.0	40.0	8.0	60	71	5
71	80.0	80.0	8.0	80.0	8.0	80.0	8.0	46	75	5
52	40.0	80.0	8.0	80.0	8.0			43	76	5
27	20.0	20.0	8.0	10.0	5.0	10.0	5.0	32	74	5
61	80.0	40.0	8.0	40.0	8.0			94	72	5
15	10.0	20.0	8.0	5.0	5.0	5.0	5.0	20	82	5
12	10.0	20.0	8.0	10.0	5.0			27	85	6
56	40.0	80.0	8.0	80.0	5.0	40.0	5.0	36	81	6
49	40.0	40.0	8.0	40.0	5.0	20.0	5.0	48	77	6
33	20.0	40.0	8.0	40.0	5.0			36	79	6
48	40.0	40.0	8.0	20.0	8.0	20.0	8.0	40	77	6
37	40.0	20.0	8.0	20.0	5.0			81	75	6
5	10.0	10.0	8.0	3.0	3.0	3.0	3.0	29	82	6
3	10.0	10.0	8.0	5.0	5.0			34	80	6
47	40.0	40.0	8.0	20.0	5.0	20.0	5.0	45	80	6
81	80.0	20.0	8.0	10.0	8.0			156	78	7
20	20.0	10.0	8.0	3.0	3.0	3.0	3.0	63	80	8
66	80.0	80.0	8.0	80.0	8.0			71	81	8
70	80.0	80.0	8.0	80.0	5.0	40.0	5.0	62	83	8
23	20.0	20.0	8.0	10.0	5.0			53	86	8
43	40.0	40.0	8.0	40.0	5.0			63	82	8

Table 11.4 Effect of room geometry and window size and location on the exposure of exterior walls above venting window openings from 10-lb/ft² fuel-load compartment fires (sorted on equivalent exposure) (*Continued*)

Room case	Room size, ft			Window size, ft				Fire duration, min	Incident flux, Btu/ft² · min	Equivalent exposure, min
	D	W	H	w_1	h_1	w_2	h_2			
2	10.0	10.0	8.0	3.0	5.0			39	88	8
55	40.0	80.0	8.0	40.0	5.0	20.0	5.0	42	92	8
14	10.0	20.0	8.0	3.0	3.0	3.0	3.0	32	97	8
62	80.0	40.0	8.0	3.0	3.0	3.0	3.0	841	83	8
32	20.0	40.0	8.0	20.0	5.0			41	92	8
38	40.0	20.0	8.0	3.0	3.0	3.0	3.0	211	83	8
11	10.0	20.0	8.0	5.0	5.0			32	97	8
36	40.0	20.0	8.0	10.0	5.0			108	85	9
69	80.0	80.0	8.0	40.0	5.0	40.0	5.0	63	88	9
26	20.0	20.0	8.0	3.0	3.0	3.0	3.0	78	89	10
42	40.0	40.0	8.0	20.0	5.0			81	92	10
59	80.0	40.0	8.0	10.0	5.0			335	91	10
10	10.0	20.0	8.0	3.0	5.0			43	104	10
60	80.0	40.0	8.0	20.0	5.0			183	89	10
46	40.0	40.0	8.0	10.0	5.0	10.0	5.0	58	90	10
67	80.0	80.0	8.0	3.0	3.0	3.0	3.0	1190	91	10
58	80.0	40.0	8.0	5.0	5.0			664	91	10
54	40.0	80.0	8.0	20.0	5.0	20.0	5.0	42	101	10
45	40.0	40.0	8.0	3.0	3.0	3.0	3.0	297	91	10
31	20.0	40.0	8.0	10.0	5.0			54	103	11
41	40.0	40.0	8.0	10.0	5.0			127	98	11
68	80.0	80.0	8.0	20.0	5.0	20.0	5.0	93	97	11
64	80.0	80.0	8.0	3.0	3.0			3365	104	12
65	80.0	80.0	8.0	10.0	5.0			469	100	12
40	40.0	40.0	8.0	3.0	3.0			841	104	12
57	80.0	40.0	8.0	3.0	3.0			2380	104	12
22	20.0	20.0	8.0	3.0	3.0			211	104	12
18	20.0	10.0	8.0	3.0	3.0			153	104	12
1	10.0	10.0	8.0	3.0	3.0			60	103	12
35	40.0	20.0	8.0	3.0	3.0			595	104	12
74	80.0	10.0	8.0	3.0	3.0			1190	101	12
53	40.0	80.0	8.0	3.0	3.0	3.0	3.0	421	104	12
50	40.0	80.0	8.0	3.0	3.0			1190	109	13
29	20.0	40.0	8.0	3.0	3.0			297	109	13
80	80.0	20.0	8.0	3.0	3.0			1683	106	13
9	10.0	20.0	8.0	3.0	3.0			76	108	13
30	20.0	40.0	8.0	3.0	8.0			75	122	15
77	20.0	80.0	8.0	3.0	3.0			421	119	15
51	40.0	80.0	8.0	3.0	8.0			273	127	16
72	10.0	80.0	8.0	3.0	3.0			149	137	18
78	20.0	80.0	8.0	3.0	8.0			98	148	19

Table 11.5   Effect of room geometry and window size and location on the exposure of exterior walls above venting window openings from 20-lb/ft^2 fuel-load compartment fires (sorted on fire duration)

Room case	Room size, ft			Window size, ft				Fire duration, min	Incident flux, Btu/ft^2 · min	Equivalent exposure, min
	D	W	H	w_1	h_1	w_2	h_2			
573	10.0	80.0	8.0	80.0	8.0			31	62	0
516	10.0	20.0	8.0	10.0	5.0	10.0	5.0	36	73	4
517	10.0	20.0	8.0	10.0	8.0	10.0	8.0	37	67	2
515	10.0	20.0	8.0	5.0	5.0	5.0	5.0	38	83	6
506	10.0	10.0	8.0	5.0	5.0	5.0	5.0	43	72	4
507	10.0	10.0	8.0	10.0	5.0	10.0	5.0	45	61	0
513	10.0	20.0	8.0	10.0	8.0			50	70	4
508	10.0	10.0	8.0	10.0	8.0	10.0	8.0	51	36	0
579	20.0	80.0	8.0	80.0	8.0			52	68	2
512	10.0	20.0	8.0	10.0	5.0			54	85	8
505	10.0	10.0	8.0	3.0	3.0	3.0	3.0	57	82	6
504	10.0	10.0	8.0	10.0	5.0			61	59	0
528	20.0	20.0	8.0	10.0	8.0	10.0	8.0	61	70	4
514	10.0	20.0	8.0	3.0	3.0	3.0	3.0	63	97	10
527	20.0	20.0	8.0	10.0	5.0	10.0	5.0	64	74	6
511	10.0	20.0	8.0	5.0	5.0			66	97	10
503	10.0	10.0	8.0	5.0	5.0			67	80	8
534	20.0	40.0	8.0	40.0	8.0			69	58	0
533	20.0	40.0	8.0	40.0	5.0			71	79	6
556	40.0	80.0	8.0	80.0	5.0	40.0	5.0	72	81	8
521	20.0	10.0	8.0	10.0	8.0	10.0	8.0	74	48	0
502	10.0	10.0	8.0	3.0	5.0			78	88	10
548	40.0	40.0	8.0	20.0	8.0	20.0	8.0	79	77	6
532	20.0	40.0	8.0	20.0	5.0			81	92	12
555	40.0	80.0	8.0	40.0	5.0	20.0	5.0	84	92	12
510	10.0	20.0	8.0	3.0	5.0			85	104	14
554	40.0	80.0	8.0	20.0	5.0	20.0	5.0	85	101	14
552	40.0	80.0	8.0	80.0	8.0			85	76	6
525	20.0	20.0	8.0	20.0	8.0			86	48	0
547	40.0	40.0	8.0	20.0	5.0	20.0	5.0	90	80	8
524	20.0	20.0	8.0	20.0	5.0			91	71	4
571	80.0	80.0	8.0	80.0	8.0	80.0	8.0	92	75	6
549	40.0	40.0	8.0	40.0	5.0	20.0	5.0	96	77	8
539	40.0	20.0	8.0	20.0	8.0	20.0	8.0	98	61	0
519	20.0	10.0	8.0	10.0	8.0			100	42	0
523	20.0	20.0	8.0	10.0	5.0			106	86	12
531	20.0	40.0	8.0	10.0	5.0			108	103	16
544	40.0	40.0	8.0	40.0	8.0			112	66	0
546	40.0	40.0	8.0	10.0	5.0	10.0	5.0	115	90	14
501	10.0	10.0	8.0	3.0	3.0			119	103	18
563	80.0	40.0	8.0	40.0	8.0	40.0	8.0	120	71	6

Table 11.5　Effect of room geometry and window size and location on the exposure of exterior walls above venting window openings from 20-lb/ft² fuel-load compartment fires (sorted on fire duration) (*Continued*)

Room case	Room size, ft			Window size, ft				Fire duration, min	Incident flux, Btu/ft² · min	Equivalent exposure, min
	D	W	H	w_1	h_1	w_2	h_2			
570	80.0	80.0	8.0	80.0	5.0	40.0	5.0	125	83	10
569	80.0	80.0	8.0	40.0	5.0	40.0	5.0	125	88	12
520	20.0	10.0	8.0	3.0	3.0	3.0	3.0	126	80	10
543	40.0	40.0	8.0	40.0	5.0			127	82	10
566	80.0	80.0	8.0	80.0	8.0			141	81	10
582	80.0	20.0	8.0	20.0	8.0	20.0	8.0	149	67	2
530	20.0	40.0	8.0	3.0	8.0			150	122	22
509	10.0	20.0	8.0	3.0	3.0			153	108	18
526	20.0	20.0	8.0	3.0	3.0	3.0	3.0	157	89	14
537	40.0	20.0	8.0	20.0	5.0			162	75	8
542	40.0	40.0	8.0	20.0	5.0			163	92	14
576	80.0	10.0	8.0	10.0	8.0	10.0	8.0	171	64	0
568	80.0	80.0	8.0	20.0	5.0	20.0	5.0	186	97	16
561	80.0	40.0	8.0	40.0	8.0			188	72	6
578	20.0	80.0	8.0	3.0	8.0			195	148	30
536	40.0	20.0	8.0	10.0	5.0			216	85	12
541	40.0	40.0	8.0	10.0	5.0			254	98	16
575	80.0	10.0	8.0	10.0	8.0			286	56	0
572	10.0	80.0	8.0	3.0	3.0			297	137	26
518	20.0	10.0	8.0	3.0	3.0			306	104	18
581	80.0	20.0	8.0	10.0	8.0			312	78	8
560	80.0	40.0	8.0	20.0	5.0			365	89	14
522	20.0	20.0	8.0	3.0	3.0			422	104	18
538	40.0	20.0	8.0	3.0	3.0	3.0	3.0	423	83	10
551	40.0	80.0	8.0	3.0	8.0			547	127	24
545	40.0	40.0	8.0	3.0	3.0	3.0	3.0	595	91	14
529	20.0	40.0	8.0	3.0	3.0			595	109	20
559	80.0	40.0	8.0	10.0	5.0			669	91	14
577	20.0	80.0	8.0	3.0	3.0			841	119	22
553	40.0	80.0	8.0	3.0	3.0	3.0	3.0	841	104	18
565	80.0	80.0	8.0	10.0	5.0			939	100	16
535	40.0	20.0	8.0	3.0	3.0			1190	104	18
558	80.0	40.0	8.0	5.0	5.0			1327	91	14
540	40.0	40.0	8.0	3.0	3.0			1683	104	18
562	80.0	40.0	8.0	3.0	3.0	3.0	3.0	1683	83	10
550	40.0	80.0	8.0	3.0	3.0			2380	109	20
574	80.0	10.0	8.0	3.0	3.0			2380	101	16
567	80.0	80.0	8.0	3.0	3.0	3.0	3.0	2380	91	14
580	80.0	20.0	8.0	3.0	3.0			3365	106	18
557	80.0	40.0	8.0	3.0	3.0			4759	104	18
564	80.0	80.0	8.0	3.0	3.0			6730	104	18

Table 11.6   Effect of room geometry and window size and location on the exposure of exterior walls above venting window openings from 20-lb/ft² fuel-load compartment fires (sorted on equivalent exposure)

Room case	Room size, ft			Window size, ft				Fire duration, min	Incident flux, Btu/ft² · min	Equivalent exposure, min
	D	W	H	w_1	h_1	w_2	h_2			
521	20.0	10.0	8.0	10.0	8.0	10.0	8.0	74	48	0
573	10.0	80.0	8.0	80.0	8.0			31	62	0
544	40.0	40.0	8.0	40.0	8.0			112	66	0
504	10.0	10.0	8.0	10.0	5.0			61	59	0
525	20.0	20.0	8.0	20.0	8.0			86	48	0
508	10.0	10.0	8.0	10.0	8.0	10.0	8.0	51	36	0
507	10.0	10.0	8.0	10.0	5.0	10.0	5.0	45	61	0
576	80.0	10.0	8.0	10.0	8.0	10.0	8.0	171	64	0
519	20.0	10.0	8.0	10.0	8.0			100	42	0
534	20.0	40.0	8.0	40.0	8.0			69	58	0
575	80.0	10.0	8.0	10.0	8.0			286	56	0
539	40.0	20.0	8.0	20.0	8.0	20.0	8.0	98	61	0
582	80.0	20.0	8.0	20.0	8.0	20.0	8.0	149	67	2
517	10.0	20.0	8.0	10.0	8.0	10.0	8.0	37	67	2
579	20.0	80.0	8.0	80.0	8.0			52	68	2
516	10.0	20.0	8.0	10.0	5.0	10.0	5.0	36	73	4
506	10.0	10.0	8.0	5.0	5.0	5.0	5.0	43	72	4
513	10.0	20.0	8.0	10.0	8.0			50	70	4
528	20.0	20.0	8.0	10.0	8.0	10.0	8.0	61	70	4
524	20.0	20.0	8.0	20.0	5.0			91	71	4
563	80.0	40.0	8.0	40.0	8.0	40.0	8.0	120	71	6
571	80.0	80.0	8.0	80.0	8.0	80.0	8.0	92	75	6
552	40.0	80.0	8.0	80.0	8.0			85	76	6
527	20.0	20.0	8.0	10.0	5.0	10.0	5.0	64	74	6
561	80.0	40.0	8.0	40.0	8.0			188	72	6
533	20.0	40.0	8.0	40.0	5.0			71	79	6
548	40.0	40.0	8.0	20.0	8.0	20.0	8.0	79	77	6
505	10.0	10.0	8.0	3.0	3.0	3.0	3.0	57	82	6
515	10.0	20.0	8.0	5.0	5.0	5.0	5.0	38	83	6
556	40.0	80.0	8.0	80.0	5.0	40.0	5.0	72	81	8
512	10.0	20.0	8.0	10.0	5.0			54	85	8
537	40.0	20.0	8.0	20.0	5.0			162	75	8
549	40.0	40.0	8.0	40.0	5.0	20.0	5.0	96	77	8
503	10.0	10.0	8.0	5.0	5.0			67	80	8
547	40.0	40.0	8.0	20.0	5.0	20.0	5.0	90	80	8
581	80.0	20.0	8.0	10.0	8.0			312	78	8
520	20.0	10.0	8.0	3.0	3.0	3.0	3.0	126	80	10
566	80.0	80.0	8.0	80.0	8.0			141	81	10
570	80.0	80.0	8.0	80.0	5.0	40.0	5.0	125	83	10
543	40.0	40.0	8.0	40.0	5.0			127	82	10
562	80.0	40.0	8.0	3.0	3.0	3.0	3.0	1683	83	10

Table 11.6   Effect of room geometry and window size and location on the exposure of exterior walls above venting window openings from 20-lb/ft² fuel-load compartment fires (sorted on equivalent exposure) (Continued)

Room case	Room size, ft			Window size, ft				Fire duration, min	Incident flux, Btu/ft² · min	Equivalent exposure, min
	D	W	H	w_1	h_1	w_2	h_2			
502	10.0	10.0	8.0	3.0	5.0			78	88	10
538	40.0	20.0	8.0	3.0	3.0	3.0	3.0	423	83	10
514	10.0	20.0	8.0	3.0	3.0	3.0	3.0	63	97	10
511	10.0	20.0	8.0	5.0	5.0			66	97	10
536	40.0	20.0	8.0	10.0	5.0			216	85	12
523	20.0	20.0	8.0	10.0	5.0			106	86	12
532	20.0	40.0	8.0	20.0	5.0			81	92	12
555	40.0	80.0	8.0	40.0	5.0	20.0	5.0	84	92	12
569	80.0	80.0	8.0	40.0	5.0	40.0	5.0	125	88	12
526	20.0	20.0	8.0	3.0	3.0	3.0	3.0	157	89	14
542	40.0	40.0	8.0	20.0	5.0			163	92	14
559	80.0	40.0	8.0	10.0	5.0			669	91	14
510	10.0	20.0	8.0	3.0	5.0			85	104	14
560	80.0	40.0	8.0	20.0	5.0			365	89	14
546	40.0	40.0	8.0	10.0	5.0	10.0	5.0	115	90	14
567	80.0	80.0	8.0	3.0	3.0	3.0	3.0	2380	91	14
558	80.0	40.0	8.0	5.0	5.0			1327	91	14
554	40.0	80.0	8.0	20.0	5.0	20.0	5.0	85	101	14
545	40.0	40.0	8.0	3.0	3.0	3.0	3.0	595	91	14
531	20.0	40.0	8.0	10.0	5.0			108	103	16
541	40.0	40.0	8.0	10.0	5.0			254	98	16
568	80.0	80.0	8.0	20.0	5.0	20.0	5.0	186	97	16
574	80.0	10.0	8.0	3.0	3.0			2380	101	16
565	80.0	80.0	8.0	10.0	5.0			939	100	16
540	40.0	40.0	8.0	3.0	3.0			1683	104	18
557	80.0	40.0	8.0	3.0	3.0			4759	104	18
522	20.0	20.0	8.0	3.0	3.0			422	104	18
509	10.0	20.0	8.0	3.0	3.0			153	108	18
580	80.0	20.0	8.0	3.0	3.0			3365	106	18
535	40.0	20.0	8.0	3.0	3.0			1190	104	18
501	10.0	10.0	8.0	3.0	3.0			119	103	18
553	40.0	80.0	8.0	3.0	3.0	3.0	3.0	841	104	18
564	80.0	80.0	8.0	3.0	3.0			6730	104	18
518	20.0	10.0	8.0	3.0	3.0			306	104	18
529	20.0	40.0	8.0	3.0	3.0			595	109	20
550	40.0	80.0	8.0	3.0	3.0			2380	109	20
530	20.0	40.0	8.0	3.0	8.0			150	122	22
577	20.0	80.0	8.0	3.0	3.0			841	119	22
551	40.0	80.0	8.0	3.0	8.0			547	127	24
572	10.0	80.0	8.0	3.0	3.0			297	137	26
578	20.0	80.0	8.0	3.0	8.0			195	148	30

story to story through the floor assemblies. If no exterior wall resistance is required, fire will be able to spread readily from story to story via the exterior wall. The technique illustrated in this study, though originally developed in the late 1970s, nonetheless can be used today to determine the level of fire exposure on an exterior wall for any specific combination of room configuration, window geometry, and fuel load. The analytical technique discussed herein could be used to assess potential fire spread via the windows above a venting window. Such an analysis would indicate the need for fire-resistive spandrel walls, flame deflectors, or other protective measures. The technique can also be used to select a minimum level of exposure that could be the basis of exterior wall design. Studies along these lines continue to be pursued, so that building regulations might reflect, more realistically, the exposures that are to be expected from room-fire circumstances.

11.3 CONDENSED REFERENCES/BIBLIOGRAPHY

AISI 1979, *Fire-Safe Structural Steel—A Design Guide*
ASTM E-119 1988, *Standard Test Methods for Fire Tests of Building Construction and Ma-*
Ingberg 1928, *Tests of Severity of Building Fires*
ISO 1975, *Fire Resistance Tests—Elements of Building Construction*
Oberg 1977, *Development of an Evaluative Technique for the Fire Performance of*
Ove Arup & Partners 1977, *Design Guide for Fire Safety of Bare Structural Steel*

12

Fire-Endurance Testing and Developments in Fireproofing Technology

Empirical methods have been used to determine the fire performance of sample construction elements used in tall buildings for many years. In addition, recently the results of such testing have provided checks for analytical methods which reduce testing and complexity while increasing design flexibility. In this chapter performance as well as test and analytical methods to model performance of specific systems are discussed.

12.1 LARGE-SCALE FIRE-ENDURANCE TESTING

The classical practice of testing to determine the fire-related behavior of building components has consisted of exposing such parts or relatively small models to a "standard fire." This type of testing, currently practiced by all testing laboratories, implies two types of uncertainties: those originating from results based on structural parts and systems of unrealistic dimensions, and uncertainties because of the application of thermal programs which may be nonrepresentative of actual fires.

This consideration led to the beginning of testing as early as in the 1960s to reduce uncertainties and inaccuracies due to specimen size and thermal programs. Because of concerns that the standard fire today does not resemble a real fire and because supporting parts or structures should be submitted to static loads during fire tests, the experimental approach followed was felt to allow a more reliable and accurate evaluation of the real structural behavior of elements during actual fire situations.

The testing station where this was occurring, built at the Centre Technique Industriel de la Construction Métallique (CTICM) in Maizières-lès-Metz, France, contained a larger test furnace and has proved to respond to research needs by considerably extending the range of test object size and by improved control capability during testing. The experience of 20 years of service there has shown its value as a source of advanced experimental structural engineering information. The following discussion describes that station and the specific types of tests carried out there.

1 The Fire-Testing Installation

The interior dimensions of the test furnace shown in Fig. 12.1 are 5.75 by 7.75 by 3.00 m (19 by 25 by 100 ft). A height increase of an additional 3 m (10 ft) is possible. The walls consist of three layers [with a total thickness of 35 cm (14 in.)]; namely, one layer of ordinary brick masonry toward the exterior, then a layer of insulation, and, on the inner side, a layer of light terra-cotta brick (density = 1). The front wall is formed of two doors, which facilitate access and assembly of specimens in the interior of the furnace. The whole installation is covered with removable light-concrete slabs having a sufficient fire resistance.

The base of the furnace lies on a slab which consists of several equal layers [with a total thickness of 46 cm (18 in.)], the first (interior) layer lightweight ceramic insulation, the following layer built of light insulation brick, and the upper layer of hard terra-cotta bricks.

The interior room dimensions can be reduced by inserting a removable transverse wall into the fire room. The resulting dimensions are 4.60 by 5.75 m (15 by 19 ft). Besides the two doors, the walls of the fire room are equipped with nine small observation windows.

The heating of the furnace allows reproduction of the standardized time-temperature (ISO) curve or any other thermal program with peak temperatures in the range of 1200°C (2200°F). These can be reached in 25 to 30 min. The installation consists of 16 high-pressure burners, the fuel being supplied with an adjustable flow of between 25 and 90 l/hr (6.5 and 24.0 gal/hr), corresponding to 13×10^6 cal/hr. Removal of hot gases is carried out by means of two lateral chimneys, as illustrated in Fig. 12.2.

The test-specimen loading device comprises six double-acting wide-course jacks with a power of 30 tons each and six jacks of 150 tons each. The jacks are

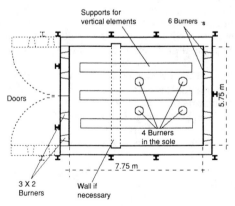

Fig. 12.1 Test furnace.

connected by pairs to six groups of pumps, which automatically regulate the loading level and maintain a constant load during testing.

2 Experience Derived from Large-Scale Testing

The tests, which have been carried out since the construction of the large furnace at CTICM, have been valuable in describing the behavior of many types of materials under realistic test conditions. These experiments can be grouped into four categories: structural elements, partition elements, miscellaneous mechanical systems, and compartments.

Structural Elements. The furnace was initially conceived for carrying out tests on steel constructions. The first tests were made with steel beams 8 m (26 ft) long, which were placed on three supports; the second tests were made with frames of two bays of 3-m (10-ft) span each; and the third tests were with two-story-high frames having a height of 6 m (20 ft). These elements were submitted to static loads of several hundred tons, which permitted observation of plastic-hinge formations in relationship to the longitudinal and cross-sectional temperature gradients.
It was thus possible to elaborate and verify a method of plastic design at high temperatures. Later this research became the reference source for official calculations of structural fire resistance in France (CTICM, 1975).
Research was also conducted with reinforced-concrete slabs. Such specimens were 9 m (30 ft) long and placed on three supports with or without restraint. Hence it was possible to accurately observe the collapse of steel reinforcement associated with support locations because of the sagging of a slab under normally occurring thermal gradients between the exposed and unexposed sides of the

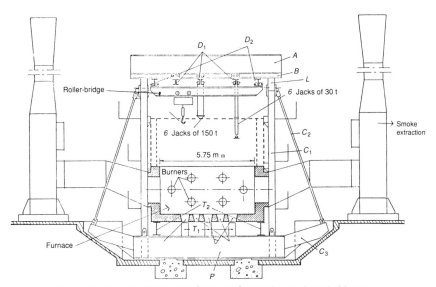

Fig. 12.2   Removal of hot gases from test furnace by two lateral chimneys.

slab. Previously this phenomenon could not be established with tests of isostatic elements of small dimensions. These results have led to a modification of design calculations for concrete structures which now take into account the rotational capacity of steel reinforcement.

Composite structures have also been the subject of instructive tests. Such structures have included steel beams and concrete slabs connected by shear connectors, or concrete beams with steel sheets. Statically indeterminate elements protected or unprotected by thermal insulation products have also been tested. A phenomenon similar to steel-reinforcement collapse was verified at support locations, and adhesion qualities of certain steel sheets were described.

For hollow steel columns filled with concrete, one particular case tested was a 3.7-m (12-ft)-high column with a diameter of 400 mm (16 in.). It was submitted to a load of 450 tons, and fire stability of 36 min was observed.

Wood structures, where the fire performances are briefer than for noncombustible elements, have not been extensively tested in this research program. However, one test was performed on two half-arches of laminated wood with a height of 4 m (13 ft) and a span of 6.75 m (22 ft). The arches were joined by wooden purlins and fixed by metallic connectors. The failure mechanism of these elements was precisely observed. The testing confirmed that it is not sufficient to increase the section of the arches in order to obtain a greater fire resistance. Instead, the fire resistance was influenced by the behavior of the stabilizing, smaller section of the purlins.

Partition Elements. Tests have been carried out on 4-m (13-ft)-high walls which have shown a need to include horizontal joints. The latter are generally not necessary for partitions shorter than 2.70 m (9 ft). In several cases the connections of partition assemblies were tested. Standardized tests usually check single elements only, independent of their wall-system connection details. The interaction of *all* elements of a wall can have a great influence on the final test result, and thus such testing is appropriate.

For example, a plaster partition may have a fire resistance of 2 to 3 hr. However, when equipped with a metallic door frame, the expansion of the frame in the first 30 min of the fire can cause the wall to crack and very rapidly will lead to the wall's collapse. Likewise, when non-load-bearing partitions are situated under a slab directly exposed to fire, deformation of the floor (due primarily to the thermal gradient between the interior and the exterior faces of the slab) can lead to a premature collapse of the partition elements.

Some tests conducted have been used to estimate the thermal insulation needed for various elements of steel structures. These have included, for example, a 3-m (10-ft)-high column connected to a 3-m (10-ft)-long beam, protected on both sides by cladding elements. CTICM also conducted tests on closing elements such as a horizontal metallic curtain, 5.7 m (19 ft) long and 6 m (20 ft) wide, on a door of 4 by 3.6 m (13 by 12 ft).

Thus it has been possible to examine real construction conditions and materials and assure their reliability if subjected to fire. More importantly, because of some of these tests, the manufacturers have been able to improve their products in order to meet code requirements or obtain other specific desired results.

Miscellaneous Mechanical Equipment. Intrafloor HVAC ducts which pass through different rooms are potential paths for fire spread. In France, therefore, ducts are subjected to specific fire-resistance tests, which require minimum specimen lengths of 6 m (20 ft). This length simulates as closely as possible the internal stresses which could be created due to a restrained elongation. Tests on sections varying between 200 by 200 mm (62 in.[2]) and 1500 by 100 mm (2325 in.[2]) have demonstrated the importance of this parameter for the internal heating of the duct as well as for the risks of cracking.

Recently the center evaluated a duct section 3000 by 3000 mm (10 by 10 ft). When duct sections are this large, it is imperative that the fire rating of the wall across which the duct passes be assured, and it is necessary to install dampers at the wall. In this case a group of eight dampers with a total dimension of 4 by 2 m (13 by 6½ ft) were tested, and the quality of bonding between the dampers and the buckling conditions of the entire system were verified.

Obviously, when the sections of such ducts are very large, fans which extract gases must have the corresponding power. Fan tests have been carried out with a 45-kW blower of 2000-mm (6-ft 6-in.) diameter. For such large elements it is impossible to apply the results from small testing apparatus because they are not systematically interchangeable and because it is difficult to provide for scaleup of all the components.

Compartment Fire-Endurance Testing. The large internal volume of CTICM's furnace allows the full-scale construction of test units, or complete cells, such as the following:

1. A passageway 1.5 m (5 ft) wide by 3 m (10 ft) high by 7 m (23 ft) long with a suspended ceiling for testing the fire developed in an adjoining space while the aspiration of hot gases was carried by a duct located in the technical casing.
2. A room 4.9 m (16 ft) wide by 3.6 m (12 ft) high by 5.75 m (19 ft) long adjacent to a 2.85-m (9-ft)-long passageway. Construction details included steel columns, a concrete floor and beams, suspended vermiculite ceiling, and plaster partition walls with a door.
3. A three-dimensional structure of reinforced-concrete panels, 5 m (16 ft) high, with a concrete floor of 7.50 by 5.50 m (24½ by 18 ft). This was exposed to fire on half its surface.

These tests were conducted for research purposes to better understand the interaction of various elements and, at the same time, to get a more accurate insight into the temperature distributions in the case of a localized fire in a building.

3 Impact of Large-Scale Fire Testing

Testing of building elements of full-scale or large dimensions should not be an unusual practice since ISO 834 requires that the test specimen be of full size. However, there is an obvious limitation to such testing due to furnace dimensions. The size of the installation at CTICM was chosen to allow full-size testing of typical structural members and one- or possibly two-story frames or subassemblies in buildings (see Fig. 12.3).

Tests of steel structures during the station's first years provided general observations in the following areas:

1. Effects of specimen dimensions (standard-size sample versus relatively small sample versus one-story-high sample of large cross section with significant transverse thermal gradient)

2. Effects of continuity in continuous beams or of the degree of fixation in beam-to-column connections

3. Effects of static loads on structural elements exposed to fire

This experimental research has been closely associated with theoretical studies that lead to practical findings regarding fire resistance of unprotected steel elements and structures in particular.

The dimensions chosen for specimen-handling facilities at this installation are justified on the following grounds:

1. Results are obtained on normally sized samples for buildings.

2. Field assemblies of larger or unusual dimension may show a response beyond the range normally encountered in standard testing.

The 20-year experience with this installation (which initially was built mainly for the study of steel structures) has led to important and practical insights into the behavior of many types of structural and nonstructural building components exposed to fire.

Furthermore, such testing need not be limited to determining a fire classification. Instead, all possible measures and observations which contribute to the un-

Fig. 12.3   Furnace with both front doors open and all burners working at full power—for demonstration only.

derstanding and evaluation of the behavior of building components in fire situations can be pursued and researched. To acquire a clear view on the behavior of large or full-scale elements, the following procedures appear to be minimum test requirements:

1. Assess the available knowledge of the element to be tested and provide an initial model of its possible behavior during fire.
2. Conduct standard tests or specific tests to verify the origins of certain hypotheses.
3. Perform tests in large dimensions to determine and control the effects of fire on a realistic scale and revise, if necessary, the initial model.

Thus tested and elaborated, the limits of a model should be valid, and can be specified and included in national regulations.

12.2 FIREPROOFING TECHNOLOGY FOR TALL BUILDINGS

A technical project sponsored by Japan's Ministry of Construction from 1982 through 1987 led to the development of a new technique to evaluate building-fire safety (Wakamatsu, 1988). The technique was not based on conventional specifications conforming to the Japanese Building Standards Act and other regulations, but rather on probable building-fire characteristics, design conditions in the project, and the performance of materials used. The fire-resistant steel (FR steel) described here is a product developed in response to this new evaluation technique.

Compared with ordinary structural steel for buildings, FR steel has the characteristic of being more resistant to high temperatures than ordinary grades of steel. By using FR steel, it is possible not only to reduce the amount of fire-resistant covering required on a steel-frame structure, but also, on occasion, to eliminate this covering completely, depending on the fire and design conditions of the structure. Several buildings using FR steel have been evaluated recently and the results of those evaluations are also presented here.

The enhanced high-temperature yield strength of FR steel is obtained by adding chromium, molybdenum, and other alloying elements to ordinary steel. The manufacturing processes (after rolling) are the same as for ordinary steel.

Under high temperatures it is well known that ordinary steel shows a decrease in yield strength. Figure 12.4 compares the change in strength (yield point) of FR steel and ordinary steel under high temperatures. FR steel (designated as SM50A-NFR) and ordinary steel (designated as SM50A) both meet the Japanese building standard specified in JIS G 3106 (1987). Rolled steel for welded structures has a yield point ≥ 324 N/mm^2 (≥ 0.5 lb/in.^{2}) and has a tensile strength of 490 to 608 N/mm^2 (0.07 to 0.09 lb/in.^{2}).

The Japanese Ministry of Construction specifies 350°C (660°F) as the average allowable temperature and 450°C (840°F) as the maximum allowable temperature of the steel members in a building during a fire, and stipulates that any steel-frame building shall be provided with fire-resistant covering to ensure that the specified temperature is not exceeded. This is because, as shown in Fig. 12.4, at a temperature of around 350°C, the yield point of ordinary steel declines to two-thirds of the standard yield point at normal temperature [allowable stress 216 N/

mm^2 (0.03 lb/in.^{2}) for SM50A]. Hence the yield point required for the steel structure in a building during a fire cannot be maintained.

In contrast, FR steel will retain the required stress at temperatures up to 600°C (1100°F). Since the normal temperature of a fire is approximately 1000°C (1800°F, ordinary steel requires heat-resistant covering which keeps the steel temperature below 350°C (660°F). This allows the heat-resistant covering for FR steel required to keep the steel temperature below 600°C (1100°F) to be significantly reduced in size. In addition, depending on the fire and design conditions of a building, FR steel may be used without heat-resistant covering when steel temperatures will not exceed 600°C (1100°F) during a fire.

1 Physical Properties of FR Steel

The high-temperature yield strength is much higher than that of ordinary steel. [The yield point of FR steel at 600°C (1100°F) is two-thirds higher than the standard yield point at normal temperature.] The performance at normal temperature conforms to Japanese standard JIS G 3106. (Design requirements at normal temperature are the same as those for ordinary steel.) The workability is comparable to that of ordinary steel. (FR steel can be cut and welded in the same way as ordinary steel.) *However,* welding materials and high-strength bolts used with FR steel must have a fire resistance comparable or superior to that of the FR steel itself. At present, FR steel is available in plates and shapes of 400 N/mm^2 (0.06-lb/in.^{2}) class (SM41) and 500 N/mm^2 (0.07 lb/in.^{2}) class (SM50). In the following, the material characteristics of 500 N/mm^2 class (SM50) plate are presented.

Normal-Temperature Characteristics.   The chemical composition and mechanical properties at normal temperature are shown in Tables 12.1 and 12.2, respectively. Both conform to JIS G 3106.

High-Temperature Characteristics.   The temperature dependence of high-temperature yield strength is shown in Fig. 12.5, and that of the high-temperature Young's modulus is shown in Fig. 12.6. While the yield point of ordinary steel declines to approximately two-thirds of the standard value at normal temperature

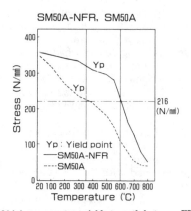

Fig. 12.4   Comparison of high-temperature yield strength between FR steel and ordinary steel.

around 350°C (660°F), FR steel retains its yield point at temperatures up to 600°C (1100°F), which is two-thirds greater than the standard value.

With respect to Young's modulus, FR steel sharply declines at temperatures above 700°C (1300°F), whereas ordinary steel sharply declines at temperatures above 600°C (1100°F).

Figure 12.7 compares the high-temperature creep characteristics of FR steel

Table 12.1 Chemical composition

Symbol	Thickness (mm)	Chemical Composition (%)						
		C	Si	Mn	P	S	C_{eq}	P_{CM}
SM50A-NFR	32	0.10	0.20	1.11	0.019	0.003	0.42	0.20
SM50A	32	0.16	0.36	1.45	0.019	0.006	0.40	0.25
SM50A Specification[a]	—	≦0.20	≦0.55	≦1.60	≦0.035	≦0.035	—	—

$C_{eq} = C + Mn/6 + Si/24 + Ni/40 + Cr/5 + Mo/4 + V/14$ (%)
$P_{CM} = C + Si/30 + Mn/20 + Cu/20 + Ni/60 + Cr/20 + Mo/15 + V/10 + 5B$ (%)
a) JIS G 3106

Table 12.2 Mechanical properties

Symbol	Thickness (mm)	Yield point (N/mm²)	Tensile strenght (N/mm²)	Elongation (%)	Yield ratio (%)
SM50A-NFR	32	410	550	31	75
SM50A	32	380	550	31	69
SM50A Specification[a]	≦16	≧324	490~608	≧17	—
	>16,≦40	≧314		≧21	
	>40	≧294		≧23	

a) JIS G 3106

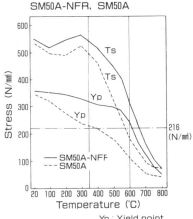

Yp : Yield point
Ts : Tensile strength

Fig. 12.5 Temperature dependence of high-temperature yield strength.

with those of ordinary steel. In the case of ordinary steel, high-temperature creep strain develops in a short time even under a small load of 100 N/mm² (0.015 lb/in.²). In contrast, FR steel shows little creep strain under a load of 150 N/mm² (0.02 lb/in.²). Thus by using FR steel, it is possible to reduce the deformation of a building during a fire.

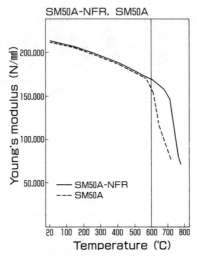

Fig. 12.6   Temperature dependence of high-temperature Young's modulus.

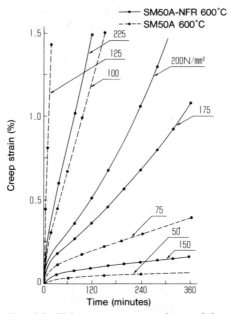

Fig. 12.7   High-temperature creep characteristics.

Normal-Temperature Characteristics after Thermal History.   Figure 12.8 shows the mechanical properties of FR steel at normal temperature after a thermal history of up to 700°C (1300°F). In fact, FR steel shows reproducibility in its mechanical properties at normal temperature after it has cooled. Hence FR steel can be reused after it has undergone a fire.

Weldability.   FR steel contains elements which improve high-temperature yield strength. In addition, the contents of carbon, manganese, and other elements, which are detrimental to weldability, are small. Therefore it is comparable in weldability to ordinary steel. Weldability test results are shown in Figs. 12.9 and 12.10, and in both tests FR steel had better results than ordinary steel.

Figure 12.11 depicts a welding test on a column and beam joint of actual size. Yield strength, hardness, impact strength, and other characteristics of joints obtained by CO_2 welding, automatic welding, electroslag welding, and other welding procedures were tested.

High-Strength Bolts.   The high-strength bolts for FR steel are F10T torque-control bolts [tensile strength 100 N/mm² (0.015 lb/in.²)], which are used in the same way as conventional high-strength bolts. The performance of this bolt at

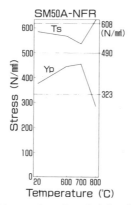

Fig. 12.8   Normal-temperature characteristics after thermal history.

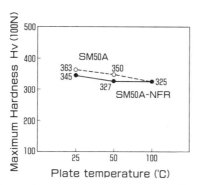

Fig. 12.9   Maximum-hardness test in weld-heat-affected zone.

normal temperatures has been approved by the Ministry of Construction as conforming to Japanese standard JSS II 09, High-Strength Bolts of Torque Control Type, Hexagon Nuts, and Flat Washer Sets for Structural Use.

Material Identification.   In fabricating and erecting FR steel frames it is necessary to accurately distinguish between FR steel and ordinary steel. Because FR steel

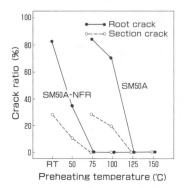

Fig. 12.10   Y-groove cracking test.

Fig. 12.11   Weldability test on full-scale column-beam joint.

is difficult to identify once it is cut, it is given the mill mark shown in Fig. 12.12 on its entire surface. Welding materials and high-strength bolts are identified by a brand name, indicated on each type of material and bolt. (An example of a high-strength-bolt identification mark is shown in Fig. 12.13.)

2 Testing Fire-Resistant Steel

The high-temperature characteristics of FR steel have been confirmed by heating tests under load (Yamaguchi et al., 1990). The tests were carried out on beams by the Japan Testing Center for Construction Materials and on columns by the General Building Research Corporation.

The column test is depicted in Fig. 12.14, where the size of the test piece was H-300 by 300 by 10 by 15 mm [length 3.5 m (11.5 ft)]. The point of buckling in the test piece—when it was subjected to a load corresponding to the allowable stress [approximately 2.10 kN (472 lb)] and heated as specified according to standard practice—was assumed as the fire-resisting time of the test piece. Wet mineral fiber was used as the fire-resistant covering. Figure 12.15 shows the heating temperature (furnace temperature) and the steel temperature, and Fig. 12.16 shows the column elongation in the vertical direction. It was confirmed from these tests that FR steel does not buckle at temperatures up to 600°C (1100°F) and that it has a fire resistance corresponding to 1 hr with 10 mm (½ in.) of fire-resistant covering. [Ordinary fire-resistant covering is typically 30 mm (1¼ in.).]

Column deformation and buckling temperature (time) can be simulated using the high-temperature characteristics of FR steel and an elastoplastic thermal deformation analysis program (Okabe et al., 1988). The results of a simulation are shown in Fig. 12.16. From the test results the following was confirmed:

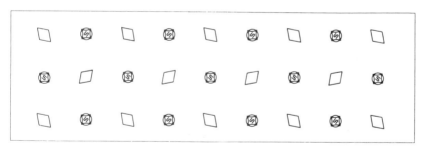

Fig. 12.12   Mill mark on FR steel plate.

Fig. 12.13   Impression on high-strength bolt for FR steel.

1. At 600°C (1100°F) FR steel retains a yield strength (yield point) higher than the allowable stress (two-thirds of the standard yield point at normal temperature).
2. Similarly, at 600°C (1100°F) FR steel members (columns, beams, and the like) can bear a load corresponding to the allowable stress.

When FR steel is used in a building frame, it is necessary to confirm the fire-resistant performance of the entire frame as well as the fire-resistant performance of the FR steel itself and the FR steel members. This is because the thermal de-

Fig. 12.14   Heating test under load.

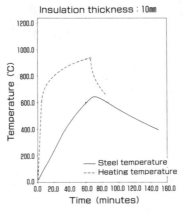

Fig. 12.15   Column heating temperature and steel temperature.

formation of the FR steel frame is greater than that of a conventional steel frame due to the higher steel temperature of FR steel [maximum 600°C (1100°F)] compared with that of ordinary steel [maximum 350°C (660°F)].

In particular, there is fear that the thermal expansion of beams during a fire causes considerable thermal deformation of the outer columns, which in turn causes local buckling of the columns, resulting in a decline of the load-bearing capacity of the columns. The load-bearing capacity of FR steel columns subjected to this kind of deformation was confirmed by a high-temperature bending and buckling test conducted at Chiba University, Japan.

Figure 12.17 shows a general view of the testing equipment, and Fig. 12.18 depicts the appearance of the test piece after the test. The test was carried out on various steel shapes. The test results from steel pipes are shown in Fig. 12.19. It was confirmed that an FR steel pipe column can support a vertical load up to approximately ½₅ (interstory deflection angle) of deformation at a steel temperature of 600°C (1100°F) with a width-to-thickness ratio of 30 and an axial force ratio of 0.30.

3 Case Studies

Under Japanese building regulations, buildings using FR steel fall into a different category than those of conventional steel structures. Therefore FR steel buildings must be evaluated individually on their fire-safety factors. The structural ability of FR steel frames during a fire must also be evaluated.

The results of some appraisals are shown in Table 12.3. The appraisal results for the Shinkawa building (presently the NSC no. 2 building) and the Yawata Works no. 2 building (presently the Tobihata building) are presented in this section.

Shinkawa Building.   This building, shown in Figs. 12.20 and 12.21, was the first to use FR steel. Its physical data are presented in Table 12.4. Its construction

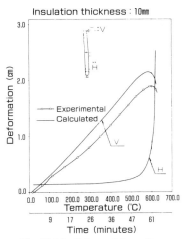

Fig. 12.16 Column deformation.

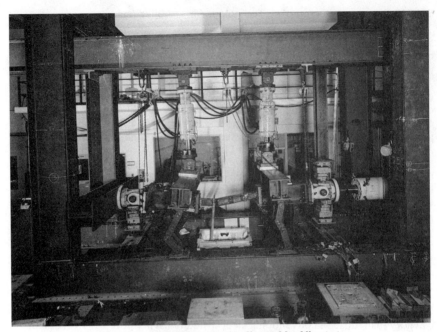

Fig. 12.17   High-temperature bending and buckling test.

Fig. 12.18   Test piece after having been subjected to high-temperature bending and buckling test.

216

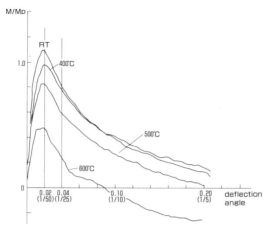

Fig. 12.19 High-temperature bending and buckling test.

Table 12.3 Results of the appraisal of buildings using FR steel

Date of appraisal	Name	Number of stories	Total floor area	FR steel frame weight	Specifications
April 1989	Shinkawa Building	15	23.500 ㎡	3,000ton	Reduce fire-resisting covering
September 1989	Yawata No.2 Building	8	15.800 ㎡	1,700ton	External frame : no covering
December 1989	Yokohama Sogo Building	6	28.900 ㎡	2,200ton	Reduce fire-resisting covering
Jung 1990	Tokoname Park Gymnasium	4	8.700 ㎡	800ton	External frame : no covering
July 1990	Kamiogi-1-chome Building	18	27.100 ㎡	700ton	External frame : no covering
July 1990	Tokyo Metropolitan Health plaza	17	83,700 ㎡	1,500ton	Reduce fire-resisting covering
November 1990	BBKK Hiroo Building	3	550 ㎡	20ton	Reduce fire-resisting covering
December 1990	Procter & Gamble Far East. Inc. Japan Headquarters and Technical Center	31	46,100 ㎡	4,700ton	External frame : no covering

Fig. 12.20 View of Shinkawa building.

economics were enhanced, in part, by reducing the thickness of the required fire-proofing for structural members.

In the building's fire-safety evaluation its performance was verified on the basis of experimental results on FR steel columns and beams, provided with a light fire-resistant covering, and the thermal deformation analysis (Okabe, 1988) results on FR steel frames. The results of the thermal deformation analysis of the frame are shown in Fig. 12.22. The projected fire conditions conformed to the required fire-resisting period (1 to 3 hr).

The amount of reduction in the fire-resistant covering is shown in Table 12.5. It can be seen that the covering thickness is one-half to one-third of the ordinary covering thickness.

Yawata No. 2 Building.   This building, shown in Figs. 12.23 and 12.24, used FR

Fig. 12.21   Steel frame of Shinkawa building.

Table 12.4   Outline of Shinkawa building

	A ward	B ward
Building area	906 m²	776 m²
Floor area	15,418 m²	8,560 m²
No. of stories	15	10
Building height	64.9 m	45.6 m
Story height	3.8 m	3.8 m

steel in its external steel frames, in which the columns and beams are arranged outside the buildings. The frame was not provided with any fireproofing since the exterior-frame steel temperature would not exceed 600°C (1100°F).

The purpose of the external steel frame is to expose the columns, beams, and other structural members outside the building to achieve an aesthetic effect, and

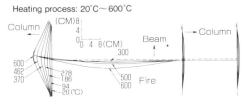

Fig. 12.22   Simulated thermal deformation of building.

Table 12.5   Reduction of fire-resistant covering in Shinkawa building

Member	Fire-resisting hours	Member size	Covering thickness (mm)	
			Ordinary steel	FR steel
Column	1	□ − 600 × 600 × 16	30	10
	2	□ − 600 × 600 × 25	40	10
	3	□ − 600 × 600 × 32	50	15
Beam	1	H − 850 × 300 × 12 × 22	25	10
	2	H − 850 × 300 × 14 × 25	35	20
	3	H − 900 × 350 × 14 × 32	45	30

Fig. 12.23   View of unfinished Yawata no. 2 building.

Fig. 12.24   View of finished Yawata no. 2 building.

Table 12.6   Outline of Yawata no. 2 building

Building area	2,318 m²
Floor area	15,834 m²
No. of stories	8
Buijding height	32.9 m
Story height	3.7 m

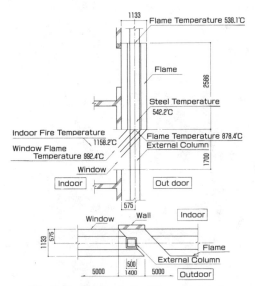

Fig. 12.25   Temperatures of flames and steel.

220

to obtain larger effective floor space by eliminating internal columns. Frames of the design are popular both at home and abroad.

The physical data of this building are given in Table 12.6. Figure 12.25 describes the fire-exposure geometry related to window openings and the associated steel frame of the building during a simulated fire exposure. The maximum calculated steel temperature is 542.5°C (1000°F), which is considerably less than 600°C (1100°F).

Thus depending on the fire and design conditions, it is possible to design a steel-frame building without a fire-resistant covering.

12.3 ROBOTIC SYSTEMS FOR SPRAY FIREPROOFING

The application of fireproofing to structural steel, regardless of type, remains critical to the fire performance of tall buildings. It is of interest, then, that at an international industrial robot exhibition held in Tokyo in 1987, a new category of construction robots was introduced by Shimizu and other Japanese companies. Research and development of construction robots began in Japan about 10 years ago. Studies and designs of robotic systems for various kinds of construction work were executed through projects funded by the Japan Industrial Robot Association. Since these projects were initiated, construction robots have been under development by general contractors, and some are available on the open market. This was a response in part to Japan's national policy which led to an increase in construction projects with a resultant shortage of skilled construction workers. As a result, expectations for construction robots are quite high in the construction industry.

Fireproofing work is considered one of the most unpleasant and hazardous tasks in the construction trades, in part because the working environment includes copious quantities of small rock wool particles. As a result, prototype robotic systems were developed to investigate the possibility of using robots that could perform the skilled construction work required for the application of sprayed fireproof systems.

The Shimizu SSR-3 robot is designed to spray fireproofing material onto structural steel frames and is illustrated in Figs. 12.26 and 12.27. This is the third model in its series. The first robot model to demonstrate successfully the feasibility of using robots on a construction site was developed in June 1986, and was used at two construction sites for trial applications. However, the quality of robotic spray work was not comparable to that of a human worker because the sensor systems required improvement. The problem lay with the distance sensor of the prototype robot, which was set beside the spray nozzle and could not function while the robot was moving. Thus the robot sprayed material using a wrist swing motion. As a result, the finished surface was not flat enough, especially at the overlapping zones of each spray pattern. To overcome this problem, the sensor system was improved by using a position sensor to detect the distance from the robot arm to the steel beam during spraying. A second model of robot was used at two demonstration construction sites where finished surfaces were flat, and the appearance of the surfaces was almost as good as that of a skilled worker. In addition, the uniformity of the fireproofing thickness was improved by the new sensor system.

Although both robots were successful prototypes, a number of problems re-

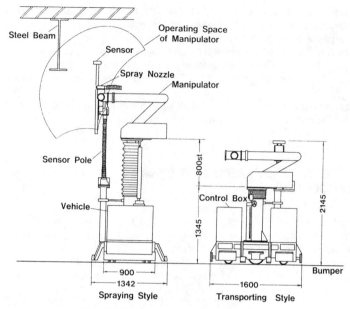

Fig. 12.26   Components of Shimizu robot.

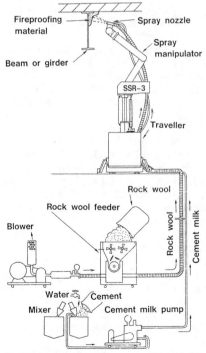

Fig. 12.27   Rock-wool spray system of Shimizu robot.

mained from the standpoint of practical applications. These included the following:

1. The robots were too large to be transported by a construction lift and had to be disassembled.
2. The operator had to be thoroughly trained to instruct the robot, as direct computer instruction could not correct previous mistakes.
3. The cost of the robots was high.

A third model was developed to overcome these liabilities. For instance, the second model was driven by hydraulic power, but the third is driven by d-c power, which reduced the cost of the robot. In addition controllers, which were separated in the case of the second model, are mounted on the body of the third. As a result, control cables are eliminated and only the power cable is necessary. A view of the third robot is shown in Figs. 12.28 and 12.29 and specifications are listed in Table 12.7.

The robot consists of a vehicle, a manipulator, a distance sensor assembly, and a control box. The manipulator has two degrees of freedom for arm motion and two degrees of freedom for wrist motion. A spray nozzle is attached to the top of the manipulator that moves the nozzle. The vehicle has two drive wheels at the center and four casters at each corner of the body. It moves back and forth, left and right, using a powered steering mechanism.

The robot has an ultrasonic sensor system which measures its distance from beams and adjusts its position using these data. Two ultrasonic sensors are attached at the top of a telescopic pole, which can be adjusted according to the height of the beam to be sprayed. Using this sensor system, the robot can spray while it is traveling along the beam. Because the height and size of steel beams to be sprayed are different from building to building, the height of the robot's manipulator arm is adjusted manually using a screw jack. Similarly the height of the robot itself can be reduced to about 2 m (7 ft) when it is being transported using this jack.

The robot is numerically controlled and can be operated remotely. However, the nozzle (attached to the manipulator) is positioned away from the beams being sprayed, making it difficult to instruct. In response to this problem, a robot programming system called "off-line teaching system" was developed, consisting of a personal computer and a digital cassette recorder. The action of the robot is simulated on the screen of a personal computer, from which a program of instruction is created. As the robot utilizes a digital cassette recorder for its memory, program data can be easily transferred to another cassette recorder. Furthermore, with this system, errors in the program can easily be corrected without changing entire programs.

Initially it took too much time to teach different programs to a robot. Because of this, the off-line teaching system was also designed to reduce teaching time. Data used for programming the robot are generated automatically by the system when the dimensions of a beam are input through the computer keyboard. Figure 12.30 is an example of typical computer programming. In addition, the system can be programmed to account for particular characteristics of structural members. For example, if a beam has a series of holes or openings, the robot can be programmed to avoid spraying such openings with fireproofing. Construction machines, often called construction robots, are not always real robots. At present, however, simple construction machines are more practical from the standpoint of low-cost, easy operation and maintenance. Yet in spite of technical and economic

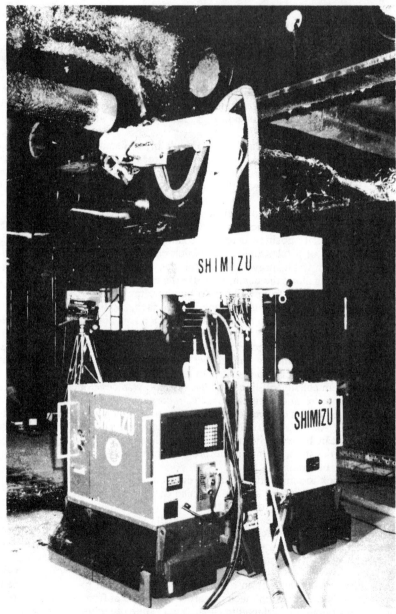

Fig. 12.28   Shimizu fireproofing robot.

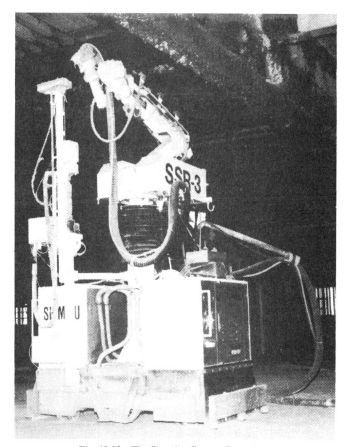

Fig. 12.29   The Shumizu fireproofing robot.

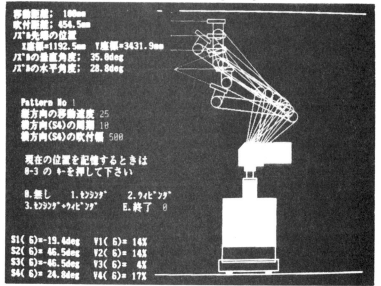

移動距離；　100mm
吹付距離；　454.5mm
ﾉｽﾞﾙ先端の位置
　　Ｘ座標＝1192.5mm　Ｙ座標＝3431.9mm
ﾉｽﾞﾙの垂直角度；　35.0deg
ﾉｽﾞﾙの水平角度；　28.8deg

Pattern No 1
縦方向の移動速度　25
横方向(S4)の周期　10
横方向(S4)の吹付幅　500

現在の位置を記憶するときは
0-3 の ｷｰを押して下さい

0.無し　　　1.ｾﾝｼﾞﾝｸﾞ　　　2.ｳｨﾋﾞﾝｸﾞ
3.ｾﾝｼﾞﾝｸﾞ＋ｳｨﾋﾞﾝｸﾞ　　　E.終了　0

S1(6)=-19.4deg　　V1(6)= 14%
S2(6)= 46.5deg　　V2(6)= 14%
S3(6)=-46.5deg　　V3(6)= 4%
S4(6)= 24.8deg　　V4(6)= 17%

Fig. 12.30   Off-line teaching system, arm-motion data.

Table 12.7   Specifications of the Shimizu robot

Traveling device		Degrees of freedom Positioning Traveling speed Positioning precision	2 (traveling, revolution) DC servo motor max. 20 m/min ± 5 mm
Manipulator		Degrees of freedom	4	
	Speed	Forward-backward turning Up-down turning Right-left swing Up-down swing	max. 18°/sec max. 18°/sec max. 60°/sec max. 60°/sec
		Load capacity Positioning Positioning precision Variable height	max. 5 kgf DC servo motor ± 3 mm max. 800 mm (manual)
Control functions		Teaching functions	• PTP • Manual data input • Data input from cassette tape
		Display Memory capacity	9" CRT and LED monitor master program 100 steps block program 20 steps pattern program 30 point 40 patterns
		Position correcting function External signal	position correction by ultrasonic sensor 6 points
Others		Safety function Dimensions Weight Temperature Power source	falldown prevention switch×4 bumper switch×2 emergency switch×3 length 1,600 mm width 1,050 mm height 2,145 mm 1,000 kg 0–40°C AC100V,50/60 Hz, 20A

concerns, it has been demonstrated that construction robots can be practical and can decrease the shortage of skilled workers. Therefore research and development in robotic technology for construction work are important for future applications and should be divided as follows:

1. Real robotic systems using high technology for future applications
2. Robots for existing, attainable applications

Universities and national laboratories are expected to pursue the first type of research, while construction manufacturers and general contractors will most likely conduct the second area of development.

12.4 CONDENSED REFERENCES/BIBLIOGRAPHY

CTICM 1975, *Forecasting Fire Effects on Steel Structures*
JIS G 3106 1987, *Rolled Steels for Welded Structure*
Okabe 1988, *Elasto-Plastic Creep Thermal Deformation Behavior of Multi-Story*
Ueno 1986, *Construction Robots for Site Automation*
Wakamatsu 1988, *Development of Design System for Building Fire Safety*
Yamaguchi 1990, *Full Scale Fire Test of Steel Column with Excellent Mechanical*
Yoshida 1984, *Development of Sprayrobot for Fireproof Cover Work*

13

Computer Program for Engineering the Fire Performance of Steel Structures

During the 1970s it was suggested in European countries that the fire resistance of steel structures is inferior to that of concrete structures. For that reason, research has been under way in several western European countries to address this concept. Research begun in 1981 in Luxembourg, which utilized computer-assisted analyses of the fire resistance of steel structures (Schleich, 1987e), is only a fraction of the overall research, but its results are useful in that they are clearly oriented to structural shapes, lead to a general approach for composite and steel structures, and will produce a useful computer code.

Given the common research effort, calculation results tend now to be accepted in place of fire tests throughout Europe. In that respect, useful work is being done by the European Convention for Constructional Steelwork (ECCS) through the publication of its European Technical Notes, such as the document Calculation of the Fire Resistance of Centrally Loaded Composite Steel Concrete Columns Exposed to the Standard Fire (ECCS, 1988).

This tendency to replace fire tests with fire-safety engineering calculations is now being accepted on official and legal levels. In fact, the 1992 Eurocodes contain clearly separated chapters on structural fire design where calculation procedures are accepted (Commission of the European Community, 1990).

13.1 THEORETICAL BACKGROUND

At the Bridges and Structural Engineering Department of the University of Liège, Belgium, there have been new developments to be used with steel and composite structures within the Council of European Countries (Schleich, 1987e; Franssen, 1987). The first aim of this research was to establish a computer program for the analysis of steel and composite structures during fire conditions. A flowchart of the relevant computer program subdivides a typical explicit problem as follows:

Structural analysis of a test specimen at ambient temperature with a step-by-step increase of the static load

A similar analysis during fire with alternative thermal and static calculations as a function of the fire-duration time

This procedure takes into account the geometric effects and the thermal material laws and stress-strain relationships for steel and concrete (included in the computer program), given as a function of temperature, and it also considers the material nonlinear effects.

The program, called CEFICOSS (Computer Engineering of the FIre resistance for Composite and Steel Structures), is a general thermomechanical numerical computer code which allows behavior prediction of structural building parts such as columns, beams, and frames during a fire. Structural elements can be either bare steel profiles or steel sections protected by insulation of any type.

The numerical code is based on the finite-element method using beam elements (Franssen, 1987). Therefore any frame structure composed of interconnected columns and beams can be divided into discrete elements, as illustrated in Fig. 13.1.

The beam element chosen has two nodes and three degrees of freedom at each node, as shown in Fig. 13.2. The midplane axial displacement x is linear, whereas the lateral displacement y is a cubic function of x. The cross section of these beam elements is divided into a rectangular mesh (Fig. 13.3), which makes it possible to know exactly the stress and strain history of each mesh element as a

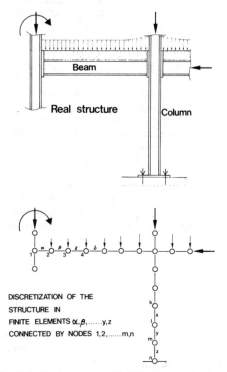

Fig. 13.1 Real structure with finite-element division.

function of time. Furthermore, this subdivision of the cross section can analyze any type of column or beam.

The same rectangular mesh is used when calculating the evolution of the differential cross-section temperature field as a function of time. Temperatures are obtained by a finite-difference method, which is the second main numerical procedure used in the computer code. This method is used to account for the heat balance between adjacent mesh elements.

13.2 FULL-SCALE FIRE TESTS

To verify simulation results given by CEFICOSS and to estimate the fundamental physical parameters with greater accuracy, a new series of full-scale fire tests based on the ISO 834 heating curve were performed (Minne et al., 1985; Kordina and Hass, 1985a; Kordina et al., 1985b). This research provided a better comparison between test and simulation results, and new, interesting information became available. The CEFICOSS numerical model was validated by these full-scale fire tests in which an acceptable correspondence was found to exist between test and simulation results for measurable physical parameters, including temperature, deformation, and fire-resistance time.

13.3 THERMOMECHANICAL ANALYSIS

CEFICOSS is a new tool which makes many previously difficult investigations feasible, greatly improving the knowledge attainable of fire-safety levels for tall

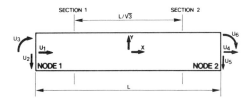

Fig. 13.2   Finite beam element.

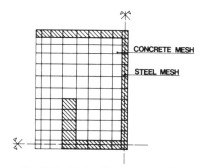

Fig. 13.3   Mesh of beam cross section.

buildings. First, since internal temperature and stress fields can be established for any structural component, optimum fire design is provided without requiring excessive structural fire proofing. Steel reinforcement can be located at more convenient places since temperature fields can be predicted (Fig. 13.4). Internal stress fields can give correct physical explanations for certain failure behaviors. Moreover, the deformation of structures can be calculated in order to show either their configuration at various times during fire exposure or their situation just before failure, as illustrated in Fig. 13.5 (see also Fig. 13.9) (Schleich, 1987).

Of course, the most important prediction is that of the actual failure time. In this respect it is worthwhile to stress the correspondence that exists with CEFICOSS between the mathematical failure expression and its physical meaning. Indeed, failure is reached mathematically when the value of the determinant related to the structure stiffness matrix (DSSM) falls to zero. Physically this means that static equilibrium can no longer be obtained because either a column is buckling or a plastic hinge has formed in a statically determinate beam. Both situations give rise to rapidly growing deformations, which will lead to the collapse of a structural element.

In real structures the understanding of failure becomes more difficult. Nevertheless, failure occurs when the value of the matrix determinant or its minimum

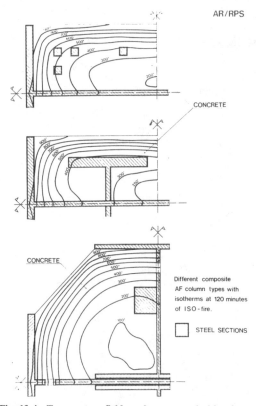

Fig. 13.4   Temperature fields under symmetrical heating.

proper value (MPV), which is equivalent, falls to zero. Physically the failure of a global frame corresponds to the successive formation of plastic hinges which, with or without a column buckling, lead to so-called structural mechanism (Schleich, 1987–1989a).

13.4 MOMENT INTERACTION DIAGRAMS FOR COLUMNS

Practical design tools for use by engineers and architects are being established for four different composite column types (Schleich, 1987–1989b). Vertical loads and bending moments are considered simultaneously, allowing the elaboration of moment diagrams which cover the entire static field from pure axial loads to pure bending moments. This is the first time that such diagrams are being elaborated for fire conditions while considering a column's buckling effect (Fig. 13.6).

Because of the good correspondence between actual testing and CEFICOSS, CEFICOSS gives the entire moment field a clear definition of mathematical failure. The failure corresponds to either a column buckling in the realm of axial loads and higher eccentricities [60 to 180 cm (2 to 6 ft)], or a plastic hinge failure in the realm of highest eccentricities and pure moments.

As global deformations are computed up to the real failure time, however, different service criteria could also be activated, such as deformation $D \le L/30$ or $L/10$. This should be done, for instance, for the composite-beam failure diagrams (Schleich, 1987–1989b). In this case the vertical midspan deformation D and its corresponding horizontal shortening can be extremely high at the failure

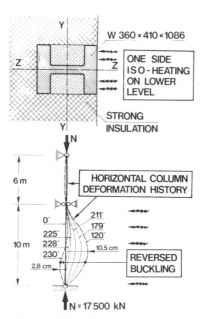

Fig. 13.5 Column deformation under asymmetrical heating.

time of the plastic hinge. Therefore a deformation limitation criterion ought to be considered. However, the conventional criterion $D \leq L/30$, chosen in the past for furnace conditions, underestimates the real fire resistance of composite beams. The proposed solution may be found in the minimum failure time given by the new deformation criterion $D \leq L/10$ combined with the plastic-hinge criterion.

It should be noted that the Ryan Robertson failure criterion [corresponding to the deflection limit $D = (L^2/800)h$, combined with the deflection-rate limitation $dD/dt = (L^2/9000)h$] cannot be considered for composite beams. In fact, this criterion gives a substantially reduced fire-resistance time compared with the plastic-hinge failure time.

13.5   STRUCTURAL CONTINUITY

Comparisons could be made of the efficiency of beams with different cross sections as well as of the beneficial effect of continuous beams. Indeed, when testing four composite beams based on the same profile (HE 300 AA), it was found that the fire resistance is clearly increased by measures improving monolithic behav-

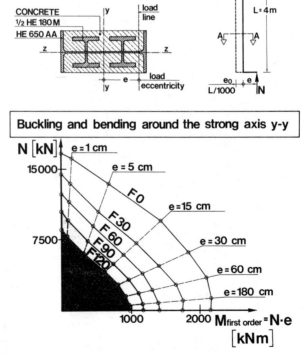

Fig. 13.6   Moment diagram for service condition $F0$ and ISO fire classes F30 to F120.

ior (Minne et al., 1985). This was noticed for an improved composite collaboration in the cross section itself, as well as for a real profile and slab continuity. This last continuity effect is, of course, enlarged by the application of fire to one span only.

A similar beneficial effect was found when analyzing continuous columns heated on only one intermediate level. For instance, a wide-flange column, W14 × 16 × 342, with a length of 3000 mm (10 ft) protected by 15 mm (½ in.) of insulation and supporting a full axial load of 14,000 kN (315,000 lb), is given a fire classification of F90 when calculated as a double-hinged single column. However, if this same column is considered as a continuous column, fire being applied only on one intermediate level, the fire class F120 can be guaranteed.

13.6 COLLAPSE OF STRUCTURES DURING FIRE

The previous considerations have shown that static continuity of beams or columns has an explicitly favorable effect on the fire-resistance time of these structural elements. This effect should increase for beams and columns which are more or less rigidly interconnected, since this actually occurs in real frames.

With CEFICOSS it has become possible to study the global behavior of real structures during fires. This was of particular interest because fires should remain localized through the use of building compartments and because separately heated structural elements, as parts of a global frame, should probably not induce global collapse at an early stage of the fire.

This was the first time that the effect of a local ISO fire on a global steel structure had been analyzed in a credible, reproducible manner. This study illustrated that even unprotected steel beams have a rather high fire resistance (Schleich, 1987–1989c).

The structure that was analyzed was a two-bay sway frame with four stories. The two upper levels were considered only for their load effect on the lower two-story frame. Figure 13.7 illustrates this structure with all its load, geometrical, and cross-sectional characteristics. Wind and snow loads are assumed *not* to act simultaneously during fire. For the service conditions defined in Fig. 13.7, for which this frame is designed, it can be demonstrated that the global safety factor seems not to exceed 1.9. Figure 13.8 shows the result of the nonlinear failure analysis by the CEFICOSS program. All service loads have been progressively increased to 1.9 before global equilibrium failure occurred by the formation of a structural mechanism to the right.

Regarding the local fire simulation, it was found that [provided the (HE 300 A) unprotected steel beams were connected to the concrete slabs by correctly designed shear studs, and provided the beams had convenient connections to the composite columns] the heated beam was able to transmit loads to the neighboring columns for the first 100 min of fire. After this exposure, the global equilibrium failure occurred because of the formation of a structural mechanism to the left, as illustrated in Fig. 13.9.

During fire exposure, the global deformation of the total frame as well as the internal cross-sectional forces m, n, and t were followed step by step. The first two plastic hinges appeared at ends 1 and 2 of the heated beam at 5 min, whereas a midspan hinge 3 was created at 29 min. These three plastic hinges deformed permanently following the heating period.

It is interesting to follow the horizontal deformations δ_h of the top of the heated column (see Fig. 13.10). This deformation was governed by the heated-beam behavior which, up to 27 min, gave way to total thermal elongation but presented an ever-growing shortening after that time period. This beam shortening induced high bending moments in the right, unheated column, which led to plastic

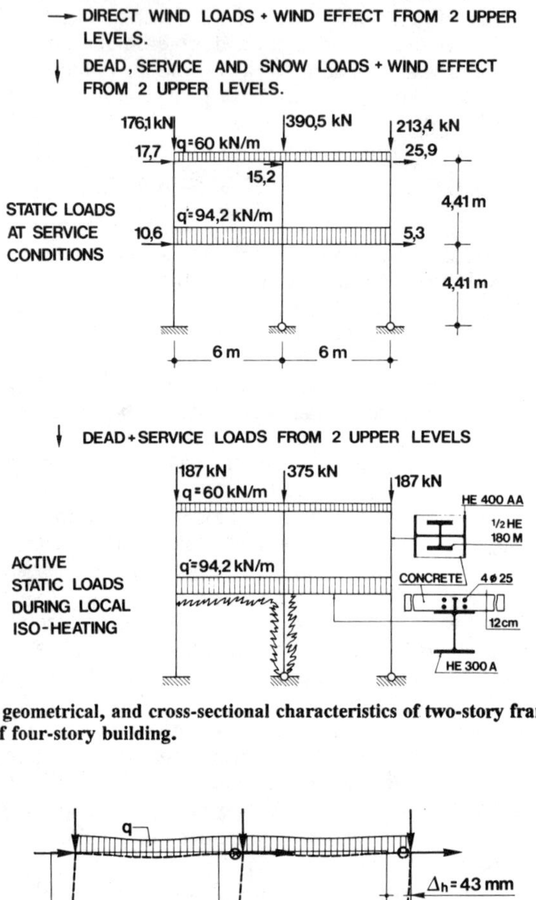

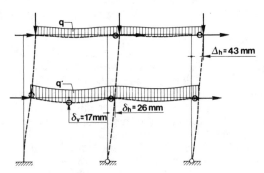

Fig. 13.7   Load, geometrical, and cross-sectional characteristics of two-story frame, forming lower part of four-story building.

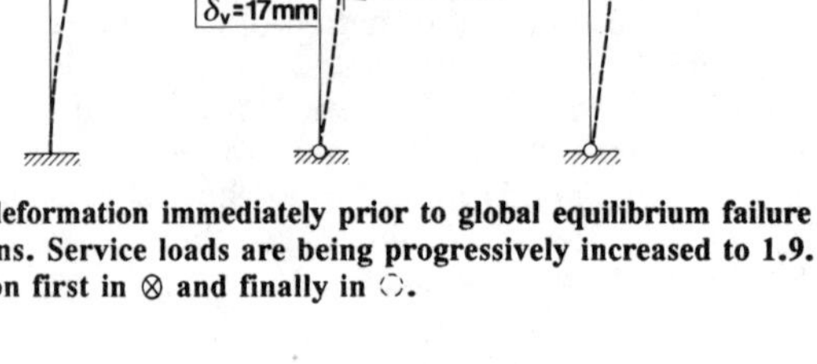

Fig. 13.8   Frame deformation immediately prior to global equilibrium failure mechanism in cold conditions. Service loads are being progressively increased to 1.9. Note plastic hinge formation first in ⊗ and finally in ○.

hinge 4 at 80 min. At 87 min, plastic hinge 5 formed at the top of the heated column; however, no buckling occurred, as the core of this composite column was still relatively cool. Subsequently the heated beam shortening was accelerated, which finally induced the formation of plastic hinges 6 and 7. These final plastic hinges provoked the total collapse, as a structural mechanism had been formed.

This physical failure sequence is supported by the mathematical evaluation of the minimum proper value (MPV), which clearly shows critical downfalls at each weakening of the structural stiffness due to load plastic hinges. As long as these plastic hinges permit equilibrium with a new internal load redistribution, the MPV value remains positive; it falls to zero when equilibrium becomes impossible, which corresponds physically to the previously quoted structural mechanism.

Failure of the specimen in the ISO test occurred only 100 min after heating was started, in spite of the presence of composite beams with unprotected structural shapes. This means that consideration of the actual interior frame continuity has a highly favorable influence on fire resistance. Indeed, the unprotected composite beam considered as a single element would have a fire resistance of only 20 min.

In conclusion, it can be stated that the fire resistance of steel structures can be

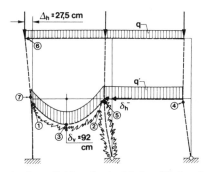

Fig. 13.9   Frame deformation immediately prior to global equilibrium failure mechanism at 100 min of local ISO heating. Applied static loads are remaining constant. Note successive plastic hinge formation at one to ① to ⑦.

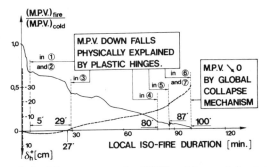

Fig. 13.10   Mathematical minimum proper value (MPV) and its correlation to heated column-top deformation δ_h.

increased substantially if the actual structural continuity is taken into account by fire-engineering design.

13.7 ARCHITECTURAL IMPLICATIONS

The CEFICOSS computer code will contribute to improving the efficiency of steel construction since structural fire-safety levels will be reached at substantial cost savings without paying for excessive fire protection. Furthermore, unsafe structural elements will be avoided, thus reducing fire losses due to structural collapse.

In addition, architects will be able to choose composite structural sections of any shape and construction elements, with visible steel surfaces available for all desired fire-safety levels. This latter advantage is illustrated in Figs. 13.11 through 13.14 which show some of the possible composite-column shapes. They are based on rolled H profiles and present a systematic variation from steel to concrete surfaces. This system, the so-called universal composite construction, allows the creation of many aesthetic building elements with a wide range of ar-chitectural possibilities (Schleich, 1987).

The following characteristics make this composite construction system economical:

High flexibility because of the great number of structural connections avail-able which offer practical solutions (Fig. 13.15)

High construction speed because of prefabrication, allowing early project completion (Fig. 13.16)

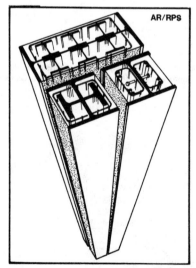

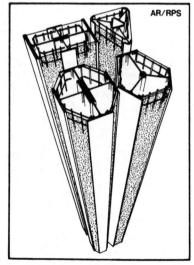

Fig. 13.11 Possible column cross sections con-sidered by universal composite construction.

Fig. 13.12 Composite beams connected to composite column.

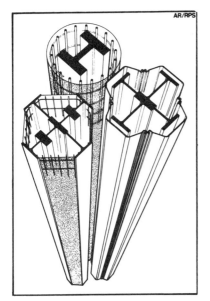

Fig. 13.13   Connection type allowing full prefabrication.

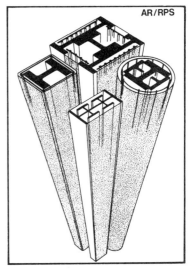

Fig. 13.14   Evolution of load-bearing capacity.

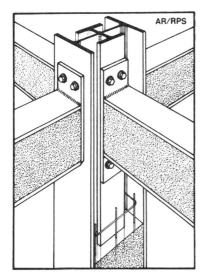

Fig. 13.15   Composite space frame with exposed steel beams; fire resistant to
F120 under localized fire.

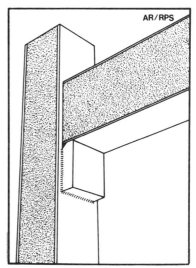

Fig. 13.16   Connection type allowing full prefabrication.

Very small building elements, which lead to slender construction and offer more space for the building

These considerations explain why numerous buildings have been erected in Europe using these composite construction elements (Schleich, 1987). With the numerical computer code CEFICOSS, however, an even wider application of these construction elements is possible. These elements are universally adaptable to practical engineering configurations, and are highly competitive in comparison with concrete construction (Baus and Schleich, 1987).

In this respect, two case studies are cited where composite-column solutions have been used with great success. The first example is the Ko-Gallery in Düsseldorf, Germany, erected between 1985 and 1986. This central shopping complex was built under difficult conditions on a site located in a built up, inner-city area. As a result, state-of-the-art construction methods were used and the "lid-cover construction method" (a method where work proceeds simultaneously from an intermediate floor above ground) was carried out. Use was also made of prefabricated composite AF columns, which served as primary columns and were founded on large-diameter bored piles. The use of these two construction methods shortened the total construction period by 35%.

The choice of this composite-column type allowed the primary column to be prefabricated in one piece of 14.1-m (46-ft) length. This covered the four underground levels. The wide flange shape W40 × 16 × 249, having a weight that allowed easy erection, was used with concrete infill. Another important feature was that the visible steel flanges guaranteed mechanical protection during excavation.

The composite column chosen corresponded to fire class F90 without any additional fire protection. It supported a service load of 21,000 kN (2000 tons), and with its small cross section allowed space savings, which were greatly needed for the underground parking levels.

A second building example is the new Saab factory in Malmö, Sweden, erected between 1987 and 1989, where old shipyard buildings were renovated to provide car manufacturing facilities. For the new floors, added within the existing building, octagonal composite columns were chosen. This column type was selected by the design engineer for the following reasons:

A fire resistance of F60 was required, but ordinary fire protection, for example, by gypsum wallboard, was not acceptable for the new industrial purpose.

High loads and bending moments had to be transmitted whereas the space available for columns was reduced to a strict minimum.

It should be mentioned that these composite columns, including the concrete, were prefabricated in the steel contractor's workshop and transported over a distance of 483 km (300 mi) to the site in Malmö.

13.8 FUTURE DEVELOPMENTS

Various studies have been undertaken to increase applications of this universal composite construction. As a result of initial studies, it can be concluded that steel structures have a promising behavior in fires (Schleich, 1987). It can be demonstrated that the inner part of composite components heats up to a low maximum temperature and that a relatively high critical load level should exist below which the structural composite element would no longer fail. This effect is less important for unprotected steel, in which case the total steel cross section quickly heats up to 600°C (1100°F), but this temperature clearly decreases after a period of about 45 min.

Furthermore, an initial attempt was successful at establishing the equivalent time of an ISO fire exposure t_{eg}^{ISO} for a given structure for which the load-bearing capacity under an ISO fire is identical to the minimum load-bearing capacity under a given natural fire (Schleich, 1987). This load-bearing equivalence is more accurate and useful than the temperature equivalence proposed up to now, which is completely inadequate for composite construction (Figs. 13.17 and 13.18).

Finally, studies will clarify the conditions under which even unprotected steel structures, such as the one shown in Fig. 13.19, are fire-resistant. In fact, heavy steel profile columns could be classified in fire class F60 (Schleich, 1987–1989), whereas steel sections, as a part of complete structures, could have even higher fire resistance (see Section 13.6).

The research of the Commission of the European Community (1990) has begun to adapt this universal composite construction to earthquake conditions. The intent of this research is to combine the ductility of rolled sections with the damping qualities of concrete and to take advantage of the concrete filling between the profile flanges in order to improve local steel buckling. The challenge of the studies is to develop prefabricated beam-column connections with high bending-moment capacity.

The new computer code CEFICOSS precisely determines structural fire safety and advances the application of computers in the design of fire-safety sys-

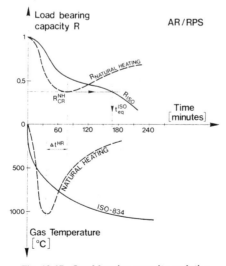

Fig. 13.17   Load-bearing capacity evolution.

tems by predicting fire resistance. This tool has become available at an appropriate time so that fire testing can occur without actual loading tests and with structural fire safety being provided to a predetermined level without the need of paying for excessive fire-protection requirements.

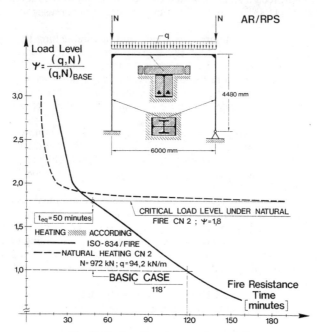

Fig. 13.18   Load-bearing capacity evolution.

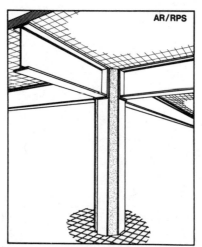

Fig. 13.19   Composite space frame with exposed steel beams, fire resistant up to F120 under localized fire.

13.9 CONDENSED REFERENCES/BIBLIOGRAPHY

ARBED Research Centre 1987–1990, *Seismic Resistance of Composite Structures*
Baus 1987, *Prédétermination de la Résistance au Feu des Constructions Mixtes*
Commission of the European Community 1990, *Eurocodes—Part 10, Structural Fire Design*
ECCS-TC3 1988, *Calculation of the Fire Resistance of Centrally Loaded Composite*
Franssen 1987, *Etude du Comportement au Feu des Structures Mixtes Acier-Béton*
Kordina 1985a, *Untersuchungsbericht Nr. 85636*
Kordina 1985b, *Untersuchungsbericht Nr. 85833*
Minne 1985, *Fire Test Reports Nr. 5091 to 5099*
Pederson 1989, *Structural Fire Protection in Multi-Storey Open Car-Park Buildings*
Schleich 1987a, *CEFICOSS: A Computer Program for the Fire Engineering of Steel*
Schleich 1987b, *L'Acier Face aux Incendies*
Schleich 1987c, *Numerical Simulations, the Forthcoming Approach in Fire*
Schleich 1987d, *Numerical Simulation—Zukunftsorientierte Vorgehensweise zur*
Schleich 1987e, *REFAO-CAFIR, Computer Assisted Analysis of the Fire Resistance of*
Schleich 1987–1989a, *Global Behaviour of Steel Structures under Local Fires*
Schleich 1987–1989b, *Practical Design Tools for Composite Steel-Concrete*
Schleich 1987–1989c, *Practical Design Tools for Unprotected Steel Columns*

14

Optimizing the Behavior of Concrete Structures

In general the conventional way to determine structural fire resistance is based on two main assumptions. First, tests or other evaluations are done with individual structural members, normally without taking into account restraint which, in reality, may be caused by adjoining members. Nevertheless, it is supposed that the provision of sufficient fire resistance of all members provides adequate fire resistance for the whole building. Second, national standards, as well as ISO 834 (1975), which specify how to evaluate and determine the fire resistance of structural elements, allow failure after the required fire-resistance time has been reached. No requirements, or sometimes minimal ones, concerning residual load-bearing capacity are raised, and minimizing damage is not considered.

If considering the fire behavior of complete concrete structures, two main questions should be asked as a result of the foregoing assumptions:

1. Does the design of individual concrete members for specified fire-resistance ratings guarantee adequate fire behavior of an entire system?
2. What measures can be taken to optimize the fire behavior of an entire system with respect to fire exposure *and* the condition after a fire?

The answers to these questions overlap, and investigations regarding typical structural elements are discussed in this chapter.

14.1 FLEXURAL MEMBERS

1 Beams, One-Way Slabs

Fire tests carried out on simply supported, unrestrained members have classically provided the basis for assessing the behavior of concrete flexural structures under fire. In practice, however, the actual structural system is often different because it allows changes in stress distribution and, thus, can affect the actual available fire resistance.

Continuity. The monolithic connection of continuous beams and slabs results in favorable behavior under fire conditions. The effect may be illustrated by the example given in Fig. 14.1 (ACI, 1981; Anderberg et al., 1975; Comité Euro-International du Béton, 1982; Deutsche Normen DIN 4102, 1981; Institution of Structural Engineers, 1978). With a three-span reinforced concrete beam, the bending moment under service load may follow the upper part of the figure. If the beam is heated from below, it tends to deflect, but is restrained by the monolithic connection over the supports. As a result, an additional moment is produced which relieves the midspan regions, while the support regions are stressed to a higher degree.

Normally the increase of moment is limited when the yield point of the upper (still cool) reinforcement in the support region is reached. The third moment line then applies. It is evident that the midspan bottom reinforcement, which is heated more intensely, though relieved of stress, cannot resist deflection indefinitely. The system fails when plastic hinges occur in the span in addition to those plastic hinges which occur over the supports.

The mechanism described requires an appropriate length of the upper (support) reinforcement, whereas the concrete cover of the lower (span) reinforcement can be reduced, as shown in Fig. 14.2. The increase of stresses in the concrete compressive zone near the supports must be mentioned, but it does not become decisive for the load-bearing capacity of the system in the majority of cases.

Some national standards, and international recommendations as well, take into account the effect of continuity. Thus the application of results from an in-

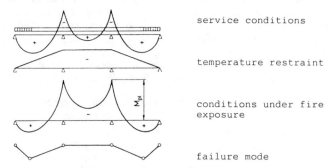

service conditions

temperature restraint

conditions under fire exposure

failure mode

Fig. 14.1   Moment redistribution and failure mode of three-span continuous concrete element.

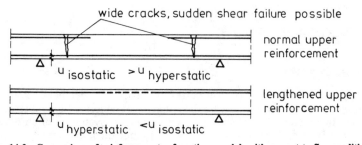

wide cracks, sudden shear failure possible

normal upper reinforcement

u isostatic $> u$ hyperstatic

lengthened upper reinforcement

u hyperstatic $< u$ isostatic

Fig. 14.2   Comparison of reinforcements of continuous slab with respect to fire conditions.

dividual load-bearing member can be applied to the entire structure (CEB, 1982; Deutsche Normen, 1981). If continuous systems are designed without taking into account temperature restraint effects—but do use design rules for isostatic members—a premature collapse must be prevented. Therefore some designers tend to exchange the loss of effective depth (caused by increased concrete cover required for the lower reinforcement) for shorter upper reinforcement over the supports and enlarged areas without any upper reinforcement being used, especially with one-way slabs. These may be troublesome when the slab is under construction (see Fig. 14.2).

End restraint. Longitudinal end restraint against thermal elongation often develops in members during fire and amplifies the effects of continuity. When a fire occurs beneath an interior structural member, the heated portion tends to expand, but it is resisted by adjoining members, which push against the heated portion and, in effect, prestress it. When the fire starts, with *monolithic* structures the thermal thrust normally acts near the bottom of the member and the effect on the load-bearing capacity is similar to that of "fictitious reinforcement," located at the line of action of the thrust (American Concrete Institute, 1981).

The favorable effect of longitudinal and rotational end restraint provided by adjoining members on the fire-exposed member can be demonstrated very clearly. Figure 14.3 shows test results (ISO 834 conditions) for reinforced concrete double-T-beams under a full service load, which were expected to have a fire resistance of about 90 min if no thermal restraint occurred. During the tests, however, the free elongation of the specimens was hindered by hydraulic jacks acting near the bottom of the beams. Thus compressive forces and relieving moments developed, which elongated the fire resistance of the members considerably. It is obvious, therefore, that the restraint force and the fire-resistance time depend on the degree to which the free elongation is hindered (Abrams and Lin, 1974; Kordina et al., 1977). It can be seen from Fig. 14.3 that the deflection of end-restrained members remains low during the fire, depending on the level of restraint.

To a limited degree, the effect of longitudinal restraint can be used even for precast flexural members. This may be explained by the example shown in Fig. 14.4. If the support of a T- or double-T-beam is constructed as shown in Fig. 14.4a, during a fire the lower part of the beam will contact the ledge (caused by simultaneous deflection and dilation), thus causing restraint as described previously. In later stages of a fire, if contact between slab and wall occurs, the force in the lower part will prevail and keep the negative (favorable) moment dominant. It is evident that these favorable effects will not occur if the beam support is constructed as shown in Fig. 14.4b.

Until now it was not possible to state the magnitude of restraint forces, that is, the rigidity of the adjoining structure and the joint width a in Fig. 14.4, which could guarantee a favorable influence on the fire behavior of such precast beams. Early evaluations resulted in joint widths of $a \leq l/500$ (l being the span of the beam), which is sufficient to produce a favorable effect. This is true if the restraining force acts considerably below the neutral axis and if sufficient load-bearing capacity of the supporting structural members is provided.

Horizontal stressing. Industrial buildings are often roofed by bare profiled steel sheets. This kind of construction is economical if no fire resistance is needed. However, in case of a fully developed fire, such roofs fail quickly, showing extreme deformations. Consequently not only does the horizontal stiffening of some roof girders fail, but external tension forces from membrane action also occur, which lead to horizontal bending and torsion in the remaining roof girders, as can be seen in Fig. 14.5. Normally such lightweight roofs and their supporting structures are not designed to withstand the horizontal stressing produced by fire, and

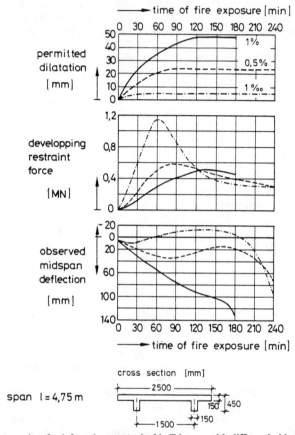

Fig. 14.3   Test results of reinforced concrete double-T-beams with differently hindered horizontal deformation (ISO 834 fire conditions). (*From Kordina et al., 1977, 1980, 1983.*)

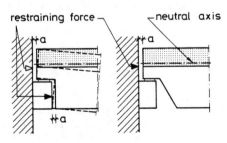

Fig. 14.4   Support of precast T- or double-T-beam.

it does not seem rational to do so. However, the designer should be aware of this risk, which does not occur with stiffer kinds of roofing. The danger of progressive collapse of long-span roofs can be decreased by proper detailing of supports; examples are given in Fig. 14.6.

2 Two-Way Slabs

When exposed to fire, two-way concrete slabs undergo stress redistributions, which have a favorable influence on their load-bearing behavior (Wesche, 1985). This fact is accounted for in a number of standards by allowing a decrease in the required concrete coverage of the lower reinforcement even when effects on continuity are considered.

Similar to one-way slabs, the development of thermal restraint in two-way slabs can again be explained by the action of load-bearing members. If a fire is limited in size (which is true of the typical fire), the heated area of a floor slab is surrounded by cooler regions which restrict the thermal elongation. This initiates compressive stresses in the heated area and tensile, as well as compressive, stresses in the cold area, as illustrated in Fig. 14.7 (Kordina et al., 1977; Walter, 1981).

The stress distribution illustrated in Fig. 14.7 has been evaluated by a theoretical disk model. For clear demonstration, the uncracked state of concrete has been assumed, although concrete tensile stresses exceed the tensile strength in some regions. The disk stresses have to be superimposed onto those coming from the mechanical (vertical) loading of the slab.

Fig. 14.5(a)   Strong horizontal deformations of roof girders caused by stresses from light roofing.

In general, the results are favorable, since in the slab area affected directly (being heated from beneath) the additional compressive stresses relieve the span (lower) reinforcement, thus increasing its failure temperature. However, the compressive zone in the support regions, if also heated, becomes stressed to a higher level. This may lead to damage if, under typical loading conditions, these

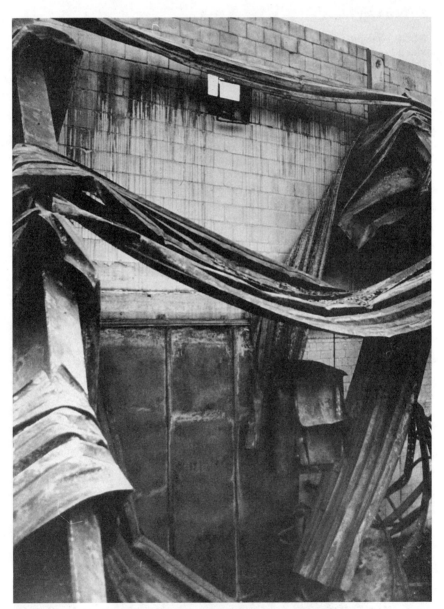

Fig. 14.5(b) Failure of bare profiled steel-sheet roof.

regions are stressed heavily. Nevertheless, failure due to support deterioration is not normal. Tensile stresses forced upon the cool, surrounding concrete may lead to cracks, but in the case of fire, these are of minor importance.

The disk model used earlier may also serve to give an impression of the reduction in thermal dilatations coming from a heated floor area, which is caused by heat transfer to cool surroundings. It can be seen from Fig. 14.8 that the ratio between the horizontal deformation of the outer, cool border u_o and that of the inner, heated area u_i of the disk is strongly influenced by the cool surroundings and the specimen stiffness. The impact of columns and walls has not been accounted for in the example.

14.2 COMPRESSIVE MEMBERS

The traditional way to evaluate the fire resistance of compressive members (especially columns) is to test them under constant symmetrical loading. Unfortunately this method does not correspond well to actual conditions in a building. The influence of adjoining structural members on the fire performance of col-

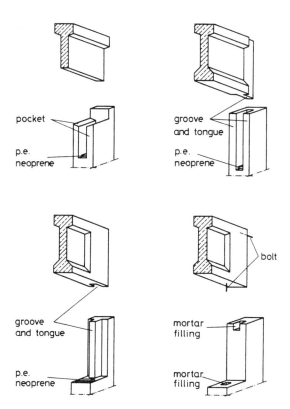

Fig. 14.6 Recommended supports for precast concrete girders.

umns is less favorable than with flexural members, and this influence must be considered.

If several neighboring columns are affected differently by a fire, their thermal elongation will be different and restraint to elongation will occur. Thus additional axial forces will arise primarily in the column that is exposed to the most severe fire (Kordina et al., 1977). Besides the thermal effect, the degree of additional compressive forces depends on the entire system. These forces especially depend on the stiffness of the area above the columns, which will transfer additional loads to them. High-temperature creep and relaxation effects will reduce the forces in later stages of fire exposure. Figure 14.9 shows that the effect of elongation restraint on the fire performance of concrete columns is less severe than might be expected.

If the test results given in Fig. 14.9 are to be of value to actual conditions, the simply supported ends of specimens must be considered. This support mechanism, in combination with induced-load eccentricity, favors the increase of hor-

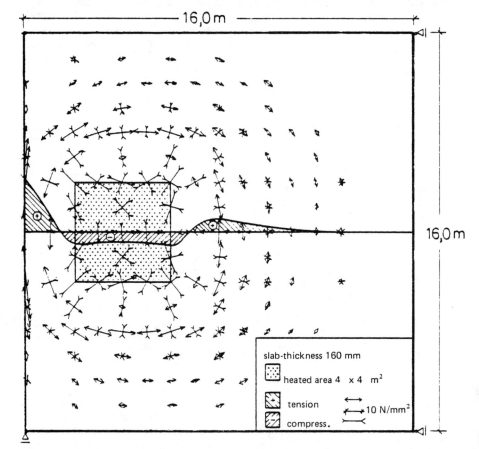

Fig. 14.7 Stresses in partially heated concrete slab (90 min of ISO standard fire). *Note:* **To demonstrate the behavior, the uncracked state has been assumed.** (*From Kordina et al., 1977, 1980, 1983.*)

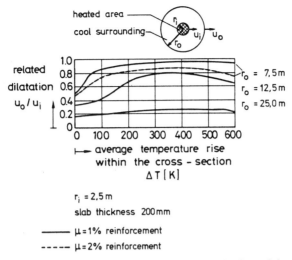

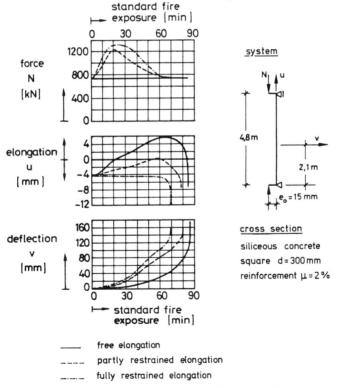

Fig. 14.8 Relation between horizontal movement u_o of outer cool border and thermal dilatation u_i of inner heated area of reinforced concrete floor (theoretical disk model). (*From Walter, 1981*.)

Fig. 14.9 Test results of reinforced concrete columns with free or restrained elongation (ISO 834 fire conditions).

253

izontal deflection. Consequently the hinges further the development and increasing influence of second-order moments.

With normal precast columns, and even more so with poured-in-place columns, more favorable performance is to be expected. In practice, there is no danger of premature failure of such columns due to axial forces increased by longitudinal restraint.

Contrary to this, dilatations of a beam-floor system adjoining columns can have a negative impact on their load-bearing capacity during a fire because these dilatations introduce horizontal forces to the columns (see Fig. 14.8). Failure due to the combined action of moment, shear, and axial forces can result if large areas of floors are affected by a fire. It is questionable whether columns should be designed to withstand such restraint stresses. However, proper anchorage of reinforcing bars and arrangement of an adequate number of stirrups improve a column's resistance capacity.

Most importantly, the dilatations expected must be kept small by preventing a fire from spreading over too large an area, providing proper compartmentation of the building, and arranging dilatation joints with the shortest possible distances. In addition, the dilatation joints mut be tight to prevent gas or flame penetration (see Fig. 14.11).

14.3 STRUCTURAL ASSEMBLIES—ENTIRE STRUCTURES

The change from analyzing individual structural members for fire resistance to analyzing entire systems can be done on a qualitative level using similar concepts. However, it is extremely difficult to quantitatively estimate thermal influences on entire monolithic systems and their structural response. This is especially so because a building typically consists of multidirectional frames connected by slabs. Even if the building is reduced to two-dimensional multistory, multibay frames, quantitative statements are difficult. These are complex not only with respect to performing calculations and testing, but also because the restraining influence of the adjoined slabs and their effective width under fire conditions is more or less unknown. Recent investigations have not yet solved such analytical problems satisfactorily and do not provide practical guidance for design (Becker and Bresler, 1977; Haksver, 1977; Kordina et al., 1977).

It can be expected (and has been proven) that, during a localized fire, the monolithic behavior of load-bearing elements normally results in increased fire resistance. However, the interaction of the individual elements may lead to additional stresses, far beyond the immediately affected area. Inversion of the moment sign may even occur, especially in the corners of multistory frames at the ends of continuous girders, and this should be taken into account by adequate detailing.

Consideration of the large thermal deformations to which a structure is subjected during a fire leads to the conclusion that stiff elements used as stabilizing cores, such as staircases and shafts, should be concentrated in the central part of a building (or located in the middle of two expansion joints). These locations give the possibility of balanced dilatation in all directions.

The interaction of individual elements within total structures, as explained, occurs in situ with concrete buildings. Although appropriate structural design provides increased load-bearing capacity in case of fire (even columns will not fail prematurely with a rational overall design), the possibility of severe damage cannot be prevented even in areas outside immediate fire attack. With properly designed precast systems having flexible joints between individual elements, the monolithic effects of concrete during a fire will not normally occur (see Fig. 14.4 for an exception). Failure in areas of extreme fire therefore becomes more probable. However, it can be expected that damage will remain local and limited.

14.4 JOINTS

The significance of expansion joints has already been introduced and a detailed explanation of their importance follows. To minimize the impact of fire, the designer should adequately section the building by joints. Monolithic parts between expansion joints should be kept as small as possible so that horizontal forces introduced in columns and walls by expanding slabs do not become too large. In addition, the width of expansion joints must be sufficient so as not to transfer high forces to adjacent parts of a building in the case of fire. However, joint sizes which are wide enough for this purpose often create design difficulties; for example, the German standard for reinforced concrete structures requires joint widths up to $l/600$, l being the distance between two joints. A designer might compromise this width where suitable compartmentation of the building can serve as tradeoff.

The current tendency toward "jointless" construction is unfavorable with respect to the fire performance of a concrete structure. However, it does not seem likely that this trend will desist in spite of its higher fire risk. As a result, compensation by tradeoff measures, such as proper compartmentation and sprinklers, will get even greater importance.

A numerical evaluation of the necessary width for expansion joints is difficult, and, until recently, no satisfactory proposals have been made. It is not sufficient to use the simple equation $\Delta l = \alpha \Delta \delta l$ because it results in two unfavorable values which have to be regarded: vertical deflection, high-temperature creep, and relaxation of the heated area; and partial elongation restraint at the connections between horizontal and vertical elements. From this it is evident that the dilatation of a building section totally or partly exposed to a fire not only depends on the time and extent of temperature rise but also, essentially, on the load-bearing system and its rigidity (see Fig. 14.8).

Multiple problems occur if joints have to withstand dilatation, vertical displacement, and end rotation of the adjoining structural members simultaneously. Additional problems occur if a joint must also prevent the unexposed surfaces of the two assemblies from being excessively heated and must hinder the passage of flames and hot gases.

At the Technical University of Braunschweig, Germany, a number of tests have been done to obtain better information on the performance of joints subjected to such conditions (Kordina et al., 1986). Before exposing them to a fire, some of the specimens had to undergo several cycles of mechanical stressing and destressing, which simulated the actual forces occurring during the lifetime of a joint. From the tests, some general recommendations can be derived:

The filling material for expansion joints must be noncombustible.

The filling material has to be compressible in order to compensate for the movements of adjoining members, and it has to be installed in a slightly compressed state to allow for contractions of the concrete structure under normal loading conditions as well as for rotational forces during fire exposure.

Sealing of the joint is necessary not only for normal conditions, but also for expansion and contraction of the structure. In addition, the seal hinders the filling material from displacement during fire exposure. The sealing material may be combustible.

If the expected movement of structural elements closes the joint, no mechanical fastening of the filling material at the two assemblies is necessary (Fig. 14.10a).

If considerable rotation is expected, the filling material must be fixed to both adjoining members with an adhesive (Fig. 14.10b and c).

Adhesive bonding has not proven to be favorable with vertical displacements (Fig. 14.10d).

By a notched joint as shown in Fig. 14.11, combined movements of the structure can, to a certain degree, be accounted for.

If expansion joints are arranged in such a way that the horizontal element expected to move is lying on a column corbel or a wall ledge, a proper gliding plane must be provided.

14.5 POST-FIRE REPAIR

Generally speaking, an increase of a building's fire-resistance time is not economically feasible. However, repairability and reusability of a fire-damaged structure

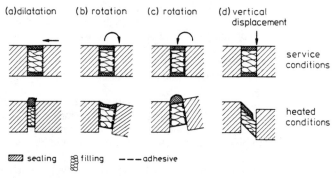

Fig. 14.10 Expansion joints suitable for different movements of adjoining structural elements.

Fig. 14.11 Notched expansion joint suitable for combined movements of adjoining structural elements.

are feasible, and this is of great importance to a designer. All measures taken to improve a structure's performance during a fire by accommodating thermal restraint also increase the likelihood that the structure can be repaired after a fire has occurred.

Reusing concrete elements often depends on the residual deformations which are revealed after a fire. Simply supported flexural elements undergo considerable thermal deflections during a fire which are not fully reversible; in contrast, the deflection will be significantly smaller with continuous and restrained systems (see Fig. 14.3). Moreover, it is evident that proper compartmenting and interior finishes increase the repairability of a building. Nevertheless, the effects of measures which reduce post-fire damage cannot be quantified.

In exceptional cases (for example, underwater tunnels or industrial buildings such as nuclear power plants), where evaluation of fire damage is necessary, or where reuse within a very short time after a fire is of overwhelming importance, the concept of adjusted building-fire protection is useful. The concept takes into account the fire expected under given circumstances and evaluates the structural response to this fire. The building's fire safety is then adjusted and quantified. Consequently, design parameters must not be exceeded either during or after fire exposure (for example, parameters with respect to material stress and strain, crack width and depth, deformation) because they are necessary for the overall stability, repairability, and reuse of a tall building. Unfortunately this adjusted-protection concept contains considerable difficulties and should only be used in special cases (Kordina, 1981).

14.6 OPTIMUM BEHAVIOR

The first question posed at the beginning of this discussion (whether fire-resistant design of individual structural elements leads to adequate performance of the entire structure) may be answered as follows: If certain rules concerning continuity of flexural systems and interaction of vertical and horizontal concrete elements are followed, it can be expected that an entire concrete structure will behave as satisfactorily as the individual members or even better.

If a designer realizes the effects of elevated temperature on reinforced or prestressed concrete structures with respect to their load-bearing capacity and their thermal deformations (including the restraint forces resulting from them), the designer will be able to optimize their behavior during fire and their condition after a fire (such as the possibilities of reuse).

14.7 CONDENSED REFERENCES/BIBLIOGRAPHY

Abrams 1974, *Simulation of Realistic Thermal Restraint during Fire Tests of Floor*
American Concrete Institute 1981, *Guide for Determining the Fire Endurance of Concrete*
Anderberg 1975, *A Differentiated Design of Fire Exposed Concrete Structures*
Becker 1977, *Reinforced Concrete Frames in Fire Environments*
Comité Euro-International du Béton 1982, *Design of Concrete Structures for Fire Resis-*
Centre Scientifique et Technique du Batiment 1980, *Méthode de Prévision par le Calcul du*

Deutsche Normen DIN 4102 1981, *Teil 4, Brandverhalten von Baustoffen und*
Haksver 1977, *Zur Frage des Trag- and Verformungsverhaltens ebener*
Institution of Structural Engineers 1978, *Design and Detailing of Concrete*
ISO 834 1975, *Fire Resistance Tests—Elements of Building Construction*
Kordina 1977, *Jahresberichte 1975–1977, 1978, and 1981–1983*
Kordina 1981, *Baulicher Brandschutz in Strassen- and U-Bahn-Tunneln*
Kordina 1984, *Empfehlungen für brandschutztechnisch richtiges Konstruieren von*
Kordina 1986, *Dehnfugen, Anforderungen und Konstruktion unter Berücksichtigung*
Salse 1971, *Structural Capacity of Concrete Beams during Fires and Affected by*
Walter 1981, *Partiell brandbeanspruchte Stahlbetondecken—Berechnung des Inneren*
Wesche 1985, *Tragverhalten von Stahlbetonplatten im baupraktischen Einbauzustand*

15

Fire-Resistant Composite Structures: Calculation and Applications

Composite steel-concrete structures ideally combine advantages of both their components: reduced overall dimensions, easy prefabrication and connections, and reduced thermal diffusivity for steel; easy shaping and reduced section factor F/V for concrete. Figure 15.1 shows some typical plan sections of structures that have been investigated.

15.1 CALCULATION PRINCIPLES

Analytical models for determining thermal material properties for concrete and steel are described in this chapter. These models lead into the application of the computer program STABA-F, which was developed to support the experimental investigations on the load-bearing and deformation behavior of uniaxial structures during fire.

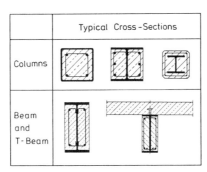

Fig. 15.1 Typical cross sections of composite structures.

Heat transfer from a fire to a structural element depends on the material and nature of the absorbing surface of the member; the color of the flames; the geometry, material, and the material properties of the test furnace walls; and the ventilation conditions in the furnace. Extensive investigations into the heating of structural columns and beams in test furnaces at the Technical University of Braunschweig, Germany, showed that there is sufficient correspondence between measured and calculated temperature distributions in a section. This correspondence assumed the coefficient of convection heat transfer to be α = 25 W/ $m^2 \cdot$ K and the resultant emissivity ε = 0.3 to 0.7 for concrete and 0.5 to 0.9 for steel.

Heat conduction is described by the well-known equation from Fourier, which is valid for homogeneous and isotropic materials. In the application of this equation to composite structures, the following simplifications are necessary:

Water vaporizes as soon as it reaches the boiling point.

Movement of steam is taken together with other effects.

Consumption of energy for vaporizing water is accounted for by using suitable values for the specific heat capacity of concrete up to 200°C (390°F)

Concrete is considered a homogeneous material, while the heterogeneous structures, as well as capillary pores and internal cracks, are grouped together.

A finite-element method in connection with a time-step integration is used to calculate the temperature distribution in a section. The time steps chosen have to be quite small (Δt = 2.5 to 5 min) because the characteristic values of the thermal conductivity λ, specific density δ, and specific heat capacity c_p are very much dependent on temperature. Figures 15.2 and 15.3 show the temperature dependence of the thermal material properties for concrete and steel. To determine the temperature distribution, a rectangular network is preferred with a maximum width of less than 20 mm (1 in.). In the area of the structural steel, it is advantageous to reduce the width of the network to the thickness of the structural steel profile. The elements of the cross-sectional discretization have corresponding thermal materials of steel or concrete.

Due to temperature effects, the thermal strains for the cross-section elements

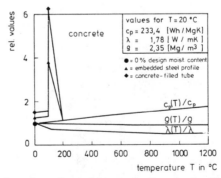

Fig. 15.2   Design values, thermal material properties for concrete with siliceous aggregates.

are derived by using the specific temperature-dependent thermal strain for concrete and steel as shown in Fig. 15.4.

15.2 STRESS-STRAIN RELATIONSHIPS

In an actual fire, material is normally subjected to a transient process with varying temperatures and stresses. To get material data of direct relevance, transient creep tests were carried out. During the tests, the specimen was subjected to a certain constant load and a constant heating rate. From these tests, uniaxial

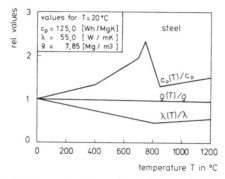

Fig. 15.3   Design values, thermal material properties for steel.

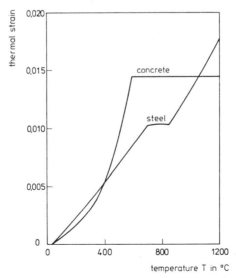

Fig. 15.4   Design values, thermal strains of concrete with siliceous aggregates and steel.

stress-strain characteristics were obtained which included the temperature-dependent elastic strains and the comparatively large transient creep strains. Figure 15.5 shows the analytical model for determining the temperature-dependent stress-strain relationships for concrete and structural steel. The model is also valid for other materials such as reinforcing and prestressing steel. A complete stress-strain curve consists of different segments described by Eq. 15.1 in Fig. 15.5. Material constants ε_0, β_0, and E_0 are used as starting points for the parameters ε_i, G_i, and $dG_i/d\varepsilon_i$. The temperature dependence is introduced by varying these parameters with respect to temperature. The coefficients a_k, b_k, and c_k for calculating the material temperature dependence are available from the Technical University of Braunschweig.

Theoretical stress-strain relationships for concrete and structural steel are shown in Figs. 15.6 and 15.7. The input parameters for the calculation are listed in Table 15.1.

The calculation of the load-bearing and deformation behavior of composite structural elements during fire is determined in four parts, each independent of the other:

1. Geometric description of the composite structural cross section
2. Determination of the temperature distribution during fire
3. Determination of the nonlinear interaction between bending moment M and curvature $1/r$, dependent on normal force N and temperature distribution

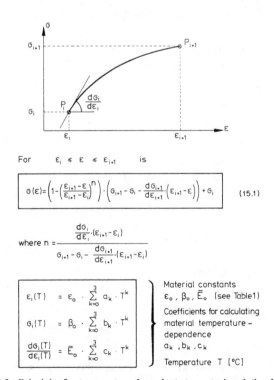

For $\varepsilon_i \leqslant \varepsilon \leqslant \varepsilon_{i+1}$ is

$$G(\varepsilon) = \left(1 - \left(\frac{\varepsilon_{i+1} - \varepsilon}{\varepsilon_{i+1} - \varepsilon_i}\right)^n\right) \cdot \left(G_{i+1} - G_i - \frac{dG_{i+1}}{d\varepsilon_{i+1}} \cdot (\varepsilon_{i+1} - \varepsilon)\right) + G_i \qquad (15.1)$$

where $n = \dfrac{\dfrac{dG_i}{d\varepsilon_i} \cdot (\varepsilon_{i+1} - \varepsilon_i)}{G_{i+1} - G_i - \dfrac{dG_{i+1}}{d\varepsilon_{i+1}} \cdot (\varepsilon_{i+1} - \varepsilon_i)}$

$$\varepsilon_i(T) = \varepsilon_0 \cdot \sum_{k=0}^{3} a_k \cdot T^k$$

$$G_i(T) = \beta_0 \cdot \sum_{k=0}^{3} b_k \cdot T^k$$

$$\frac{dG_i(T)}{d\varepsilon_i(T)} = \bar{E}_0 \cdot \sum_{k=0}^{3} c_k \cdot T^k$$

Material constants
ε_0, β_0, $\bar{E}_0$ (see Table 1)

Coefficients for calculating material temperature-dependence
a_k, b_k, c_k

Temperature T [°C]

Fig. 15.5   Principles for temperature-dependent stress-strain relationships.

4. Determination of all forces and deformations in accordance with second-order theory and arbitrary boundary conditions.

The first step to determine the properties of a composite structural element should be to separate the total plan section area into discrete elements for analysis. Figure 15.8 shows a quarter of a composite cross section composed of a rolled shape HE 240 B embedded in concrete. To ensure a detailed temperature specification of different parts of the cross section (structural steel, concrete, and reinforcement), the section is divided into rectangular and triangular elements.

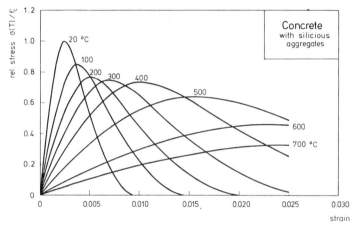

Fig. 15.6   Stress-strain relationships at elevated temperatures for concrete with siliceous aggregates.

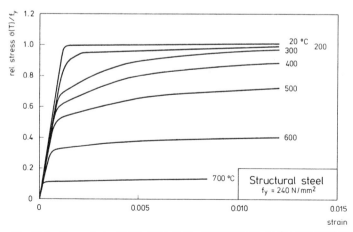

Fig. 15.7   Stress-strain relationships at elevated temperatures for structural steel.

A finite-element method in connection with a time-step integration is used to calculate the temperature distribution in the cross section (Becker et al., 1974; Wickstrom, 1979). The calculated temperature distribution of a composite cross section after 30 min of fire exposure, according to ISO 834 (1975), is shown in Fig. 15.9.

Knowing the temperature distribution and assuming the following simplifications, it is possible to determine the relation between loads and deflections of a bar:

Table 15.1   Material constants for calculating stress-strain relationships at elevated temperatures for concrete and structural steel (see Figs. 15.6 and 15.7)

Material constants	Concrete	Structural steel
ε_0, mm/m	10^{-3}	10^{-3}
β_0, N/mm^2	$f_c'{}^*$	240
$\bar{E}_0$, N/mm^2	$\beta_0 \times 10^3$	2.1×10^5

$^*f_c'$ is the specified compressive strength at $T = 20°C$.

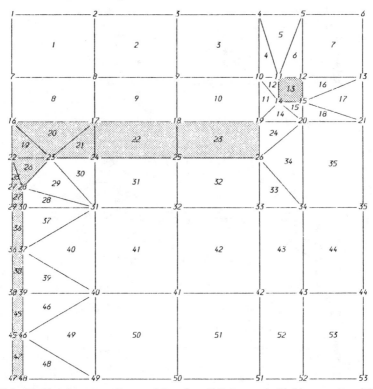

Fig. 15.8   Quarter of a cross section consisting of a rolled shape embedded in concrete and 1% reinforcement.

The Bernoulli-Navier hypothesis.

Only uniaxial stresses are taken into account, shear stresses are neglected.

There is no slip between concrete and steel.

The stress-strain relationships are nonlinear elastic.

The same network as for the calculation of temperatures is used. The stress-causing strain ε_i^σ at location i follows from

$$\varepsilon_i^\sigma = \varepsilon_0 + \frac{1}{r_z} \cdot z_1 + \frac{1}{r_y} \cdot y_i - \varepsilon_i^{th} \qquad (15.2)$$

With the actual temperature T_i the stress G_i can be evaluated by the temperature-dependent stress-strain relationships (see Figs. 15.6 and 15.7). Figure 15.10 shows the calculated temperature, strain, and stress distributions in section A-A—a composite cross section consisting of a concrete-filled hollow section with 3% reinforcement—after 30 min of fire exposure, according to ISO 834.

The integration of the stress distribution yields the following stress resultants according to the temperature:

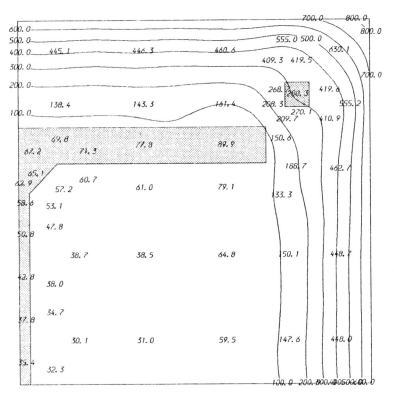

Fig. 15.9   Calculated temperature distribution of cross section of Fig. 15.8 after 30-min fire exposure according to ISO 834.

$$N = \int_A \sigma \, dA \simeq \sum_{i=1}^{\eta} \sigma_i \Delta A_i \qquad (15.3)$$

$$M_y = \int_A \sigma z \, dA \simeq \sum_{i=1}^{\eta} \sigma_i z_i \Delta A_i \qquad (15.4)$$

$$M_z = \int_A \sigma y \, dA \simeq \sum_{i=1}^{\eta} \sigma_i y_i \Delta A_i \qquad (15.5)$$

```
concrete                        f'c  =   30 MPa
reinforcement    3.0 %          fy   =  420 MPa
steel tube   QR 220 6.3         fy   =  240 MPa
```

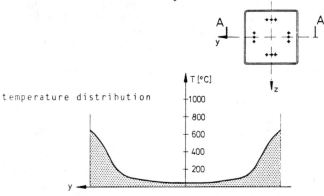

temperature distribution

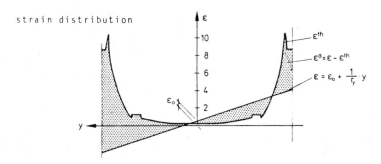

strain distribution

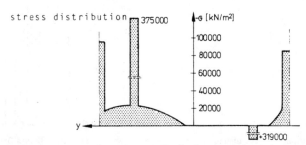

stress distribution

Fig. 15.10 Temperature, strain, and stress distributions in section A-A of composite cross section composed of a concrete-filled hollow section after 30-min fire exposure.

The relations between bending moment M_y and curvature $1/r_z$ of a composite cross section is shown in Fig. 15.11. Every curve is dependent on the actual applied normal force N_x. An accurate evaluation of the load-bearing behavior has to take into account the influence of mechanical (nonlinear moment-curvature relationship) and geometrical (second-order theory) nonlinear interaction between load and deformation. To determine the bending moment M, shear force Q, slope of the bar α, and deflection w, the method of transferring these values from one partition point to the next is used, as shown in Fig. 15.12.

The unknown forces or deformations at the beginning of the analysis of a structural element have to be determined by integration in a manner compatible with the condition at the end of the structural element. The definition of the stiffness as the gradient of the partially linear moment-curvature relationship results in a quick, converging calculation algorithm. It is possible to determine the load-bearing and deformation behavior of the structural elements with different moment-curvature relationships along the axis of bar elements. Figure 15.13 shows evaluated moments and deformations of a composite column under ultimate load after 60 min of fire exposure.

Comparison of calculation results with test observations revealed an agreement between measured and calculated values (Hass and Quast, 1985). In a research report, the calculated ultimate loads are summed up in charts and diagrams (Quast and Rudolph, 1985).

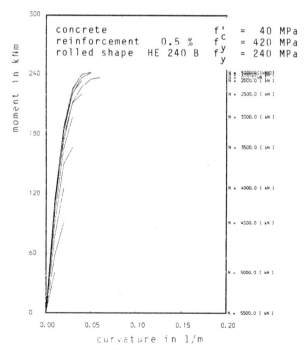

Fig. 15.11   Relationship between bending moment M_y and curvature $1/r_z$ dependent on normal force N_x. (Cross section and temperature distribution are shown in Figs. 15.8 and 15.9.)

15.3 FIRE BEHAVIOR OF COMPOSITE COLUMNS

As an example of composite members, the behavior of different types of columns during fire tests is discussed in this section. In concrete sections, temperature propagates at approximately 1/20 the rate in steel sections. This accounts for the

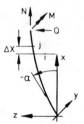

Fig. 15.12   System of bar element without intermediate supports.

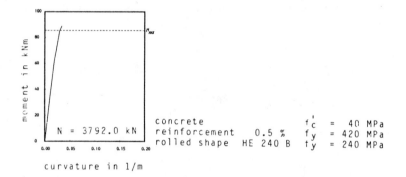

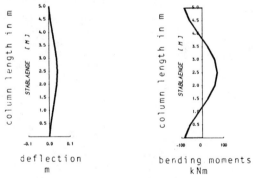

Fig. 15.13   Ultimate limit state of composite column consisting of a rolled shape HE 240 B embedded in concrete and 0.5% reinforcement after 30-min fire exposure. Determined ultimate load is 3792.0 kN.

long fire-resistance times of composite columns with embedded steel profiles, as seen in Fig. 15.14.

Due to the reduction of strength and stiffness in the outer parts of the concrete section with temperatures above 300°C (570°F), the stresses in the steel are augmented and, therefore, collapse occurs before the steel is heated up to 500°C (930°F), which is normally the critical temperature. It is preferable to have a high bearing capacity N_{st} for the steel, compared with that of concrete. This means that columns with a high ratio of N_{st}/N_{pl} are less affected by fire attack (N_{pl} being the bearing capacity of the composite cross section). Only a minimum number of longitudinal bars and stirrups are necessary to avoid loosening of the concrete cover by spalling and to provide bond between concrete and steel.

The fire resistance of concrete-filled tubes is primarily due to the fire resistance of the reinforced concrete core. The temperatures develop somewhat slower compared with concrete columns because of the greater water content of enclosed concrete. Due to restraining of the greater thermal elongation of steel, which heats up more quickly than concrete, the tube takes more of the column force at the beginning of the fire action. Because of this reduction in load, decompression of the concrete may occur. After 15 min of fire exposure under ISO conditions, yielding of the tube takes place and the total load is taken by the reinforced concrete core. Local buckling of the tube may occur, but it only seems to affect the deflections slightly and does not affect the stability of the composite column. This can be seen in Fig. 15.15. With increasing temperatures, the column stiffness decreases and, finally, the column fails due to the loss of stiffness, which causes a stability failure with remarkable deflections. It is preferable for the tube to have a low bearing capacity compared to that of the reinforced concrete core. The fire resistance increases when the ratio N_{st}/N_{pl} decreases, unlike for columns with embedded steel sections.

Figure 15.16 shows the various stresses of the tube, the concrete, and its reinforcement during fire action. A minimum amount of longitudinal reinforcement is necessary to control the crack width during the temporary decompression of the concrete. Otherwise a single widely opened crack can occur, which causes a

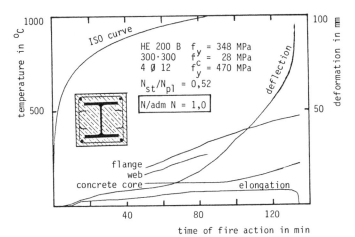

Fig. 15.14   Experimental data of 4.20m (13.7-ft)-long composite column with embedded HEB profile, tested with one end fixed and the other pin-jointed.

plastic hinge when the tube initially yields at one side only. In this case the deflection of the column becomes so large that it immediately leads to rapid failure.

Openings in the tubes are important to avoid excessive steam pressure inside the tube. To attain a fire resistance of 60 min or more, load reductions are unavoidable. Therefore N/N_{adm} ratios smaller than 1.0 have to be chosen. This is the reason why in Canada, for example, the bearing capacity is determined for the tube only, without taking into account any capacity of the concrete. In both cases the admissible load of the composite column depends on the required fire-resistance time. The temperature in the steel is not a suitable criterion for the ultimate limit state. By using a special high-temperature-resistant concrete with steel-fiber reinforcement, load reductions may be avoided as some tests have proven.

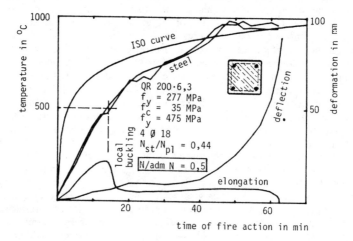

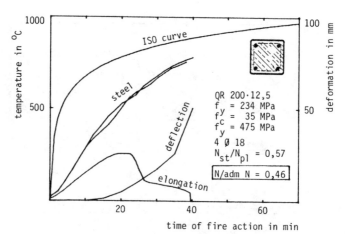

Fig. 15.15　Experimental data of 4.20-m (13.7-ft)-long composite columns with concrete-filled quadratic tubes 6.3 and 12.5 mm (0.25 and 0.50 in.) thick, tested with one end fixed and the other pin-jointed.

Braced columns in frames are generally centrically loaded, or are assumed to be, for simplification of a stability check. During an isolated fire in only one story, end rotations are significantly restrained due to the fact that the stiffness of the hot column decreases, whereas the stiffness of the connected elements beneath and above the referred column remains unchanged. Only the ceiling is heated, whereas the floor is protected by cool air, ashes, or other materials. Tests on columns were carried out by controlling end rotations in a manner that imposed equivalent or restraining actions on them. Typical results are shown in Fig. 15.17.

At the beginning of a fire test, for approximately 30 min, a nearly constant rate of dilatation of heated ceiling causes imposed bending moments, which are released when a specimen's steel tube yields. After 30 min the rate of displacement decreases due to increasing deflections of the ceiling. Therefore, and because of the continuous loss of column stiffness, the restraint decreases. It disappears completely when a certain flexural deformation of the column is attained. Then

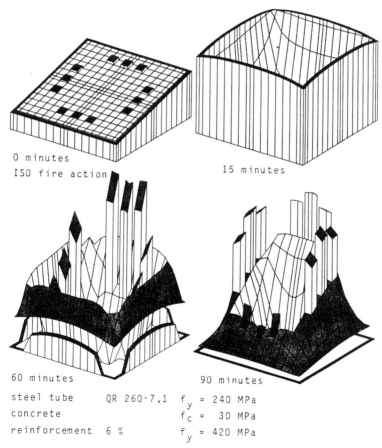

0 minutes
ISO fire action   15 minutes

60 minutes   90 minutes

steel tube   QR 260·7,1 f_y = 240 MPa
concrete f_c = 30 MPa
reinforcement   6 % f_y = 420 MPa

Fig. 15.16 Calculated stress distributions for reinforced tube section after different times of fire action. Stresses are related to corresponding strengths to allow for better comparison.

the increasing deflection is hindered by the connected beams or similar elements. Differences between stiff connections joining flanges and webs of the beams and soft connections joining only the webs were observed only at the beginning of the fire-action time. The measured fire-resistance time corresponds well with that obtained when the columns with fixed ends and without displacements were tested (Euler mode no. 4). For edge columns of a multibay frame, node rotations due to beam deformation have to be considered. The tests revealed that the fire-resistance time for this case corresponds very well with that of the columns tested with one end fixed and the other pin-jointed. At present research on composite columns is continuing.

15.4 PRACTICAL APPLICATIONS

Although research on the structural behavior of composite elements under fire conditions is not yet finished, some results of this work have already been applied in practice. Investigations are mainly concerned with the load-bearing behavior of composite slabs with profiled steel sheets, composite beams, different types of composite columns, and—as a special aspect—the connections between composite beams and composite columns. These types of construction, which are illustrated in Figs. 15.18 and 15.19, do not yet have regulated fire protection in most of the European countries, and fire assessment still has to be done for each individual building. Because of this complex process, some details of practical application are discussed below, especially connections, which need no further cladding or additional fire protection.

The application of composite structures for certain fire-resistance requirements is facilitated by recently developed tables in which the load-bearing capacity of these structural elements is given for specific times of standard fire expo-

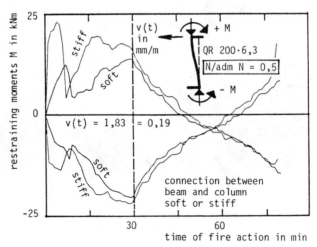

Fig. 15.17 Restraining moments due to floor dilatation in 4.20-m (13.7-ft)-long columns, tested with fixed ends.

sure. The classification of elements for various fire requirements does not cover all possible types of construction in practice. As an example of composite construction, two buildings which were constructed in 1984 with a composite construction method are presented in the following discussion.

Example 1. The first example is a four-story laboratory building in Stuttgart, Germany. Regulations required that all structures have a fire resistance of at least 90 min (ISO standard temperature-time curve). A number of different types of composite structures were constructed to demonstrate their use advantages, especially for structural fire requirements. For instance, there are:

1. Composite concrete slabs with profiled steel sheets
2. Composite beams with concrete infill in connection with reinforced concrete slabs or composite concrete slabs with profiled steel sheets
3. Different kinds of composite columns such as concrete-filled and reinforced hollow sections, hollow sections filled with refractory concrete and steel fibers, steel I sections embedded in concrete, steel I sections with concrete infill, extruded steel sections welded together and filled with concrete, and water-filled hollow sections.

The connections between columns and beams are of special interest. As the building has a central steel core, the connections should only transfer shear

Fig. 15.18   Composite slab with profiled steel sheet.

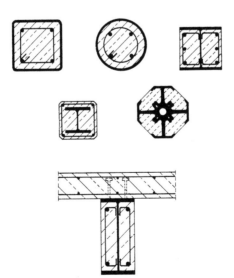

Fig. 15.19   Typical sections of composite columns and composite beams.

forces. Except for the water-filled hollow sections, which are part of the central core, all columns stand outside the building. As a result, only the beams without slabs are in connection with the columns.

Two principal types of connections are applied in this building example. The first has a vertically installed connecting plate, as shown in Fig. 15.20, which transfers shear forces from the beams to the columns. The plates are fixed to the webs of the beams by screws. If this type of connection is used to join a beam to a concrete-filled hollow section, the plate should pass through the column. It is not sufficient to fix the plate only to the hollow section because the steel sheet of the column buckles as a result of heating.

The second principal type of connection that was built uses a steel cleat, which forms a support for the beam, as shown in Fig. 15.21. To connect composite beams and columns, this steel cleat can be welded directly to I-section columns that are infilled with concrete or totally embedded in concrete. The fire resistance of this connection depends on the solidity of the steel cleat and the sizes of the welding seams. If this column-beam connection is used in combination with concrete-filled hollow sections, the steel cleat should be anchored to the concrete core of the column by studs. The reason for this is the effect of local buckling on the steel sheet of the column.

Future plans will demonstrate a real fire in the top story of a building, which simulates the fire load of furnishings by substituting them with wooden

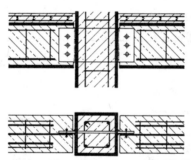

Fig. 15.20   Connection between composite column and composite beam (plate stuck through hollow-section column).

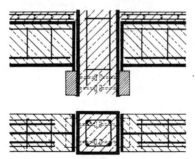

Fig. 15.21   Connection between composite column and composite beam (cleat as a support for beam).

cribs. After the fire, the damaged building will be repaired. One of the aims of this demonstration is to show the repairability of fire-damaged composite structures.

Example 2. The second building example erected utilizing the composite construction method is a seven-story laboratory and office building in Berlin. A circular-shaped testing shed with a diameter of about 60 m (180 ft) is surrounded by the laboratory section. The fire-resistance requirements are different in the two parts of the building. Because of the low fire load, there are no requirements in the testing shed so that the load-bearing steel structure can remain unprotected. The laboratory requires a fire resistance of 90 min (ISO standard temperature-time curve).

The columns in the laboratories are concrete-filled circular hollow sections with diameters ranging from about 250 to 500 mm (10 to 20 in.). The columns are reinforced in their upper stories, where the loads are less. The columns have an internal I-beam section embedded in concrete and wrapped by hollow sections in their lower stories where loads are large. These types of sections have a high fire resistance because the steel profiles are isolated by the surrounding concrete and, therefore, they maintain their load-bearing capacity for a long period of fire exposure.

The beams used in this seven-story building are I-beam sections which are infilled with concrete in connection with prefabricated concrete slabs. In case of fire, the reinforcement takes over the tensile stresses from the heated lower flange. There are holes in the webs of the steel profiles and in the concrete for the installation of supply lines running transverse to the axis of the beams. In case of fire, the areas of the beams below these holes are heated from all sides. To take over the tension force, there are T sections welded to the webs of the I sections, as shown in Fig. 15.22.

The connections between beams and columns are pinned joints. The beams are hung to welded sections which pass through the columns and are protected from heating by a concrete floor above, as shown in Fig. 15.23. Under ambient temperature, high-strength screws transfer the shear forces from the beams to the columns. There is a clear space between the end of the beams and the columns so that the connections are actually hinged. In case of fire, the beams expand and the construction turns into a continuous girder, providing the advantages of a strong beam with respect to load-bearing conditions during fire. The sections which pass through the columns take over tension forces and the compression forces are passed directly from beam to column.

These few examples show that it is frequently not advisable to use standardized constructions or tables for the design of structures with specific fire requirements. On the contrary, there is a need for individual solutions to each condition.

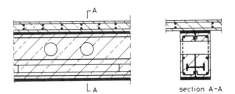

Fig. 15.22   Composite beams concreted between flanges, with holes for supply lines.

It has been proven that the design of a structure can optimize load-bearing behavior under ambient temperatures and under fire conditions.

15.5 FIRE DESIGN OF COMPOSITE CONCRETE SLABS WITH PROFILED STEEL SHEET

A composite assembly composed of a steel-reinforced concrete slab and a profiled steel sheet is the type of composite system most frequently found in buildings today. The fire resistance of such slabs is significant, even if no additional fire-safety precautions are taken. If necessary, the fire resistance can be increased to almost any desired level by simple and reliable means. Until recently, however, structural fire-engineering design of composite slab could only be based on fire-resistance tests. This procedure is time-consuming and expensive, and sometimes gives rise to anomalies due to variations in test results.

Consequently there has been a strong need for a practical design method by which the fire resistance of composite slabs can be determined analytically. This should lead to more uniform levels of safety. Furthermore, this method should be simple and systematic, thus stimulating the use of composite slabs. Such a design method was derived for normal-weight concrete and is discussed in this section. It was established by the Technical Committee 3 of the European Convention for Constructional Steelwork (ECCS) and is available now (ECCS-TC3, 1988).

1 Criteria for Fire Resistance

Fire resistance is determined under standard fire conditions, which are characterized by the so-called standard gas-temperature-time curve. This curve is shown in Fig. 15.24.

Composite steel-concrete slabs have both a load-bearing and a separating function, and the following criteria for fire resistance shall therefore be taken into account, as defined in ISO 834:

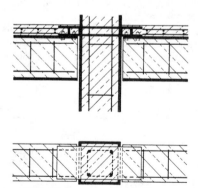

Fig. 15.23 Connection between composite column and composite beam (beam hung up to welded section).

Load-bearing capacity. Resistance to collapse or excessive deflection under structural loading

Insulation. Limitation of temperature increase on unexposed side of slab

Integrity. Ability of slab to resist penetration of flames or hot gases through formation of cracks and openings

The time taken to fail any of these three criteria is taken as the fire rating of the slab, even though failure under other criteria may not occur until much later. It is common practice to determine the fire resistance by means of standard fire-resistance tests. During such tests the test specimen is exposed to a standard fire on its underside while loaded to produce the normal maximum working stresses in its floor construction. Similar assumptions are adopted when using an analytical approach. For proper verification of the performance criteria in an analytical approach, however, some additional assumptions are necessary.

Since the following discussion uses such an analytical approach rather than actual experiments, such as were discussed earlier in this chapter, the load-bearing criterion is used here. The criterion of load-bearing capacity requires that a slab shall not, during a fire, cease to perform the load-bearing function for which it was constructed. During tests, collapse of a slab is avoided by limiting its deflection. This prevents damage to the furnace and other apparatus.

To fulfill the insulation criterion, the temperature rise of the unexposed side of a test specimen should not exceed 180°C (350°F) at any point and the average should not exceed 140°C (280°F). This criterion is applied in most national standards (ISO 834). Because of the profiled shape of the slab, care must be taken when checking that the insulation criterion is satisfied. Theoretically the temperature at the unexposed side will vary as a function of the location at which the temperature is measured. Tests show, however, that in practical cases, the temperature differences are small. In theory it is also possible that passage of heat through joints may result in a nonuniform temperature distribution at the unexposed side. However, composite steel-concrete slabs are normally manufactured in situ, and this complication does not arise. Therefore a uniform temperature distribution at the unexposed side is assumed in this discussion. A temperature increase of 140°C (280°F) at this same side is taken as the limiting insulation criterion.

Integrity is a measure of the ability of the construction to resist the passage of flames and hot gases through cracks. For composite steel-concrete floors, the in-

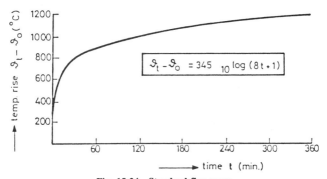

Fig. 15.24   Standard fire curve.

tegrity criterion is not difficult to fulfill. The main reason is that, as mentioned, the floor slab is cast in situ. This means that joints are adequately sealed. Any cracks which may occur in the concrete during fire exposure are unimportant because the steel sheet will prevent penetration by the flames and hot gases. Therefore it is assumed that if the insulation criterion is fulfilled, then the integrity criterion is also fulfilled.

15.6 BEHAVIOR OF FIRE-EXPOSED COMPOSITE SLABS WITHOUT FIRE PROTECTION

At ambient temperature the steel sheet of a composite slab is normally designed to transmit the tensile stresses due to positive bending moments. When exposed to fire, the temperature of the steel sheet will increase, and consequently, the mechanical properties such as yield stress and elastic modulus will decrease. At a certain temperature, which is dependent on the load level and the static system, the steel sheet is no longer able to transmit the applied tensile force, and as a result, the slab fails the criterion for load-bearing capacity. Such a failure may, depending on the thickness of the slab, be preceded by failure of the insulation criterion.

Table 15.2 gives results of fire-resistance tests on composite concrete slabs with profiled steel sheet that were conducted in various European fire test laboratories. No additional means of fire protection were present. The tests cover a practical range of applications. The table illustrates that, in all cases, the fire resistance is governed by the criterion for load-bearing capacity, and it is over 30 min. The considerable scatter in test results is caused by differences in design assumptions at room temperature. All tests, except for tests 1 and 10, were conducted on simply supported slabs. This simple static system obviously constitutes relatively unfavorable conditions, because no beneficial moment redistribution or catenary force occurs as they often do in practice.

Table 15.2 therefore suggests that the fire resistance of composite concrete slabs with profiled steel sheet without additional means of fire protection is at least 30 min when assessed under the criterion for load-bearing capacity.

The application of this principle should obviously be restricted to those cases in which the design at room temperature is based on an approved method. The European Recommendations for the Design of Composite Floors with Profiled Steel Sheet could be taken as a reference (ECCS-TC11, 1975). It is also necessary to check that the insulation criterion is fulfilled. The simple verification rules presented in Chapter 5 of the European recommendations can be used for this purpose.

As a direct consequence of this design rule, an explicit analysis of the load-bearing capacity of fire-exposed composite concrete-steel slabs is only necessary for requirements over 30 min. Additional means of fire protection may then be necessary.

15.7 ADDITIONAL MEANS OF FIRE PROTECTION FOR COMPOSITE SLABS

The following means to provide additional fire protection for composite slabs can be utilized and are discussed in this section: additional reinforcement, thermal barriers, and suspended ceilings.

Table 15.2   Fire resistance of composite concrete slabs with profiled steel sheet without additional means of fire protection

Laboratory Test + report nr. (country)	Cross section [1] (dimensions in mm)	Statical system (span in m)	Live load [N/m²]	Fire resistance criterion regarding load bearing capacity	Fire resistance time [min]
1 TNO BV-73-74 (The Netherlands)	120, 73, 63 34	restrained, 4.20	2700	$\frac{\delta}{L} = \frac{L'}{900\ H}$	40
2 EMPA 66356/1 (Switzerland)	100, 38, 133 19	2.70	4800	$\frac{\delta}{L} > \frac{1}{20}$	55
3 EMPA 66356/2 (Switzerland)	140, 38, 133 19	2.70	10100	$\frac{\delta}{L} > \frac{1}{20}$	69
4 CTICM 74V58/T41 (France)	160, 80, 40 150	4.02	2500	$\frac{\delta}{L} > \frac{1}{20}$	33
5 CTICM 74V64/T47 (France)	110, 50, 88 62	2.01	2500	$\frac{\delta}{L} > \frac{1}{30}$	35
6 CTICM 74V59/T42 (France)	160, 20, 140	4.02	2500	$\frac{\delta}{L} > \frac{1}{30}$	37
7 CSTB 66.2385D (France)	110 steel sheet, 70, 54 150	4.00	3830	$\frac{\delta}{L} > \frac{1}{30}$	38
8 CSTB 66.2478A (France)	150, 80, 54 150	3.20	3800	$\frac{\delta}{L} > \frac{1}{30}$	44
9 CSTB 70.4018 (France)	170 light weight concrete, 77, 54 150	2.85	6880	$\frac{\delta}{L} > \frac{1}{30}$	38
10 CSTB 69.3595 (France)	180 light weight concrete, 77, 54 150	55 300 55	10100	$\frac{\delta}{L} > \frac{1}{30}$	40
11 EMPA 11009 (Switzerland)	100, 58, 160 100	2.70	3300	$\frac{\delta}{L} > \frac{1}{30}$	40
12 VTT/PAL A8378 (Finland)	200, 55, 45, 37 53 93	3.70	5300	$\frac{\delta}{L} > \frac{1}{30}$	43

1) Dense concrete, unless otherwise stated.

1 Additional Reinforcement

In a composite slab a minor percentage of steel reinforcement is normally included to control shrinkage and creep of the concrete. This reinforcement may be placed directly on the steel sheet. When no coating or suspended ceiling is used, the steel sheet is directly exposed to fire. As a result, the temperature increase in the steel sheet and in the reinforcement can be expected to be approximately the same. Consequently the beneficial effect of the shrinkage reinforcement on the fire resistance of the slab may only be marginal. However, additional reinforcement placed in the center of the ribs may significantly contribute to its fire resistance. The same advantage is true of continuous slabs for the top reinforcement over intermediate supports. Figure 15.25 illustrates the various possibilities.

2 Thermal Barriers

A thermal barrier may be necessary when extremely high fire-resistance ratings are required or when deflections have to be severely limited under fire exposure. Spray-applied coatings (for example, coatings based on mineral fibers or vermiculite) are directly applied to the surface of the steel sheet. In order to achieve good adhesion, the steel surface should be properly cleaned to remove dirt and grease.

Fire-resistive slab materials can also be used (for example, vermiculite, gypsum, fiber). These are directly adhered, or mechanically fixed, to the ribs of the steel sheet. As with sprayed coatings, a thorough cleaning is necessary to ensure good adhesion. Special attention should be paid to the type of adhesive used and to ensure adequate connection of boards during fire conditions. Such construction details must be verified by experimental evidence. Only a relatively small thickness of insulation is necessary to achieve a considerable fire resistance (Muess, 1978). Nevertheless the application of fire-protective coatings will involve considerable extra cost.

3 Suspended Ceilings

A suspended ceiling specifically designed to function as a heat shield for the structural components above it can contribute to the fire resistance of the floor assembly. In the cavity above the fire-protective ceiling a time-temperature curve which is less severe than the standard fire curve can be assumed. This is subject to the condition that there is only a limited amount of combustible material in the (unventilated) cavity. The extent to which the standard fire curve is thus reduced will depend on the insulating qualities of the ceiling and on the floor above it. The behavior of a suspended ceiling during a fire is critical, but unfortunately it may

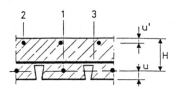

1 - additional bottom reinforcement
    in the ribs
2 - top reinforcement used over
    supports in continuous slabs
3 - reinforcement against shrinkage
H - structural height
u - concrete cover (axial)

Fig. 15.25 Reinforcement in composite concrete slab with profiled steel sheet.

vary because it depends on good detailing, workmanship, and maintenance. These design aspects have to be verified experimentally. Fire-resistance tests are therefore essential.

Thus additional steel reinforcement is a simple, reliable, and economical means for increasing the fire resistance of composite concrete-steel slabs. Moreover, the assessment of additional reinforcement is open to theoretical analysis. The reliability of the two other means of fire protection, thermal barriers, and suspended ceilings, is much more difficult to determine because they are highly dependent on factors such as detailing and workmanship. Experimental verification is necessary if these options are to be used.

15.8 CALCULATIONS OF MINIMUM SLAB THICKNESS

The insulation criterion of fire resistance is fulfilled if the average temperature increase at the unexposed side of the slab exceeds 140°C (280°F), as discussed earlier. This requires a sufficient slab thickness, which depends on the period of fire resistance required. Based on experimental data, Table 15.3 provides conservative guidelines for effective slab thickness (Kruppa, 1983).

As is seen from the equation for h_e, the effective thickness corresponds to an arithmetic average of the thickness, which accounts for the profiled shape of a slab. The calculation rule applies to standard fire exposure and normal-weight concrete.

15.9 CALCULATIONS FOR ADDITIONAL REINFORCEMENT

1 Failure Conditions

The load-bearing capacity may be analyzed on the basis of elementary plastic theory (limit-state) design. For various static systems, the failure conditions can then be easily formulated.

Table 15.3 Effective thickness of composite concrete slab with profiled steel sheet as function of fire resistance time

Required fire resistance Min.	Minium effective thickness h_e mm	Equation for effective thickness h_e	Restrictions
30	60		– for $h_2/h_1 > 1.5$
60	70	$h_e = h_1 + \dfrac{h_2}{2} \cdot \dfrac{l_1 + l_2}{l_1 + l_3}$	$h_e = h_1$
90	80		
120	100		– $h_1 > 50$ mm
180	130		
240	150		

In Table 15.4 the following notation is used: $M_{u\vartheta}^+$ $M_{u\vartheta}^-$ is the absolute value of the positive and the negative plastic bending moment, respectively, at the end of the required period of standard fire exposure, q is the load on the slab to be accounted for during fire, and L is the span of the slab.

To evaluate the failure conditions, it is necessary to quantify the plastic moments $M_{u\vartheta}^+$ and $M_{u\vartheta}^-$. Typical stress distributions over the cross section are represented in Figs. 15.26 and 15.27, respectively. Some assumptions are made for simplicity and these are discussed in the following.

General. The tensile strengths of concrete and the steel sheet do not contribute to the load-bearing capacity at elevated temperatures and thus may be ignored.

Positive Plastic Moment. The ultimate strength of concrete in the compression zone is not influenced by temperature and the room temperature values may be

Table 15.4 Failure conditions for slabs

Statical system	Failure condition
no negative reinforcement — $M_{u\vartheta}^+$	$M_{u\vartheta}^+ \leq q \cdot L^2/8$ or$\quad$ $q \geq 8 \cdot M_{u\vartheta}^+/L^2$
negative and positive reinforcement — $M_{u\vartheta}^-$; $M_{u\vartheta}^+$	$M_{u\vartheta}^+ + 0.5 \cdot M_{u\vartheta}^- \leq q \cdot L^2/8$ or$\quad$ $q \geq (8M_{u\vartheta}^+ + 4M_{u\vartheta}^-)/L^2$
negative and positive reinforcement — $M_{u\vartheta}^-$; $M_{u\vartheta}^-$; $M_{u\vartheta}^+$	$M_{u\vartheta}^+ + M_{u\vartheta}^- \leq q \cdot L^2/8$ or$\quad$ $q \geq 8 \cdot (M_{u\vartheta}^+ + M_{u\vartheta}^-)/L^2$
no positive reinforcement — $M_{u\vartheta}^-$; $M_{u\vartheta}^-$; $M_{u\vartheta}^+=0$	$M_{u\vartheta}^- \leq q \cdot L^2/8$ or$\quad$ $q \geq 8 \cdot M_{u\vartheta}^-/L^2$

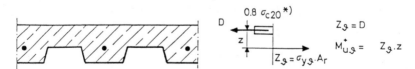

$$0.8\,\sigma_{c20}{}^{*)}$$
$$Z_\vartheta = D$$
$$M_{u\vartheta}^+ = Z_\vartheta \cdot z$$
$$Z_\vartheta = \sigma_{y\vartheta} \cdot A_r$$

Fig. 15.26 Positive plastic moment $M_{u\vartheta}^+$. *Note: The factor 0.8 is introduced to correct the assumed full-plastic stress distribution in the concrete compression zone. In the ultimate state, a nonuniform stress distribution will occur due to the limited capacity of the concrete to accept deformation.*

used. The effective yield stress of the additional reinforcement is affected by temperature.

Negative Plastic Moment. In performing calculations, the profiled concrete slab may be replaced by a slab with a uniform thickness equal to the effective thickness h_e, in accordance with Table 15.3. The ultimate strength of concrete in the compression zone (exposed side) is affected by temperature. The effective yield stress of the reinforcement (unexposed side) is not influenced by temperature, and room temperature values may be used.

The load q on a slab during a fire follows from $0.85q^*$, where q^* is the load to be used in the fire test and is chosen in accordance with ISO 834. Due to the assumptions noted, a reduced value is used because the calculation rule gives a conservative result (compared with the amount of fire resistance measured in a fire-resistance test). The value of the reduction factor (0.85) is based on comparative calculations (Pettersson and Witteveen, 1978/1980).

On the basis of the foregoing assumptions (and using the temperature distribution and mechanical properties of steel and concrete which are discussed hereafter), at elevated temperatures, the evaluation of the failure conditions can proceed in a manner similar to the case of conventional reinforced concrete slabs under ambient temperature conditions. First, however, additional design considerations should be discussed.

2 Design Considerations

In statically indeterminate slabs, a redistribution of moments will occur during a period of fire exposure. This phenomenon is now discussed for the continuous slab presented in Fig. 15.28a. The moment distribution at room temperature under working-load conditions is shown in Fig. 15.28b. Directly after commencement of the fire, a steep temperature gradient will be attained through the thickness of the slab due to the relatively low thermal conductivity of the concrete. Consequently additional negative bending moments will develop, which relieve the positive moment in the midspan region, but increase the negative moment at the supports. Since in the first stage of fire exposure the value of the full plastic moment at the supports will not be affected significantly by temperature, a moment distribution as presented in Fig. 15.28c will tend to occur. (This is a conservative assumption.) The additional negative reinforcement should then be designed to cope with such a moment distribution. This means that the additional reinforcement at the supports shall be extended at least over a distance L'. The anchorage length should be determined in accordance with room-temperature design. For other static systems, other minimum lengths will apply. As heating con-

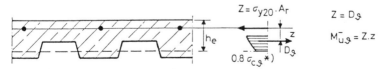

Fig. 15.27   Negative plastic moment $M_{u\vartheta}^-$. Note: *The factor 0.8 is introduced to correct the assumed full-plastic stress distribution in the concrete compression zone. In the ultimate state, a nonuniform stress distribution will occur due to the limited capacity of the concrete to accept deformation.*

tinues, both the (positive) plastic moment capacity at midspan and the (negative) plastic moment capacity at the supports will decrease, finally leading to failure.

The moment distribution at failure is presented in Fig. 15.28d and corresponds to the relevant condition given in Table 15.4. To arrive at such a moment distribution, sufficient rotation capacity is necessary, especially at the supports. The amount of negative reinforcement and its ductility are then of crucial importance. The present state of knowledge does not allow the formulation of criteria specifically derived for fire circumstances. In room-temperature design, however, a limit is normally set for both the maximum and the minimum amount of negative reinforcement in order to guarantee adequate rotation capacity at the supports. If similar limits are followed under fire conditions, the necessary moment redistribution is also possible. In addition, the ductility of reinforcement steel should meet room-temperature specifications.

3 Temperature Distribution

Concrete.   The temperature distribution in a concrete slab is assumed to be independent of the effective thickness h_e, of the slab and can be derived from Table 15.5 for various periods of standard fire exposure and normal-weight concrete (FIP, 1978).

Additional Reinforcement.   The temperature of additional reinforcement depends on the position of the reinforcement bars and the shape of the steel sheet profile. Both factors can be represented by the coefficient γ (see Fig. 15.29), which is given by the following equation (Kruppa, 1983):

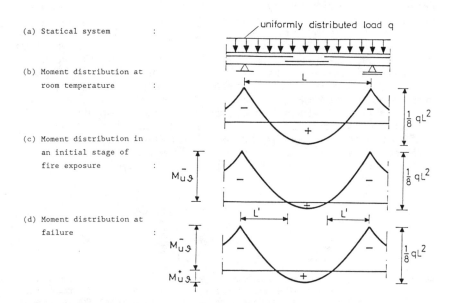

Fig. 15.28   Moment redistribution in continuous slab during fire exposure.

$$\frac{1}{\gamma} = \frac{1}{\sqrt{u_1}} + \frac{1}{\sqrt{u_2}} + \frac{1}{\sqrt{u_3}} \qquad (15.6)$$

The distances u_1, u_2, and u_3 are in millimeters.
The temperature of the reinforcement bars can be calculated using the following equations. For 60-min fire duration,

$$\sigma_s = 1175 - 350\gamma \le 810°C, \qquad \gamma \le 3.3 \qquad (15.7)$$

for 90-min fire duration,

$$\sigma_s = 1285 - 350\gamma \le 880°C, \qquad \gamma \le 3.6 \qquad (15.8)$$

and for 120-min fire duration,

$$\sigma_s = 1370 - 350\gamma \le 930°C, \qquad \gamma \le 3.8 \qquad (15.9)$$

4 Mechanical Properties at Elevated Temperature

Concrete. The following approximate relation between the ultimate compressive strength $\sigma_{c\theta}$ and the temperature θ_c of concrete may be used (Comité Euro-International du Béton, 1982). For $\theta_c < 200°C$,

Table 15.5 Temperature distribution in concrete

Depth X	Temperature in °C after a fire duration [min] of:			
mm	60	90	120	180
5	705			
10	642	738		
15	581	681	754	
20	525	627	697	
25	469	571	642	738
30	421	519	591	689
35	374	473	542	635
40	327	428	493	590
45	289	387	454	549
50	250	345	415	508
55	200	294	369	469
60		271	342	438

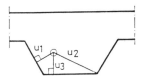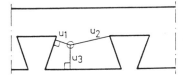

Fig. 15.29   Calculation of coefficient γ.

$$\sigma_{c\vartheta} = \sigma_{c\vartheta} \tag{15.10}$$

for $200 \le \vartheta_c \le 700°C$,

$$\sigma_{c\vartheta} = \sigma_{c20}\left(1 - 0.8 \frac{\vartheta_c - 200}{500}\right) = \alpha\sigma_{c20} \tag{15.11}$$

and for $\vartheta_c > 700°C$,

$$\sigma_{c\vartheta} = 0 \tag{15.12}$$

where σ_{c20} is the compression strength of concrete at ambient temperature.

The strength of the compressive zone of a slab with a width of 1000 mm (3.3 ft) can be calculated as follows (Fig. 15.30):

$$D_\vartheta = 0.8\Sigma(\sigma_{c\vartheta}dx \cdot 1000) = 0.8\sigma_{c20}\,\Sigma(\alpha\,dx \cdot 1000) = 0.8\sigma_{c20} \cdot A_{cr}$$

with

$$e = \frac{\Sigma\sigma_{c\vartheta}dx \cdot 1000)x}{D_\vartheta} \tag{15.13}$$

The reduced compressive area A_{cr} and the position of the compressive force e are given in Table 15.6 and Fig. 15.30.

Steel. The following approximate relation between the effective yield stress $\sigma_{y\vartheta}$, and the temperature ϑ_s of the additional reinforcement may be used (Comité Euro-International du Béton, 1982). For $\sigma \le 250°C$,

$$\sigma_{y\vartheta} = 1.0\sigma_{y20} \tag{15.14}$$

and for $250°C \le \vartheta \le 650°C$,

$$\sigma_{y\vartheta} = \sigma_{y20}\left(1 - 0.6 \frac{\vartheta - 250}{400}\right), \quad \text{for hot-rolled bar} \tag{15.15}$$

$$\sigma_{y\vartheta} = \sigma_{y20}\left(1 - 0.8 \frac{\vartheta - 250}{400}\right) \quad \text{for cold-drawn bar} \tag{15.16}$$

where σ_{y20} is the yield stress at room temperature.

15.10 APPLICATION OF METHOD

An explicit analysis of the load-bearing capacity of fire-exposed concrete slabs with profiled steel sheet is not necessary if the required fire resistance is not over 30 min and the room-temperature design complies with an approved method. To

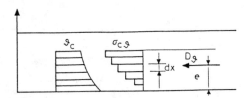

Fig. 15.30 Resultant compressive force in concrete compressive zone.

check on fire-resistance criteria, that is, insulation and integrity, the simple verification rule for minimum slab thickness presented in Chapter 5 of the European recommendations may be used.

If the required fire resistance is over 30 min, additional fire protection may be necessary. The preferred method for such protection is additional reinforcement. The practical calculation rules presented here provide a simple verification method for determining whether the amount of additional reinforcement is sufficient to meet given requirements of fire resistance. This method of assessment is derived for normal-weight concrete and gives conservative solutions.

15.11 ANALYSIS OF A PARTICULAR STEEL-CORE COLUMN

In the case reported in this section, research has been aimed at developing a steel-concrete composite column that could be delivered as a premanufactured "steel column." Ideally, the column would have a slender steel profile and a short erection time. In addition it would maintain the high load-bearing capacity of steel while providing satisfactory fire protection, reduced physical dimensions, and a visible metallic surface. As part of the research, a series of "Geilinger" column cross sections were tested. The most efficient section proved to be one that consists of a circular steel core, a steel tube jacket, and concrete infill, as illustrated in Fig. 15.31.

Table 15.6   Strength of concrete compressive zone at elevated temperatures

Total depth of compressive zone, mm	Fire duration							
	60 min		90 min		120 min		180 min	
	$A_{cr} \times 10^{-2}$, mm^2	e, mm	$A_{cr} \times 10^{-2}$, mm^2	e, mm	$A_{cr} \times 10^{-2}$, mm^2	e, mm	$A_{cr} \times 10^{-2}$, mm^2	e, mm
10	12.1	7.5						
15	29.1	10.4						
20	50.9	13.5	13.7	17.5				
25	77.1	16.5	31.8	20.3	12.4	22.5		
30	107.5	19.6	54.2	23.3	29.0	25.4		
35	141.7	22.7	80.5	26.3	49.7	28.3	13.0	32.5
40	179.6	25.9	110.4	29.3	74.2	31.4	30.0	35.3
45	221.0	29.0	143.8	32.4	102.3	34.4	50.4	38.2
50	265.4	32.1	180.5	35.5	133.5	37.5	74.1	41.2
55	313.4	35.2	220.9	38.6	168.2	40.6	101.0	44.2
60	363.4	38.3	264.2	41.7	205.7	43.7	130.7	47.2
65			309.6	44.7	245.4	46.7	162.8	50.2
70			357.0	47.8	287.2	49.7	197.0	53.2
75			405.9	50.7	331.4	52.8	233.4	56.2
80			455.9	53.7	376.7	55.7	271.4	59.2
85							311.1	62.2
90							352.3	65.1

The load-bearing properties of such a steel-core column cannot be analyzed using conventional calculation methods for ambient- and fire-temperature conditions. Therefore to carry out an analysis, different considerations needed to be addressed. The following discussion reviews some of these considerations.

1 Testing under Ambient Conditions

Compression tests were carried out (EMPA, 1980) for material testing of sample columns in two cases. In both, the load was centrally applied to the steel core, which extended 25 m (1 in.) over the concrete and the steel jacket. Except for the spacer (which was not welded to the jacket), there was no shear connection. Horizontal and vertical sections of the column are shown in Fig. 15.32.

The test results led to the following observations and conclusions.

1. In spite of the centrally applied load, a uniform response of all the components of the column could be observed through composite action of contact surfaces and transverse elongation of the materials. In Fig. 15.33 the action of the jacket is shown by the distribution of crushings in both the jacket and the core.

2. Local buckling phenomena were observed only in column 1 close to its collapse load (see Fig. 15.33).

3. A comparison between observed load at failure loads and the calculated failure load (according to the German norm DIN 18 806) versus a numerical analysis (with geometrical and mechanical nonlinearities being taken into account)

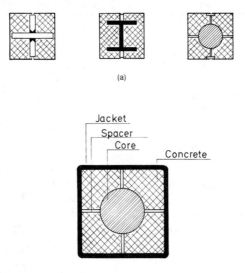

(a)

(b)

Fig. 15.31 Evolution of cross section of the Geilinger structural column. (a) Alternative column variations. (b) Current section.

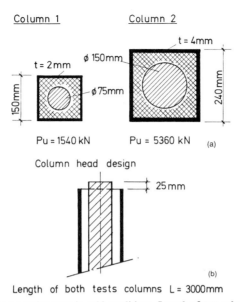

Column 1   Column 2

t = 2mm   ⌀ 150mm   t = 4mm

150mm   ⌀75mm   240mm

Pu = 1540 kN   Pu = 5360 kN   (a)

Column head design

25mm

(b)

Length of both tests columns L = 3000mm

Fig. 15.32 Static compression tests in cold conditions. Length of test columns *L* = **3000** mm. (*a*) Horizontal plan sections of columns head design. (*b*) Vertical cross section.

Column - upper part   Crushings in (⁰/oo)

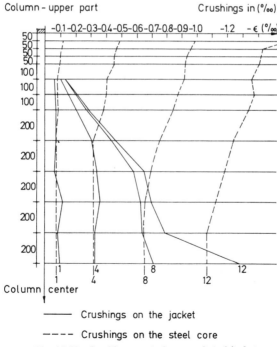

-0.1 -0.2 -0.3 -0.4 -0.5 -0.6 -0.7 -0.8 -0.9 -1.0   -1.2   - ∈ (⁰/oo)

Column center

———— Crushings on the jacket

– – – – Crushings on the steel core

Fig. 15.33 Crushings on steel core and steel jacket.

289

shows that the calculation generating the buckling curve provides the best agreement with this experiment.

2 Testing under Fire Conditions

Fire testing of similar Geilinger columns was conducted at the University of Wuppertal, Germany (Klingsch, 1984a). As part of that program, a numerical analysis of the fire behavior of this column type was carried out. Dimensions and loading data from these tests are summarized in Fig. 15.34.
Observations and conclusions from that program follow.

1. According to preliminary calculations, the applied axial load should have resulted in a fire resistance of about 90 min of the concrete-encased core. The collapse occurred 128 min after the beginning of the test. Nothing abnormal was observed during the test and the behavior of the column was similar to that of other composite columns.

2. Figure 15.35 shows the evolution of both test and calculated temperatures of the steel core. The small increase in temperature led to only a minimal de-

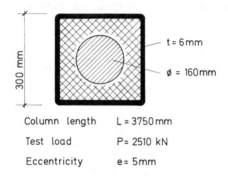

Column length   L = 3750 mm

Test load   P = 2510 kN

Eccentricity   e = 5 mm

Experimental collapse: 128 min

Fig. 15.34   Fire test of Geilinger structural column.

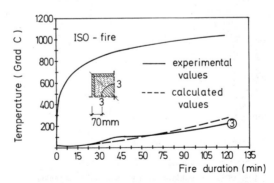

Fig. 15.35   Heating of core under ISO-curve conditions.

crease of core resistance while the concrete coating lost its resistance after only 80 min of fire duration. Because the resistance of the concrete was lost, only the steel core remained to resist the axial load. It was this loading configuration that finally induced collapse of the steel core.

3. Even during fire, the composite action between core, concrete coating, and steel jacket was maintained. As can be seen in Fig. 15.36, the test results agree with the recalculation results for this composite column.

3 Design Concept

Based on the preceding test results, a calculation procedure to describe the performance of these columns has been developed which addresses both ambient and fire conditions.

For normal, ambient temperatures, calculations to size this steel-core column may be carried out in the same way as for typical composite columns.

The current regulations for the calculation of the Geilinger structural columns are not comprehensive enough for an adequate description of the column behavior when submitted to the effects of fire. For this condition, a special numerical procedure has been developed (Bryl and Keller, 1982). The calculation, which is carried out by considering the Geilinger structural columns as composite columns, consists of the following three stages:

1. Calculation of temperature distribution across the column by the finite-element method. To obtain an accurate description of the two-dimensional transitory heating process, the two dimensions of the plan section as well as the exposure time are separated into finite elements.
2. Applying the characteristic properties of concrete and steel during elevated temperatures.
3. Predicting load-bearing capacity during elevated temperatures.

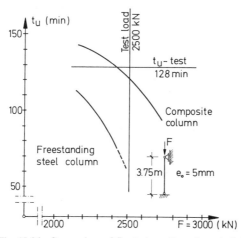

Fig. 15.36 Comparison of fire-test results and recalculation.

After having calculated the size of a cross section on the basis of given temperatures and resistance distribution, a composite column can be designed which consists of all the required component elements.

4 Applications

Because of the slenderness of the Geilinger columns and their load-bearing capacity, they are efficient structural members. Their efficiency may be illustrated by the following examples.

1. A column whose plan dimensions are 200 by 200 mm (8 by 8 in.) carries an axial force of 140 t when submitted to fire action for 90 min (F90).

2. A 250- by 250-mm (10- by 10-in.) Gilinger column submitted to the same fire action and duration carries an axial force of 300 t.

Small column sections are important in many applications where the restriction of floor area plays an important role, such as in corridors and parking places. The main fields of application of the Geilinger structural columns are stores, office buildings, hospitals, underground car parks, and schools. Figure 15.37 shows the slenderness of this steel-core column. The two columns carry the same load under the same conditions. The left column is a concrete-encased steel profile, the right a Geilinger steel-core column.

Short erection time is another important aspect of structural steel. In addition, Geilinger steel-core columns need no supplementary work after installation because their steel finish may be exposed.

Fig. 15.37   Steel-core columns.

15.12 CONDENSED REFERENCES/BIBLIOGRAPHY

Becker 1974, *FIRES-T, A Computer Program for the Fire Response of Structures*

Bryl 1982, *Über die Berechnung des Feuerwiderstandes von Verbundstützen mit*

Comité Euro-International du Béton 1982, *Design of Concrete Structures for Fire*

Dorn, *Brandverhalten von Riegelanschlüssen an Verbundstützen aus einbetonierten*

ECCS-TC3 1984, *Calculation of the Fire Resistance of Composite Concrete*

ECCS-TC11 1975, *European Recommendation for the Design of Composite Floors*

EMPA 1980, *Statische Druckversuche an zwei Geilinger-Baustützen*

FIP 1978, *FIP/CEP Report on Methods of Assessment of the Fire Resistance of*

Hass 1984, *STABA-F, A Computer Program for the Determination of Load-Bearing and*

Hass 1985, *Brandverhalten von Verbundstützen mit Berücksichtigung der*

Herschelmann 1983, *Berücksichtigung der Betonfeuchtigkeit auf Erwarmungsvorgänge*

ISO 834 1975, *Fire Resistance Tests—Elements of Building Constructions*

Jungbluth, *Feuerwiderstandsfähige Konstrucktionen durch Profilverbund*

Jungbluth 1980, *Verbundprofilkonstruktionen mit erhöhter Feuerwiderstandsdauer*

Jungbluth, *Feuerwiderstandsfähige Spezial-Verbundprofile*

Klingsch 1984a, *Analyse des Tragverhaltens von Geilinger-Baustützen bei Normal*

Klingsch 1984b, *Brandverhalten von Hohlprofil-Verbundstützen*

Klingsch 1984c, *Brandverhalten von Verbundstützen aus vollständig einbetonierten*

Kordina 1981, *Beton-Brandschutz-Handbuch*

Kordina 1982, *Zur Frage des Brandverhaltens von Stahlverbundkonstruktionen*

Kordina 1984, *Parameterstudie für Verbundträger der Feuerwiderstandsklasse F 90*

Kordina 1984, *Brandverhalten von Stahlstützen im Verbund mit Beton und von*

Kruppa 1983, *Echauffement des Planchers Béton à Bac Acier Soumis à l'Incendie*

Muess 1978, *Brandverhalten von bekleideten Stahlbauteilen*

Muess 1982, *Neubau einer zweigschossigen Fertigungshalle mit neuartiger*

Pettersson 1978/1980, *On the Critical Temperatures of Steel Elements Derived for*

Quast, *Auswirkungen der Stützen/Riegel-Verbindung auf das Brandverhalten von*

Quast 1983, *Verbundstützen unter Brandbeanspruchung*

Quast 1985, *Bemessungshilfen für Stahlverbundstützen mit definierten*

Schleich 1983, *Garantierter Feuerwiderstand in Stahlbau, eine neue Technologie*

Schmidt 1984, *Stahltrapezprofildecken—Bemessung and Brandschutz*

Schmidt, *Stahlprofildecken mit Aufbeton—Bemessung and Feuerwiderstand*

Wickstrom 1979, *TASEF, A Computer Program for Temperature Analysis of Structures*

Current Questions, Problems, and Research Needs

INTRODUCTION

The original fire monograph, prepared in the 1970s, contained a substantial number of "current questions, problems, and research needs." In reviewing these in conjunction with the present text, many of the items have been addressed while some have not. The following list incorporates or modifies some of those research needs to reflect progress made and includes additions based upon advances in engineering and construction technology since the original monograph.

POINTS OF INTEREST

1. Improved techniques for dealing with stack effect in tall buildings.
2. Further studies to detail smoke movement in [specific] tall building designs, including improved assessment technology to practically determine how smoke movement occurs in both "as built" and newly constructed tall buildings.
3. Further development of technology to reduce the risk of smoke movement into exit stairs during mass evacuation or relocation in a tall building during a fire incident.
4. Effect of fire spread on pressurization of buildings and impact of mechanical pressurization systems in preventing fire spread.
5. Incorporation and combination of rational design and fire growth and spread modeling to determine how fire development in a building influences structural element protection requirements.
6. Development of systems approaches to fire safety for use as legitimate substitutes for [the present] specification-type legal requirements.
7. Improved human factors training related to fire detection and control equipment.
8. Incorporation of improved design factors relating to partial or complete disablement of fire protection systems in tall buildings through occurrence of earthquakes, floods, or hurricanes.

9. Research into the impact of furnishings on fires in tall buildings, possible test-ing, and development of standards for furnishings.

10. Standardization "research" to develop unified building nomenclature and tall building construction terminology. This work would be akin to current work being conducted between ISO and ASTM [at the time of publication of this monograph] to unify standards in fire safety fields and could be undertaken jointly with that working group.

11. Further research with elevator systems in consideration of their possible use during fires in tall buildings for evacuation, relocation, and fire fighting.

12. Improve intra-building communication for use during fire incidents.

13. Conduct research to assess the utility of messaging systems for occupants of tall buildings during fire emergencies.

14. Further research on area of refuge concepts to ensure that such areas remain uncontaminated by smoke.

15. Further planning for the use of helicopters in fire fighting and tall building evacuation efforts.

16. Research on approaches to fire drills and dissemination of emergency plan-ning information to building occupants to develop information on effective-ness and standard of care in various countries.

17. Development of a comprehensive tall building fire incident database system describing in unified terminology causes of incidents and construction details of involved buildings, as well as mitigating or exacerbating factors effecting fires.

18. Development of improved evacuation methods for building occupants.

19. Development of improved techniques for individuals to protect themselves from exposure to smoke in tall building fire incidents.

20. Reliability studies of high-rise sprinkler systems.

21. Clarification of cost/benefit values of sprinkler systems based on data for new and existing buildings, which integrate a scale for type of use of building, level of supervision in building, and national approaches to tall building fire safety regulation.

22. Use of new sprinkler materials such as plastic pipe in tall buildings.

Nomenclature

Access door. A fire door smaller than conventional doors which provides access to utility shafts, chases, manways, plumbing, and various other concealed spaces and equipment.

Airflow. The movement of air within a building; it is especially important on the fire floor, when airflow often follows a fire fighter's advance toward the fire; for purposes of ventilation, it is usually measured in cubic feet per minute.

Air handling system. All the fans, ducts, controls, dampers, filters, intakes, and heating and cooling apparatus that remove, clean, and recirculate the air in a structure.

Arcing. The flashing occurring at electrical terminals when the circuit has been opened or closed.

Authority having jurisdiction. A term used in many standards and codes to refer to the organization, office, or individual responsible for "approving" equipment, procedures, and construction in a town, county, city, or state.

Autoignition temperature. The lowest temperature at which a flammable gas or vapor-air mixture will ignite without a spark or flame. Vapors and gases will spontaneously ignite at a lower temperature in oxygen than in air, and their autoignition temperature may be influenced by the presence of catalytic substances. (*See also* Spontaneous ignition.)

Automatic closing device. A mechanism that can be fitted to a door which will cause the door to close if there is a fire. (*See also* Automatic fire door; Self-closing device.)

Automatic dry sprinkler system. A sprinkler system that has air under pressure in the pipes; when the sprinkler head fuses from the fire, the air escapes and allows water into the pipes and through the open sprinkler heads. (*Also called* dry pipe system.)

Automatic fire door. A door designed so that it will close automatically if there is a fire; the closing device is often a fusible link which melts in the presence of heat or activation of a smoke detector.

Automatic sprinkler system. A system of pipes with water under pressure that allows water to be discharged immediately when a sprinkler head operates. (*See also* Wet-pipe sprinkler system.)

Backdraft. The explosion or rapid burning of heated gases that occurs when oxygen is introduced into a building that has not been properly ventilated and has a depleted supply of oxygen due to fire. (*See also* Hot air explosion; Smoke explosion.)

Bearing wall. A wall that supports floor or roof beams, girders, or other structural loads.

BLEVE. Acronym for Boiling liquid-expanding vapor explosion.

Boiling liquid-expanding vapor explosion (BLEVE). A major container failure, in two or more pieces, at a moment when the contained liquid is at a temperature well above its boiling point at normal atmospheric pressure. If the liquid is flammable, an enormous explosion may result.

Boiling point (BP). The temperature (which varies with the pressure and nature of the liquid) at which a liquid is rapidly converted to a vapor; normally it is reported at a pressure of one atmosphere.

Building standard. A document defining minimum standards for design.

Busbar. A short conductor forming a common junction between two or more electrical circuits. (*Also called* bus.)

Char. 1. Carbonaceous matter formed by incomplete combustion of organic material such as wood. 2. To change into charcoal or carbon by pyrolysis. 3. To burn or scorch.

Circuit. 1. An electrical path of conducting components through which electrical current flows. 2. The conductor, or radio channel, and associated equipment used to transmit a fire alarm.

Circuit breaker. A device which opens a circuit when an electrical overload occurs.

Coefficient of variation. The ratio of the standard deviation to the mean of a random variable.

Common wall. *See* Panel wall.

Compartmentation. A type of building design in which a building is divided into sections that can be closed off from each other so that there is resistance to fire spread beyond the area of origin; it is most common in high-rise buildings and health care facilities.

Conduction. 1. The transmission of heat through or by means of a conductor by direct contact with a heated element. 2. The transmission of an electrical current.

Conductor. A substance that transmits electrical or thermal energy.

Conduit. 1. A trough or pipe containing and protecting electrical wires or cables. 2. A pipe or channel for conveying water.

Convection. The transfer of heat that occurs because of the mixing or circulation of heated fluid.

Convection column. *See* Thermal column.

Curtain wall. An exterior nonloadbearing prefabricated wall, usually more than one story high supported by the structural frame, which protects the building's interior from weather, noise, or fire. (*See also* Panel wall.)

Damage. The total loss caused by fire, including indirect losses such as business interruption, loss of future production, and loss of grazing, wood products, wildlife habitat, recreation, and watershed values in forest, brush, or grass fires.

Decomposition. 1. Slow oxidation. 2. Burning without noticeable heat or light. 3. Chemical change of a single substance into two or more different substances.

Deflagration. 1. Thermal decomposition that proceeds at less than some velocity, and may or may not develop hazardous pressures. 2. A rapid combustion that does not generate shock waves. 3. A burning that takes place at a flame speed below the velocity of sound in the unburned medium.

Deluge sprinkler system. A type of automatic sprinkler system, used to protect special risks. It consists of pipes and open sprinkler heads supplied through a valve connected to a detection system.

Detonation. 1. A thermal decomposition that occurs at supersonic velocity, accompanied by a shock wave in the decomposing material. 2. A burning that takes place at a flame speed above the velocity of sound in the unburned medium.

Dry pipe system. *See* Automatic dry sprinkler system.

Dry system. A dry-pipe automatic sprinkler system that has air instead of water under pressure in its piping; dry systems are often installed in areas subject of freezing.

Egress. A way out or exit. (*See also* Means of egress.)

Electrical grounding. A connection between a conductive body and the earth that eliminates the difference in potential between the object and ground.

Electrical raceways. Trenches in concrete floors, fitted with removable metal covers, where electrical connections and junction boxes are concealed.

Endothermic reaction. A process or change that absorbs heat and requires it for initiation and maintenance.

Energy dissipation. The process of reducing kinetic energy by converting it into other forms of energy (e.g., work done by friction, plastic strain energy).

Exit. 1. The portion of the means of egress that leads from the interior of a building or structure to the outside at ground level. 2. An area of refuge. (*See also* Egress.)

Exit access. Any portion of an evacuation path that leads to an exit.

Exit discharge. That portion of a means of egress between the termination of the exit and the exterior of the building at ground level.

Explosion. A sudden and violent release of energy from a material or compound as it decomposes, undergoes rapid chemical reaction, or changes from a solid to a liquid.

Explosive atmosphere. An atmosphere containing a mixture of a vapor or gas in any concentration within the explosive range.

Explosive limits. The range of concentration of a flammable gas or vapor (percent by volume in air) in which explosion can occur upon ignition in a confined area.

Express elevators. Elevators that operate nonstop to sky lobbies or observation floors.

Extra hazard occupancies. Properties in which flash fires that open all the sprinklers in the area are a problem. In such occupancies, sprinkler spacing must be closer and pipe sizes larger than in other occupancies. (*See also* Light hazard industrial occupancy; Ordinary hazard industrial occupancy.)

Facade. The face and especially the principal elevation of a building.

Faced wall. A wall comprised of two different masonry material, called wythes. The wythes are bonded together to act as one unit under load.

Firebreak. 1. A natural or constructed barrier that stops the spread of a fire or provides a control line from which to work. (*Also called* Fire lane.) 2. Urban areas in which decrepit buildings have been destroyed in order to reduce the chance of a conflagration.

Fire code. A set of rules and requirements whose purpose is to establish levels of fire protection considered adequate for procedures, practices, and equipment.

Fire damper. A device arranged to interrupt airflow automatically through a duct system, so as to restrict the passage of heat.

Fire department connection. *See* Sprinkler connection.

Fire door. A tested, listed, or approved door and frame assembly which prevents the spread of fire through a vertical or horizontal opening. (*See also* Automatic fire door.)

Fire lane. *See* Firebreak.

Fire resistance rating. The time, in minutes or hours, that materials or assemblies have withstood exposure to a fire, as established in accordance with the procedures of NFPA 251 and ASTM E119.

Fire resistive construction. Construction in which the structural members, including walls, columns, floors, and roofs, are of noncombustible or limited-combustible materials, and have fire resistance ratings not less than those specified in NFPA 220; fire resistive construction has more ability to resist structural damage from fire than any other construction type.

Fire retardant. A surface or construction that will slow or limit the spread of fire.

Fire separation. 1. A floor or wall that meets specified fire endurance requirements as a barrier which prevents or retards fire spread.   2. A space or aisle between objects, such as goods in storage, buildings, or structures, that serves as a fire break and also as an area for fire fighting operations. (*See also* Separation.)

Fire signature. Any product of combustion from a fire that can be used to detect or identify the fire.

Fire stop. An obstruction across an air passage or concealed space in a building to prevent fire from spreading.

Firestopping. The blocking off of concealed spaces in structures to prevent fire spread through walls and ceilings. (*Also called* stopping.)

Fire tower. 1. A smokeproof stair designed to limit or prevent penetration of heat, smoke, and gases into any part of a building.   2. A fire department training tower.   3. A fire department communications center.   4. A fire department water tower truck or an aerial ladder employed as a water tower.   5. A forest fire lookout tower. (*See also* Smokeproof tower.)

Fire wall. 1. A wall constructed of solid masonry units faced on each side with brick or reinforced concrete used to subdivide a building or separate buildings to restrict the spread of fire; it begins at the foundation and extends through all stories to and above the roof, unless the roof is of fire-resistive or semi-fire-resistive construction, in which case the wall is carried up tightly against the underside of the roof slab.   2. A wall with adequate fire resistance used to subdivide buildings to restrict the spread of fire.

Flame over. The rapid spread of flame over one or more surfaces.

Flame resistant. A material or surface that does not propagate flame once the external source of flame is removed.

Flame shield. A nonstructural element that directs flames away from an exposed structural element to protect it.

Flame spread. The propagation of a flame away from the source of ignition in a gas or across the surface of a liquid or a solid.

Flammability. The relative ease with which a fuel ignites and burns.

Flammable. Capable of being readily ignited.

Flammable limit. The highest or lowest concentration of a flammable gas or vapor in air that will explode or ignite. (*See also* Explosive limits; Upper flammable limit.)

Flammable liquid. Any liquid with a flash point below 100°F and a vapor pressure not exceeding 40 psi at 100°F. Class IA includes those with a flash point below 73°, and a boiling point below 100°F. Class IB includes those with a flash point below 73°F and a boiling point at or above 100°F. Class IC includes those with a flash point at or above 70°F and below 100°F.

Flammable range. The range between the upper and lower explosive or flammable limits of a flammable vapor or gas.

Flashback. The jump of a flame from an ignition source across a distance to a supply of flammable liquid.

Flashover. The stage of a fire at which all surfaces and objects are heated to their ignition temperatures and flame breaks out almost all at once over the entire surface.

Flash point. The minimum temperature at which a liquid gives off sufficient vapor to form an ignitable mixture with air; it is usually determined by a closed cup test. (*See also* Flammable liquid.)

Floor populations. Number of potential elevator passengers per floor.

Flue gas. Combustion products vented up a chimney from a fireplace or other combustion chamber.

Fly ash. The finely divided residue resulting from the combustion of ground or powdered coal and which is transported from the firebox through the boiler by flue gases.

Fuel. 1. Any substance that produces heat through combustion. 2. A substance that reacts with oxygen, or with the oxygen yielded by an oxidizer, to produce combustion.

Fuel gas. 1. Acetylene, hydrogen, natural gas, LP-gas, methylacetylene-propadiene, and some other liquefied or nonliquefied flammable gases that are stable because of composition, or because of the conditions of storage and use. 2. Manufactured gas, natural gas, undiluted liquefied petroleum gas (vapor phase only), liquefied petroleum gas–air mixture, or mixtures of these gases that would ignite in the presence of oxygen.

Fuel load. The expected maximum amount of combustible material in a given fire area, usually expressed as weight of combustible material per square foot.

Fuse. 1. An electrical safety device consisting of or including a wire or strip of fusible metal that melts and interrupts a circuit when the current exceeds a particular amperage. 2. To reduce to a liquid or plastic state by heat.

Fusible link. A connecting link of a low-melting alloy that holds an automatic sprinkler head in the closed position and melts at a predetermined temperature; it may also be used to hold a fire door or fire damper in the open position.

Gas. 1. The state of matter characterized by very low density and viscosity, comparatively great fluctuation as pressures and temperatures change, the ability to diffuse readily into other gases, and the ability to occupy with almost complete uniformity the whole of any container. 2. Gasoline.

Gasoline. A common flammable liquid mixture of volatile hydrocarbons used in internal combustion engines; it has an octane number of at least 56, and an explosive range of about 1.4 percent to 7.6 percent. The flash point and ignition temperature vary with the octane number.

Gravity. 1. The force of mutual attraction between masses. 2. The force that causes a body to accelerate while falling, usually expressed as 32.2 ft/sec^2. (*See also* Specific gravity.)

Ground. A conducting connection, whether intentional or accidental, between an electrical circuit or equipment and the earth, or some conducting body that serves in place of the earth.

Halon. Any one of several halogenated hydrocarbon compounds, two of which (bromotrifluoromethane and bromochlorodifluoromethane) are commonly used as extinguishing agents; they are inert to almost all chemicals, and resistant to both high and low temperatures.

Halon extinguisher. A fire extinguisher charged with a halogenated agent and an expellant; it is rated for use on flammable liquid and live electrical equipment fires.

Health care occupancies. Those facilities used for the medical treatment or care of persons suffering from physical or mental illness, disease, and infirmity, and for the care of infants, convalescents, or aged persons.

Heat. Energy that is associated with the proportional to molecular motion, and that can be transferred from one body to another by radiation, conduction, and convection. (*See also* Spontaneous heating.)

Heat exhaustion. Distress characterized by headache, nausea, heavy perspiration, pale clammy skin, abnormal temperature, dilated pupils, and a general weakness brought on by overexposure to high temperature and a deficiency of salt. (*See also* Heat stroke.)

Heat flux. The intensity of heat transfer across a surface expressed in watts/cm^2, joules/m^2, or Btu/in.^{2}/sec.

Heat of combustion. The amount of heat released during a substance's complete oxidation (combustion). Heat of combustion, commonly referred to as calorific or fuel value, depends upon the kinds and numbers of atoms in the molecule as well as upon their arrangement.

Heat of vaporization. The quantity of heat required to change a unit quantity of a liquid into a vapor. One pound of water at 212°F requires 972 Btu to vaporize. (*Also called* latent heat of vaporization.)

Heat stroke. A condition more serious than heat exhaustion, characterized by dry skin, high body temperature, and collapse due to long exposure to high temperature; if untreated, it may lead to delirium, convulsions, coma, and even death.

Horizontal exit. A protected way of travel from one area of a building to another area in the same building or in an adjoining building on approximately the same level.

Hot air explosion. A backdraft explosion. (*See also* Backdraft; Smoke explosion.)

Hydrocarbon. An organic compound that consists exclusively of carbon and hydrogen and is derived principally from petroleum, coal, tar, and vegetable sources.

Ignite. To initiate combustion.

Ignitable mixture. 1. A vapor-air or dust-air mixture that can be ignited by a static spark. 2. A mixture that is capable of the propagation of flame away from the source of ignition when ignited.

Ignition. The point at which the heating of something becomes self-perpetuating.

Ignition temperature. The minimum temperature required to initiate self-sustained combustion of a material or compound.

Ignition time. The time, in seconds, between the application of an ignition source to material and the instant self-sustained combustion begins.

Incendiary. 1. A fire believed to have been deliberately set. 2. An arsonist. 3. Of or relating to a device used to set an arson fire.

Incomplete combustion. A process in which there is insufficient oxygen available for complete combustion; the result is a smoldering fire that produces smoke and carbon monoxide, and may be in danger of a backdraft.

Industrial occupancy. Factories of all kinds and properties devoted to operations such as processing, assembling, mixing, packaging, finishing, or decorating, and repairing.

Inflammable. A material that will burn. (*See also* Flammable; Nonflammable.)

Ingress. An entrance or the act of entering.

Insulator. A nonconductor of electricity, usually made of porcelain or glass, which encloses dangerous electrical equipment and prevents the leakage or passage of electricity.

Interior finish. The surface material of walls, fixed or movable partitions, ceilings, and other exposed interior surfaces; the category includes plaster paneling, wood, paint, wallpaper, floor and ceiling tiles, and the like.

Ionization detector. A detector that senses the presence of particulate combustion products and actuates an alarm.

Job, worker. *See* Working fire.

Latent heat of vaporization. *See* Heat of vaporization.

Lean mixture. A mixture of air and gas that contains too much air for the amount of gas present to cause an explosion and is thus below the lower flammable limit.

LEL. *See* Lower explosive limit.

Light hazard industrial occupancy. An occupancy in which the potential rate of heat liberation is low, areas are subdivided, and a small number of sprinklers should be able to control any fires. (*See also* Extra hazard occupancies; Ordinary hazard industrial occupancy.)

Liquefied natural gas (LNG). A cryogenic fluid composed predominantly of methane, but which may contain some butane, ethane, propane, nitrogen, or other components normally found in natural gas.

Liquefied petroleum gas (LP-Gas or LPG). A colorless, nontoxic gas liquefied by compression; it has a flash point of $-100°F$, an ignition temperature of 800 to 1000°F, and is obtained as a byproduct in petroleum refining.

Listed. A device or product that has been tested, passed, and certified by a nationally recognized testing laboratory.

LNG. *See* Liquefied natural gas.

Lower explosive limit (LEL). The minimum concentration of combustible gas or vapor in air that will ignite.

Low explosive. An explosive that deflagrates or burns rather than detonates, such as propellants, certain primer mixtures, photoflash powders, and delay compositions.

LPG (or LP-gas). *See* Liquefied petroleum gas.

Means of egress. A safe, continuous, and unobstructed way of travel out of any building or structure; this includes the exit access, exit, and exit discharge.

Metal. An element that forms positive ions when its compounds are in solution, and whose oxides form hydroxides rather than acids with water. About 75 percent of the elements are metals: most are crystalline solids with metallic luster, are conductors of electricity, and have high chemical reactivity.

Minimum hourly fire-resistance rating. That degree of fire resistance deemed necessary by the authority having jurisdiction.

Natural draft. A current of air produced by the difference in the weight of a column of flue gases inside a chimney and a corresponding column of air of equal dimensions outside the chimney or vent.

Natural gas. A naturally occurring fuel gas consisting of mostly methane but which has some ethane, butane, propane, and nitrogen. It is colorless, and almost odorless unless a warning odor has been added, and has an ignition temperature of 900–1000°F, and an explosive range of 3.8 to 17 percent. It is used for fuel and cooking, for the synthesis of ammonia, and as a raw material for the petrochemical industry; as a liquid it is called liquefied natural gas (LNG).

Nonbearing wall. A wall which supports only its own weight.

Noncombustible. 1. A material that will not ignite and burn when subjected to a fire. 2. A material defined as such in NFPA 225. 3. A material classified as such by the Standard Method of Test for Noncombustibility of Elemental Materials, ASTM E136-73.

Nonflammable. 1. A liquid or gas that will not burn under the conditions set forth in the definition of flame resistant. (*See also* Flame resistant.) 2. A liquid or gas that will not burn in 100 percent oxygen at pressures of 760 torr.

Occupancy. The use or intended use of a building, floor, or other part of a building. (*See* *also* Health care occupancies; Industrial occupancy; Special purpose industrial occupancies.)

Occupant load.   The theoretical maximum number of persons that may occupy a building or an area in it at one time.

Open circuit.   1. A break in an electrical circuit.   2. A fire alarm circuit which has no current except when a signal is being transmitted.

Open-joist construction.   Construction in which solid beams project below a ceiling surface more than 4 in. with intervals of 3 feet or less.

Ordinary construction.   Building construction in which exterior bearing walls are made of noncombustible or limited-combustible materials and have minimum fire resistance ratings and stability; nonbearing exterior walls are of noncombustible or limited-combustible materials; and roofs, floors, and framing are wholly or partly of wood of smaller dimensions than heavy timber construction.

Ordinary hazard industrial occupancy.   An industrial occupancy in which the processes, materials, and equipment are such that fires will probably burn fairly rapidly, or give off a considerable volume of smoke, but in which neither poisonous fumes nor explosions need be expected.

Oxidation.   Originally, a chemical reaction in which oxygen combines with other substances; it now indicates any chemical reaction in which electrons are transferred. Oxidation and reduction always occur simultaneously, and the substance which gains the electrons is called the oxidizing agent.

Oxidizer.   1. A substance that readily gives up oxygen without requiring an equivalent of another element in return.   2. A substance that contains an atom or atomic group that gains electrons such as oxygen, ozone, chlorine, hydrogen peroxide, nitric acid, metal oxides, chlorates, and permanganates. (*Also called* oxidizing agent; *See also* Oxidation.)

Panelboard.   1. A single panel or group of panels, accessible only from the front, from which lights, heat, or power circuits are controlled.   2. Thin laminated or pressed boards with a plastic facing or wood veneer on one side used for interior finish. (*Also called* paneling.)

Panel heating.   A heating system in which the heating elements are concealed in the walls or ceilings.

Panel wall.   A nonloadbearing wall made of panels fitted into steel or reinforced concrete framing members. (*See also* Curtain wall.)

Panic hardware.   Special latches installed on exit doors so that the latch can be released and the door opened with a force of less than 15 pounds.

Partition.   An interior space divider such as a wall.

Party wall.   A wall common to two buildings, often owned or leased by both parties. (*Also called* common wall; separating wall; zero lot line.)

Passageway.   A corridor, hallway, passage, or tunnel used for pedestrian traffic.

Photoelectric detector.   A fire detector with a photocell that either changes its electrical conductivity or produces an electrical potential when exposed to radiant energy; it also detects smoke by the reduction in the transmission of light through the smoke.

Pilot ignition.   The ignition of a material by radiation, in which a local high temperature ignition source is located in the stream of gases and volatiles issuing from the exposed material.

Preaction sprinkler system.   A sprinkler system in which piping is dry and waterflow is actuated by a separate detection system.

Propagation.   The spread of combustion through a solid, gas, or vapor, or the spread of a fire from one combustible to another combustible. (*See also* Flame spread.)

Pyrolysis.   Chemical decomposition caused by heat.

Quick response sprinkler. A sprinkler head designed for rapid operation that has increased surface-to-mass areas or special actuators, i.e., metallic vane heat collectors or electronic squibs.

Raceway. A tube for carrying and protecting electrical wires, cables, or busbars.

Radiant heat. Heat energy carried by electromagnetic waves that can pass through gases without warming them, but that increases the temperature of solid and opaque objects. (*Also called* radiated heat.)

Radiation. 1. Energy that is sent forth, e.g., light, short radio, ultraviolet, and x-ray waves. 2. The transfer of energy, including heat, through visible light by electromagnetic waves. 3. Streams of high-speed atomic particles, e.g., alpha and beta particles and neutrons.

Reflash. The reignition of a flammable fuel by a hot object after flames have been extinguished. (*See also* Flashback; Rekindle.)

Regression rate. The burning rate of a solid or liquid, usually measured in centimeters per second measured perpendicular to the surface.

Rekindle. To reignite after extinguishment. (*See also* Reflash.)

Rich mixture. A fuel and oxidizer mixture having more than the stoichiometric concentration of fuel. (*See also* Lean mixture.)

Safety glass. Glass that has thin wires or plastic embedded in it or placed between laminates; if broken, it produces no flying splinters. (*See also* Tempered glass.)

Secondary exit. An alternate exit.

Self-closing device. An approved mechanism that will ensure closing after having been opened.

Self-contained breathing apparatus. Equipment that is worn by firefighters to provide respiratory protection in a hazardous environment; it consists of a facepiece, a regulating or control device, an air or oxygen supply, a harness assembly, and, sometimes, filters.

Self-ignition. *See* Spontaneous ignition.

Separated storage. Storage in the same fire area, but separated by as much space as practicable, or by intervening storage, from incompatible materials. (*See also* Firebreak; Fire separation; Separate fire division; Separation.)

Separate fire division. A portion of a building cut off from all other portions of the building by fire walls, fire doors. (*See also* Fire separation; Separation.)

Separating wall. *See* Party wall.

Separation. 1. The spacing of buildings or materials to provide fire exposure protection. 2. A barrier to fire spread, such as a fire wall. (*See also* Firebreak; Fire separation; Separated storage; Separate fire division.)

Short circuit. Conduction of electrical current (usually inadvertent) made between points on a circuit between which the resistance is normally greater.

Simultaneous ignition. Ignition of a fire at many points at once; it is used in broadcast burning or backfiring to obtain a quick, hot, clean burn.

Slow oxidation. The oxidation of a material without the evolution of visible light, as in the rusting of iron. (*See also* Oxidation.)

Smoke. The airborne solid and liquid particles and gases evolved when a material undergoes combustion.

Smoke control door. A door designed to inhibit or act as a barrier to the spread of smoke in a building. (*Also called* smoke stop door.)

Smoke damper. A device to restrict the passage of smoke through a duct that operates automatically and is controlled by a smoke detector.

Smoke density. The proportion of solids in smoke.

Smoke detector. A detector that actuates if it senses visible or invisible particles of combustion. (*See also* Ionization detector; Photoelectric detector.)

Smoke explosion. An explosion of heated smoke and gases. (*Erroneously called* hot air explosion. *See also* Backdraft.)

Smoke inhalation. Sickness or injury caused by the respiration of smoke and other products of combustion. (*Also called* smoke poisoning.)

Smoke partition. Partitions or walls installed to divide a building into compartments and prevent the spread of smoke from a fire.

Smoke poisoning. *See* Smoke inhalation.

Smokeproof stairway. A stairway designed so that the passage of smoke and gases from a fire into it is limited.

Smokeproof tower. A continuous fire-resistive enclosure in a building that protects a stairway from fire or smoke. (*Also called* smoke tower. *See also* Fire tower.)

Smoke shaft. A continuous shaft, extending the full height of a building, with openings at each level and a fan at the top; during a fire, the dampers on the fire floor open and the fan vents the combustion products.

Smoke stop door. *See* Smoke control door.

Smoke tower. *See* Smokeproof tower.

Smolder. To burn and smoke without flame.

Space heater. A portable heating device used to heat small areas, some types of which use liquid fuel. Though these are prohibited in most jurisdictions, they are still found and have been the initial cause of many fatal fires.

Spalling. The deterioration of concrete due to fire exposure.

Spark. 1. A small, incandescent particle from burning wood or a glowing particle produced by metal grinding. 2. A localized instantaneous electrical discharge, accompanied by heat and light, between points at different voltages.

Special purpose industrial occupancies. As defined by the NFPA *Life Safety Code,* this term includes ordinary and low hazard manufacturing operations in buildings designed for and suitable only for particular types of operations, characterized by a relatively low density of employee population, with much of the area occupied by machinery or equipment.

Specific gravity (sp gr). The weight or mass of a given volume of a substance at a specified temperature, as compared to that of an equal volume of another substance. (*See also* Gravity.)

Spontaneous combustion. *See* Spontaneous heating; Spontaneous ignition.

Spontaneous heating. Heating due to chemical or bacterial action in a combustible material. (*See also* Spontaneous ignition.)

Spontaneous ignition. Ignition due to chemical reaction or bacterial action in which there is a slow oxidation of organic compounds until the material ignites; usually there is sufficient air for oxidation but insufficient ventilation to carry heat away as it is generated. (*Also called* self-ignition. *See also* Autoignition temperature.)

Spray sprinkler. An automatic sprinkler head designed to control fire by the spray principle of heat absorption. (*Also called* spray sprinkler head. *See also* Sprinkler system.)

Spread. The extension of a fire.

Sprinkler alarm. A device that sounds an audible local alarm when sprinkler heads activate.

Sprinkler block. A device used to stop the flow of water from a particular sprinkler head without shutting the system down. (*See also* Sprinkler tongs; Sprinkler wedge.)

Sprinkler connection. A connection in which hose lines from a pumper are used to increase the pressure in a sprinkler system, usually a sprinkler siamese with two pumper inlet connections. (*Also called* fire department connection.)

Sprinklered. Equipped with a sprinkler system.

Sprinkler head. A waterflow device in a sprinkler system, consisting of a threaded nipple that connects the heat to a water pipe, a fusible link or other releasing mechanism held in place, and a deflector that breaks up the water into a spray.

Sprinkler spacing. The distribution of sprinkler heads to provide adequate protection for a given hazard.

Sprinkler stopper. *See* Sprinkler tongs; Sprinkler wedge.

Sprinkler system. A system of water pipes and spaced sprinkler heads installed in a structure to control and extinguish fires; it uses a suitable water supply, such as a gravity tank, fire pump, reservoir, pressure tank, or connections to city mains and usually has a controlling valve and an alarm that signals when the system is actuated. (*See also* Automatic sprinkler system; Automatic dry sprinkler system; Deluge sprinkler system; Dry system; Preaction sprinkler system; Spray sprinkler; Sprinkler alarm; Sprinkler block; Sprinkler connection; Sprinklered; Sprinkler head; Sprinkler spacing; Sprinkler tongs; Sprinkler wedge; Wet-pipe sprinkler system.)

Sprinkler tongs. A tool used to stop the flow of water from a sprinkler head without shutting down the system. (*Also called* sprinkler block; sprinkler stopper. *See also* Sprinkler wedge.)

Sprinkler wedge. A tapered, wedge-shaped wooden block used to stop the flow of water from a sprinkler head without shutting down the system. (*Also called* sprinkler block; sprinkler stopper; sprinkler tongs.)

Stack effect. The air or smoke movement of migration through a tall building due to pressure differentials caused by temperature.

Stair pressurization. Increasing the air pressure in stair wells (usually with fan systems) to provide refuge area from fire and smoke.

Standard. A document containing requirements and specifications, such as for building construction or fire protection. (*See also* Fire code.)

Stoichiometric air. The chemically correct amount of air required for complete combustion of a given quantity of a specific fuel.

Stopping. *See* Firestopping.

Stratification. The rising or settling of layers of smoke, according to density or weight, with the heaviest layer on the bottom; smoke layers usually collect from the ceiling down.

Sublimation. Evaporation or release of vapors from a solid without going through the liquid phase.

Suppression. All actions taken to extinguish a fire, from the time of its discover; fire extinguishment.

Surface burning. *See* Flame spread.

Surface flame spread. *See* Flame spread.

Suspended maneuvering system. A helicopter-hung fire-fighting platform fully equipped to tackle the fire and rescue the occupants.

Temperature reinforcement. Reinforcement designed to carry stresses resulting from temperature changes; also the minimum reinforcement for areas of members not subjected to primary stresses or necessarily to temperature stresses.

Tempered glass. Glass heat-treated in a special process so that it has a better resistance to bending, impact, or thermal shock than ordinary glass of the same thickness; when broken it falls into small, regular fragments instead of large, random-sized shards.

Thermal column. A column of smoke and gases given off by fire because of the expansion and rise of heated gases from displacement by cooler air; from it the magnitude and intensity of a fire can often be judged. (*Also called* convection column, thermal updraft.)

Thermal conductivity. The transmission of heat through a solid or liquid.

Thermal convection. *See* Convection.

Thermal decomposition. The breakup of materials into other compounds as a result of heat. (*Also called* thermal degradation. *See also* Pyrolysis.)

Thermal radiation. The emission of radiant energy waves from a heated body.

Thermal updraft. *See* Thermal column.

Thermocouple. A temperature-measuring device composed of two dissimilar conductors connected at their ends. Heat causes a different voltage in each of the two conductors and voltage difference is proportional to the temperature of the material measured.

Thermodynamics. Study of the relation between heat and energy.

Transformer. A device that raises or lowers the voltage of alternating current.

Tunnel test. The popular name for a standard surface burning test of building materials, the details of which are given in ASTM E84.

UEL. *See* Upper explosive limit.

UFL. *See* Upper flammable limit.

Unit of exit width. The width necessary (22 in.) for the orderly movement of a single line of people along a passageway or through an exit during an emergency.

Unprotected opening. A vertical or horizontal opening through a floor, wall, or other partition in a building that allows the passage of smoke, heat, and flame.

Unstable material. Any material which will vigorously polymerize, decompose, condense, become self-reactive, or undergo other violent chemical changes.

Upper explosive limit (UEL). The maximum concentration of vapor or gas in air above which flame propagation does not occur. (*Also called* upper flammable limit. *See also* Flammable limit.)

Upper flammable limit (UFL). The highest concentration of flammable vapor in air that will burn with a flame. (*Also called* upper explosive limit. *See also* Flammable limit.)

Utility gas. Natural gas, manufactured gas, liquefied petroleum gas–air mixtures, or a mixture of any of these.

Valve. A gate in a passage used to regulate the flow of liquid, air, gas, loose material, etc.

Vapor. A substance in its gaseous state, particularly one that is liquid or solid at ordinary temperatures.

Vapor density (vd). The weight of a given volume of a gas or vapor, as compared with the weight of the same volume of dry air.

Vaporizing liquid. A liquid extinguishing agent, usually carbon tetrachloride or chlorobromomethane, used on flammable liquid and electrical fires, that vaporizes and forms a vapor blanket heavier than air that chemically extinguishes the fire. Vaporizing liquid extinguishers are no longer approved for installation. (*See also* Halon extinguisher.)

Vapor pressure. The pressure exerted at any given temperature by a vapor, either by itself or in a mixture of gases, measured at the surface of an evaporating liquid.

Vd. *See* Vapor density.

Vent. An opening for the release or dissipation of fluids, such as gases, fumes, smoke, etc.

Wall furnace. A self-contained, vented heater installed in a building, mobile home, or recreational vehicle.

Wet-pipe sprinkler system. An automatic sprinkler system in which the pipes are kept filled with water. (*See also* Automatic sprinkler system.)

Wet water. Water to which a wetting agent has been added to increase its penetration.

Wired glass. Window glass with an embedded wire mesh to improve fire resistance and make it shatter-resistant.

Wood frame construction. A type of construction in which exterior walls, bearing walls, partitions, floors, roofs, and their supports are made wholly or partly of wood and other combustible materials, when the construction does not qualify as heavy timber construction or ordinary construction.

Working fire. 1. A fire that requires fire-fighting activity by most or all of the fire department personnel assigned to the alarm. (*Also called* job, worker.) 2. A one-alarm blaze.

Zero lot line. *See* Party wall.

Zoning. The legal regulation of the use of land and buildings, in which the density of population and the height, bulk, and spacing of structures is also specified.

SYMBOLS

a	= purging rate in number of air changes per minute
A	= flow area (also called leakage area)
A	= door area
A	= number of units of exit width required
A_{cr}	= reduced compressive area
A_e	= effective area
A_F	= nominal fire compartment area = $C1.A_z$
A_{FM}	= maximum permissible fire compartment area
A_V	= maximum permitted area of openings
b	= width of the staircase between floors $r-1$ and r
b	= flow width identical to width of escape route
b	= stair width in m
b	= staircase width
B	= constant as to the construction type of building
$b(i)$	= initial width of flow
$B(i)$	= sum of widths of all partial flows
$b(i + 1)$	= width of congested flow
$b(T)$	= width of door to staircase
$b(T1)$	= width of partial flow from each floor in main flow on stairs

$b(TR)$	= stair width
$b(TR1)$	= width of partial flow from top floor n in main flow on stairs
c	= specific heat of downstream gases
C	= constant for arrangement and protection of stairs
C	= concentration of contaminant at time, t
C	= flow coefficient
$C1, C2, C3$	= trade-off coefficients
C_o	= initial concentration of contaminant
c_p	= specific heat capacity
C_w	= dimensionless pressure coefficient
d	= distance from doorknob to edge of door (knob side)
D	= constant for exposure hazard
D	= density in persons per m^2
$D(i)$	= initial flow density
$D(T)$	= density through the door to staircase (no congestion on stairs)
$D(TR; n - 1)$	= density of group emanating from overcrowded area at level $(n - 1)$
$D(TR; n)$	= flow density on stairs without congestion
$D_{\max}$	= maximum flow density
dt	= delay time
e	= compressive force
e	= constant approximately 2.718
E	= total net energy at ambient moisture content
E	= factor dependent upon height of floor above or below ground level
E_c	= energy release rate into corridor
e_F	= fire load density value
f	= perpendicular projected area of an individual
F	= rectangular radiating facade
F	= total door opening force
F_{dc}	= force to overcome the door close
fo	= flow rate/unit stair width of 0.6 m
F_v	= ventilation factor
g	= gravitational constant
G_i	= stress
h	= distance from the neutral plane
h	= distance above neutral density plane
H	= height

h_e	= effective thickness
hG	= floor height
h_v	= effective vertical opening height
I_E	= intensity of emitted radiation
I_F	= radiant energy flux
I_R	= intensity of received radiation
K	= constant (≈ 1)
$k_{10}, k_{11}, k_{12}, k_{13}$	= conversion factors used for fire zoning
K_f	= coefficient
K_v	= coefficient
k_w	= wetting down factor
K_w	= coefficient
l	= flow length
L	= span of slab
L	= limiting distance
$l(TR)$	= travel distance on stairs between adjoining stories
L_x	= protected limiting distance
$M_{uv} + M_{uv} +$	= absolute value of positive and negative plastic bending moment, respectively, at end of required period of standard fire exposure
M_y	= bending moment
n	= number of upper floors in building
n	= number of stories served by staircase
N	= number of exits required
N'	= flow rate of people per unit width down stairs
N_{pl}	= bearing capacity for composite cross-section
N_{st}	= high-bearing capacity for steel
N_x	= applied normal force
p	= evacuation population per meter effective stair width
p	= number of persons in flow
p	= number of persons in the building
P	= absolute atmospheric pressure
P	= population a staircase can accommodate
P	= flame projection distance
P_o	= outside air density
P_w	= wind pressure
q	= load on slab to be accounted for during fire
Q	= energy loss rate
Q	= volumetric airflow rate
$q(D_{\max})$	= specific flow at maximum density

$q(i)$	=	initial value of specific flow
$Q(I)$	=	sum of capacities of all partial flows
$q(T; D_{max})$	=	specific flow at maximum density through doorways
$Q(T; n - 1)$	=	flow capacity through door to staircase (each floor)
$q(T)$	=	specific flow through the door to the staircase under free flow conditions on stairs
$q(TR; n)$	=	specific flow on stairs (no congestion)
$q(TR; max)$	=	maximum flow capacity of stairs
$q(TR; n - 1)$	=	value of specific flow on stairs (after merging process)
$q(TR; D_{max})$	=	specific flow on stairs (maximum density)
$q(TR; max)$	=	stairs maximum flow capacity
$Q(TR; n)$	=	initial flow capacity stairs
$Q(TR)$	=	stairs initial flow capacity
q^*	=	load to be used in fire test
Qi	=	population of floor i
Q_{in}	=	volumetric flow rate of air into fire compartment
Q_{out}	=	volumetric flow rate of smoke out of fire compartment
Q_T	=	total flow
r	=	floor number (1 to n) which gives the max value of te
r	=	standard rate of flow
R	=	radiation distance
R	=	gas constant of air
$r(TR)$	=	evacuation time for stairs (per floor)
R_{peak}	=	maximum mass loss rate value reached during a fire
S	=	separation distance
t	=	time after doors closed in minutes
$t''(TR; STAU)$	=	length of time required for flow to leave floor level ($n - 2$)
$t(F)$	=	travel time (last evacuee along corridor)
$t(F)$	=	evacuation time of corridor (each floor)
$t(TR; STAU)$	=	length of time required for flow to leave the floor level ($n - 1$)
$t(TR)$	=	travel time (person going from top floor to adjoining story)
$t1$	=	travel time on stairs to descend one story
T_2	=	maximum fire compartment temperature
td	=	fire duration time
te	=	maximum permissible exit time from any one floor onto staircase
t_f	=	fire resistance rating (FRR)
T	=	absolute temperature of downstream mixture of air and smoke

T_F	= absolute temperature of fire compartment
T_i	= actual temperature
T_I	= absolute temperature of air inside the shaft
T_{in}	= absolute temperature of air into fire compartment
T_O	= absolute temperature of outside air
T_{out}	= absolute temperature of smoke leaving fire compartment
ts	= time for individual in unimpeded crowd to descend one story
ts	= time for person to traverse a story height of stairs at standard rate of flow
v	= flow velocity
v	= flow velocity down stairs of 0.3 m/sec
v	= crowd walking velocity (flow velocity)
V	= wind velocity
v''	= velocity of boundary between initial flow (stairs) and merged flow changing its location
$v''(T; STAU)$	= speed of congestion on corridor
$v'(TR; STAU)$	= speed of congestion on stairs
$v(T; D_{max})$	= velocity of a flow through doorway at maximum density
$v(TR; n)$	= velocity of flow at density $D(TR; n)$
$v(TR; n-1)$	= velocity of flow emanating from congested area at floor level $(n-1)$
$v(TR; STAU)$	= flow velocity on stairs at maximum density
$v(TR; n-1)$	= velocity of flow on stairs at density $D(TR; n-1)$
V_k	= critical air velocity to prevent smoke backflow
V_o	= constant velocity
w	= width of staircase
W	= door width
W	= corridor width
Z	= class of user of building
$\sigma_{c\dot{\upsilon}}$	= ultimate compressive strength
σ_{c20}	= compression strength of concrete at ambient temperature
$\sigma_{y\dot{\upsilon}}$	= effective yield stress
σ_{y20}	= yield stress at room temperature
ΔP	= pressure difference
ΔP_T	= total pressure difference
δ	= specific density
ε_i^{δ}	= stress causing strain
ε	= emissivity
λ	= thermal conductivity

ρ = density of air entering the flow path

ρ = density of upstream air

ρ_O = air density outside the shaft

ρ_I = air density inside the shaft

ρ_A = configuration factor for rectangle A

ρ_n = total configuration factor

$\dot{v}_c$ = temperature of concrete

$\dot{v}_c$ = temperature

ABBREVIATIONS

ACI	American Concrete Institute
AISI	American Iron and Steel Institute
ASET	Available safe egress time
ASHRAE	American Society of Heating, Refrigerating and Air-Conditioning Engineers
ASTM	American Society for Testing and Materials
BS	British Standard
BSI	British Standards Institute
BTR	Building Technical Regulation (Taiwan)
CEB	Comité Euro-International du Béton (European Concrete Committee)
CEFICOSS	Computer engineering of the fire resistance for composite and steel structures
CP	Code of practice
CPA	Construction and Planning Administration (Taiwan)
CTICM	Centre Technique Industriel de la Construction Métallique
ECCS	Eropean Convention for Constructional Steelwork
FFR	Fire-resistance rating
FIP	Fédération Internationale de al Précontrainte (International Federation for Prestressing)
GLC	Greater London Council
HMSO	Her Majesty's Stationary Office
HVAC	heating, ventilating, and air-conditioning
ISO	International Standards Organization
MMI	Modified Mercalli intensity
MOI	Ministry of the Interior (Taiwan)
MPV	Minimum proper value
MRT	Mass rapid transportation

NFD	Nominal fire duration
NFPA	National Fire Protection Association
NIST	National Institute of Standards and Technology (formerly National Bureau of Standards)
RHR	Rate of heat release
ROC	Republic of China
RSET	Required safe egress time

UNITS

In the table below are given conversion factors for commonly used units. The numerical values have been rounded off to the values shown. The British (Imperial) System of units is the same as the American System except where noted. Le Système International d'Unités (abbreviated "SI") is the name formally given in 1960 to the system of units partly derived from, and replacing, the old metric system.

SI	American	Old metric
	Length	
1 mm	0.03937 in.	1 mm
1 m	3.28083 ft	1 m
	1.093613 yd	
1 km	0.62137 mile	1 km
	Area	
1 mm^2	0.00155 in.^{2}	1 mm^2
1 m^2	10.76392 ft^2	1 m^2
	1.19599 yd^2	
1 km^2	247.1043 acres	1 km^2
1 hectare	2.471 acres[1]	1 hectare
	Volume	
1 cm^3	0.061023 in.^{3}	1 cc
		1 ml
1 m^3	35.3147 ft^3	1 m^3
	1.30795 yd^3	
	264.172 gal[2] liquid	
	Velocity	
1 m/sec	3.28084 ft/sec	1 m/sec
1 km/hr	0.62137 miles/hr	1 km/hr
	Acceleration	
1 m/sec^2	3.28084 ft/sec^2	1 m/sec^2

SI	American	Old metric
	Mass	
1 g	0.035274 oz	1 g
1 kg	2.2046216 lb[3]	1 kg
	Density	
1 kg/m^3	0.062428 lb/ft^3	1 kg/m^3
	Force, Weight	
1 N	0.224809 lbf	0.101972 kgf
1 kN	0.1124045 tons[4]	
1 MN	224.809 kips	
1 kN/m	0.06853 kips/ft	
1 kN/m^2	20.9 lbf/ft^2	
	Torque, Bending Moment	
1 N-m	0.73756 lbf-ft	0.101972 kgf-m
1 kN-m	0.73756 kip-ft	101.972 kgf-m
	Pressure, Stress	
1 N/m^2 = 1 Pa	0.000145038 psi	0.101972 kgf/m^2
1 kN/m^2 = 1 kPa	20.8855 psf	
1 MN/m^2 = 1 MPa	0.145038 ksi	
	Viscosity (Dynamic)	
1 N-sec/m^2	0.0208854 lbf-sec/ft^2	0.101972 kgf-sec/m^2
	Viscosity (Kinematic)	
1 m^2/sec	10.7639 ft^2/sec	1 m^2/sec
	Energy, Work	
1 J = 1 N-m	0.737562 lbf-ft	0.00027778 w-hr
1 MJ	0.37251 hp-hr	0.27778 kw-hr
	Power	
1 W = 1 J/sec	0.737562 lbf ft/sec	1 w
1 kW	1.34102 hp	1 kw
	Temperature	

$$K = 273.15 + °C \qquad °F = (°C \times 1.8) + 32 \qquad °C = (°F - 32)/1.8$$
$$K = 273.15 + 5/9(°F - 32)$$
$$K = 273.15 + 5/9(°R - 491.69)$$

(1) Hectare as an alternative for km^2 is restricted to land and water areas.
(2) 1 m^3 = 219.9693 Imperial gallons.
(3) 1 kg = 0.068522 slugs.
(4) 1 American ton = 2000 lb.　　1 kN = 0.1003612　　1 Imperial ton = 2240 lb.

Abbreviations for Units

Btu	British thermal unit	kW	kilowatt
°C	degree Celsius (centigrade)	lb	pound
cc	cubic centimeters	lbf	pound force
cm	centimeter	lb_m	pound mass
°F	degree Fahrenheit	MJ	megajoule
ft	foot	MPa	megapascal
g	gram	m	meter
gal	gallon	ml	milliliter
hp	horsepower	mm	millimeter
hr	hour	MN	meganewton
Imp	British Imperial	N	newton
in.	inch	oz	ounce
J	joule	Pa	pascal
K	Kelvin	psf	pounds per square foot
kg	kilogram	psi	pounds per square inch
kgf	kilogram-force	°R	degree Rankine
kip	1000 pound force	sec	second
km	kilometer	slug	14.594 kg
kN	kilonewton	U_o	heat transfer coefficient
kPa	kilopascal	W	watt
ksi	kips per square inch	yd	yard

References/Bibliography

Abrams, M. S., and Lin, T. D., 1974
SIMULATION OF REALISTIC THERMAL RESTRAINT DURING FIRE TESTS OF FLOOR AND ROOF ASSEMBLIES, PCA-Report, Skokie, Ill.

Alpert, R. L., 1972
CALCULATION OF RESPONSE TIME OF CEILING-MOUNTED FIRE DETECTORS, *Fire Technology*, vol. 8, no. 3, pp. 181–195.

American Concrete Institute, 1981
GUIDE FOR DETERMINING THE FIRE ENDURANCE OF CONCRETE ELEMENTS, *Concrete International*, Feb.

American Iron and Steel Institute, 1979
FIRE-SAFE STRUCTURAL STEEL—A DESIGN GUIDE, Washington, D.C., Nov.

American Society for Testing and Materials, ASTM E176, 1980
ANNUAL BOOK OF ASTM STANDARDS, PART 18, Philadelphia, Pa.

American Society of Heating, Refrigerating and Air-Conditioning Engineers, 1981
ASHRAE HANDBOOK—1981 FUNDAMENTALS, Atlanta, Ga.

Anderberg, Y., Pettersson, O., Theleandersson, S., and Wickstrom, U., 1975
A DIFFERENTIATED DESIGN OF FIRE EXPOSED CONCRETE STRUCTURES, Bull., Lund Institute of Technology, Division of Structural Mechanics and Concrete Construction, Lund, Sweden.

ARBED Research Centre, 1987–1990
SEISMIC RESISTANCE OF COMPOSITE STRUCTURES, C.E.C. Research 7210-SA/506, B-G-I-L, Luxembourg.

ASHRAE, 1985
FIRE AND SMOKE CONTROL, Technical Data Bulletin, American Society of Heating, Refrigeration and Air Conditioning Engineers, Atlanta, Ga.

Australian Fire Protection Association, 1986
FACTORY MUTUAL INTERNATIONAL ESFR UPDATE, A progress report on the Factory Mutual System's ESFR Sprinkler Programme, *Fire Journal*, vol. 11, no. 1, March.

Babrauskas, V., 1976
FIRE ENDURANCE IN BUILDINGS, Report UCB FRG 76-16, Fire Research Group, University of California, Berkeley, Nov.

Babrauskas, V., and Williamson, R. B., 1978
POST-FLASHOVER COMPARTMENT FIRES: BASIS OF A THEORETICAL MODEL, *Fire and Materials*, vol. 2, no. 2, pp. 39–53.

Barnett, C. R., 1988
FIRE SEPARATION BETWEEN EXTERNAL WALLS OF BUILDINGS, Proc. of the 2d International Symposium on Fire Safety Science, Tokyo, pp. 841–850.

Barnett, C. R., 1989
FIRE COMPARTMENT SIZING DESIGN METHOD, National Conference Publication 89/16, International Symposium on Fire Engineering for Building Structures and Safety, Institution of Engineers, Melbourne, Australia, pp. 25–31.

Baus, R., and Schleich, J. B., 1987
PREDETERMINATION DE LA RESISTANCE AU FEU DES CONSTRUCTIONS MIXTES, *Anales de L'Institut Technique du Bâtiment et des Travaux Publics*, no. 457, Sept.

Becker, J., Bizri, H., and Bresler, B., 1974
FIRES-T, A COMPUTER PROGRAM FOR THE FIRE RESPONSE OF STRUC-
TURES—THERMAL, Berkeley.

Becker, J. M., and Bresler, B., 1977
REINFORCED CONCRETE FRAMES IN FIRE ENVIRONMENTS, *Journal of the
Structural Division,* vol. 103.

Bell, J. R., 1981
137 INJURED IN NEW YORK CITY HIGH-RISE BUILDING FIRE, *Fire Journal,*
March, pp. 38–48.

Bellamy, L. J., and Geyer, T. A. W., 1988
AN EVALUATION OF THE EFFECTIVENESS OF THE COMPONENTS OF IN-
FORMATIVE FIRE WARNING SYSTEMS, International Conference on Safety in
the Built Environment, Portsmouth, July 1988, J. D. Sime (Ed.), E & F Spon, London.

Belles, D. W., and Beitel, J. J., 1988
BETWEEN THE CRACKS..., *Fire Journal,* May/June.

Berl, W. G., and Halpin, B. M., 1980
HUMAN FATALITIES FROM UNWANTED FIRES, *Johns Hopkins APL Technical
Digest,* vol. 1, no. 2, pp. 129–134.

Best, R., 1975
HIGH-RISE APARTMENT FIRE IN CHICAGO LEAVES ONE DEAD, *Fire Journal,*
Sept., pp. 38ff.

Best, R. L., 1977
RECONSTRUCTION OF A TRAGEDY: THE BEVERLY HILLS SUPPER CLUB
FIRE, SOUTHGATE KENTUCKY, MAY 28TH, LS-2, National Fire Prevention As-
sociation, Quincy, Mass.

Bimonthly Fire Record, 1980a
HIGH-RISE APARTMENT BUILDING FOR THE ELDERLY, *Fire Journal,* March,
pp. 19–21.

Bimonthly Fire Record, 1980b
DORMITORY, *Fire Journal,* Nov., p. 19.

Bryan, J. L., 1981
AN EXAMINATION AND ANALYSIS OF THE DYNAMICS OF THE HUMAN BE-
HAVIOR IN THE MGM GRAND HOTEL FIRE, Mimeo, National Fire Protection
Association, Quincy, Mass.

Bryl, S., and Keller, B., 1982
ÜBER DIE BERECHNUNG DES FEUERWIDERSTANDES VON VERBUNDSTÜ-
TZEN MIT STAHLKERN, *Ingenieur und Architekt* (Switzerland) vol. 40.

BSI BS 5588, 1978
CODE OF PRACTICE FOR FIRE PRECAUTIONS IN THE DESIGN OF HIGH-RISE
BUILDINGS, PART 4, SMOKE CONTROL IN PROTECTED ESCAPE ROUTES
USING PRESSURIZATION.

BSI BS 5588, 1983
FIRE PRECAUTIONS IN THE DESIGN AND CONSTRUCTION OF BUILDINGS,
PART 3, CODE OF PRACTICE FOR OFFICE BUILDINGS.

BSI BS 5588, 1988
FIRE PRECAUTIONS IN THE DESIGN AND CONSTRUCTION OF BUILDINGS,
PART 8, CODE OF PRACTICE FOR MEANS OF ESCAPE FOR DISABLED PEO-
PLE.

Bukowski, R., Peacock, R., Jones, W., and Forney, L., 1989
HAZARD I TECHNICAL REFERENCE GUIDE, NIST Special Publication 146, vol.
II, National Institute of Standards and Technology, Gaithersburg, Md.

Canter, D., 1985
STUDIES OF HUMAN BEHAVIOR IN FIRE: EMPIRICAL RESULTS AND THEIR
IMPLICATIONS FOR EDUCATION AND DESIGN, Report BR 61, Building Re-
search Establishment, Fire Research Station, Borehamwood, U.K.

Canter, D., Breaux, J., and Sime, J., 1980
DOMESTIC, MULTIPLE-OCCUPANCY, AND HOSPITAL FIRES, in D. Canter (Ed.), *Fires and Human Behavior,* Wiley, Chichester, pp. 117–136.

Canter, D., Powell, J., and Booker, K., 1990
THE PSYCHOLOGY OF INFORMATIVE FIRE WARNING SYSTEMS, Report, Building Research Establishment, Fire Research Station, Borehamwood, U.K.

Centre Scientifique et Technique du Bâtiment, 1980
METHODE DE PREVISION PAR LE CALCUL DU COMPORTEMENT AU FEU DES STRUCTURES EN BETON, Document Technique Unifié, Paris, April.

Cheung, K. P., 1986
STAIRCASE PRESSURIZATION—THE RATIONALE AND THE ALTERNA-TIVES, Forum of the Hong Kong Chapter of ASHRAE, Oct. 11.

Chitty, R., and Cox, G., 1988
ASKFRS, Building Research Establishment, Borehamwood, U.K.

Comité Euro-International du Béton, 1982
DESIGN OF CONCRETE STRUCTURES FOR FIRE RESISTANCE, First Draft of an Appendix to the CEB/FIP Model Code for Concrete Structures, Bulletin 145, Paris.

Commission of the European Community, 1990
EUROCODES, PART 10, STRUCTURAL FIRE DESIGN, First Draft, Brussels-Luxembourg.

Cooper, L. Y., 1982
A MATHEMATICAL MODEL FOR ESTIMATING AVAILABLE SAFE EGRESS TIME IN FIRES, *Fire and Materials,* vol. 6, pp. 135–144.

Cooper, L. Y., 1983
A CONCEPT FOR ESTIMATING SAFE AVAILABLE EGRESS TIME IN FIRES, *Fire Safety Journal,* vol. 5, pp. 135–144.

Cooper, L. Y., 1984
A BUOYANT SOURCE IN THE LOWER OF TWO, HOMOGENEOUS STABLY STRATIFIED LAYERS, in Proc. 20th Symposium (International) on Combustion, Combustion Institute, pp. 1567–1573.

Cooper, L. Y., 1986
THE NEED AND AVAILABILITY OF TEST METHODS FOR MEASURING THE SMOKE LEAKAGE CHARACTERISTICS OF DOOR ASSEMBLIES, *Fire Safety Science and Engineering,* ASTM Special Technical Publication STP 886, American Society for Testing and Materials, Philadelphia, Pa., pp. 310–329.

Cooper, L. Y., and Stroup, D. W., 1985
ASET—A COMPUTER PROGRAM FOR CALCULATING AVAILABLE SAFE EGRESS TIME, *Fire Safety Journal,* vol. 9, pp. 29–45.

Cooper, L. Y., and Davis, W. D., 1988, 1989
ESTIMATING THE ENVIRONMENT AND THE RESPONSE OF SPRINKLER LINKS IN COMPARTMENT FIRES WITH DRAFT CURTAINS AND FUSIBLE LINK-ACTUATED CEILING VENTS—PARTS I AND II, NBSIR 88-3734 and NISTIR 89-4122, National Institute of Standards and Technology, Gaithersburg, Md.

Cooper, L. Y., Forney, G. P., and Moss, W. F., 1990
THE CONSOLIDATED COMPARTMENT FIRE MODEL (CCFM) COMPUTER CODE APPLICATION CCFM. VENTS—PARTS I TO IV, NISTIR 90-4343, -4344, -4345, National Institute of Standards and Technology, Gaithersburg, Md.

Council on Tall Buildings, 1972
PLANNING AND DESIGN OF TALL BUILDINGS, Proceedings of First International Conference held in August 1972 in Bethlehem, Pennsylvania, Council on Tall Buildings, Bethlehem

Council on Tall Buildings, Committee 12A, 1992
CLADDING, McGraw-Hill, New York

Council on Tall Buildings, Committee 56, 1992
BUILDING FOR THE HANDICAPPED AND AGED: DESIGN CONCEPTS, McGraw-Hill, New York

Council on Tall Buildings, Group CL, 1980
TALL BUILDING CRITERIA AND LOADING, vol. CL, *Monograph on Planning and Design of Tall Buildings,* ASCE, New York, Chapters CL-2 and CL-4.

Cresci, R. J., 1983
SMOKE AND FIRE CONTROL IN HIGH-RISE OFFICE BUILDINGS—PART II, ANALYSIS OF STAIR PRESSURIZATION SYSTEMS, Symposium on Experience and Applications on Smoke and Fire Control, ASHRAE Annual Meeting, Atlanta, Ga., June 24–28.

CTICM, 1975
FORECASTING FIRE EFFECTS ON STEEL STRUCTURES (Prévision par le calcul du comportement au feu des structures en acier), CTICM, Pateaux, France, Sept.

Degenkolb, J. G., 1981
FIRE SAFETY REQUIREMENTS FOR EXISTING HIGH-RISE BUILDINGS, *Building Standards,* Sept.–Oct., pp. 6–12.

Deutsche Normen, 1981
DIN 4102 TEIL 4, BRANDVERHALTEN VON BAUSTOFFEN UND BAUTEILEN, ZUSAMMENSTELLUNG UND ANWENDUNG KLASSIFIZIERTER BAUS-TOFFE, *Bauteile und Sonderbauteile,* Berlin.

Dorn, T., Hass, R., and Quast, U.
BRANDVERHALTEN VON RIEGELANSCHLÜSSEN AN VERBUNDSTÜTZEN AUS EINBETONIERTEN WALZPROFILEN UND AUSBETONIERTEN HOHL-PROFILEN, Forschungsprojekt p. 86.2.10, Studiengesellschaft für Anwendung-stechnik von Eisen und Stahl e. V., in Bearbeitung.

ECCS-TC3, 1988
CALCULATION OF THE FIRE RESISTANCE OF CENTRALLY LOADED COM-POSITE STEEL-CONCRETE COLUMNS EXPOSED TO THE STANDARD FIRE, European Convention for Constructional Steelwork, Technicalk Note 55, Brussels.

ECCS-TC11, 1975
EUROPEAN RECOMMENDATION FOR THE DESIGN OF COMPOSITE FLOORS WITH PROFILED STEEL SHEET, European Convention for Constructional Steel-work, Constrado, London.

Egan, M. D., 1978
CONCEPTS IN BUILDING FIRE SAFETY, Wiley, New York.

EMPA, 1980
STATISCHE DRUCKVERSUCHE AN ZWEI GEILINGER-BAUSTÜTZEN, Eid-genössische Materialprüfungsanstalt, Dubendorf.

Engineering News Record, 1986
THE McGRAW-HILL CONSTRUCTION WEEKLY, Jan. 2.

Engineering News Record, 1990
FIRES, McGraw-Hill, New York.

Evans, D. D., 1985
CALCULATING SPRINKLER ACTUATION TIME IN COMPARTMENTS, *Fire Safety Journal,* vol. 9, pp. 147–155.

Evers, E., and Waterhouse, A., 1978
A COMPUTER MODEL FOR ANALYZING SMOKE MOVEMENT IN BUILDINGS, Building Research Establishment, Fire Research Station, Borehamwood, U.K., Nov.

Fahy, R., and Norton, H., 1989
HOW BEING POOR AFFECTS FIRE RISK, *Fire Journal,* Feb., pp. 29–36.

Fang, J. B., 1980
STATIC PRESSURES PRODUCED BY ROOM FIRES, NBSIR 80-1984, National Bu-reau of Standards, Washington, D.C.

FIP, 1978
FIP/CEB REPORT ON METHODS OF ASSESSMENT OF THE FIRE RESISTANCE OF CONCRETE STRUCTURAL MEMBERS, FIP Commission on the Fire Resistance of Prestressed Concrete Structures.

Fire Journal, 1970
BIMONTHLY FIRE RECORD, Jan., p. 47.

Fire Journal, 1975
HIGH-RISE OFFICE BUILDING, *Fire Journal*, Nov., pp. 87–89.

Fire Journal, 1977
DALLAS FIRE KILLS TWO FIREFIGHTERS, *Fire Journal*, July, p.108.

Fire Journal, 1982a
FIRE AT THE MGM GRAND, *Fire Journal,* Jan., pp. 20–37.

Fire Journal, 1982b
INVESTIGATION REPORT ON THE LAS VEGAS HILTON HOTEL FIRE, *Fire Journal*, Jan. pp. 52–57.

Fire Journal, 1983
TWELVE DIE IN FIRE AT WESTCHASE HILTON HOTEL, *Fire Journal*, Jan., pp. 11–15.

Fire Journal, 1988
CONDOMINIUM, *Fire Journal*, July/Aug., p. 14.

Fire Officers' Committee, 1973
RULES OF THE FIRE OFFICERS' COMMITTEE FOR AUTOMATIC SPRINKLER INSTALLATIONS, 29th ed., U.K., Nov.

Fire Officers' Committee, 1978
SUPPLEMENTARY HIGH RISE SPRINKLER DRAFT RULES, U.K., May.

FIRES, 1990 ENR, 4 Jan., p. 13.

Franssen, J. M., 1987
ETUDE DU COMPORTEMENT AU FEU DES STRUCTURES MIXTES ACIER-BETON, Ph.D. dissertation, University of Liege, Belgium, Feb.

Fruin, J. J., 1970
DESIGNING FOR PEDESTRIANS: A LEVEL OF SERVICE CONCEPT, Ph.D. dissertation, Polytechnic Institute of Brooklyn, N.Y.

Garkisch, R., and Heindl, W., unpublished
SIMULATION DER RÄUMUNG EINES GEBÄUDES IM GEFAHRENFALL, Forschungsarbeit im Auftrag des Bundesministeriums für Bauten und Technik.

Greater London Council, 1974
CODE OF PRACTICE, MEANS OF ESCAPE IN CASE OF FIRE, London.

Grimes, M. E., 1970
HOTEL FIRE, *Fire Journal,* May, pp. 17–20f.

Gunter, K. P., 1979
HIGH-RISE BUILDINGS, A GERMAN APPROACH, *Fire International 64,* Unisaf Publications Ltd., U.K.

Haksver, A., 1977
ZUR FRAGE DES TRAG- UND VERFORMUNGSVERHALTENS EBENER STAHLBETONRAHMEN IM BRANDFALL, Thesis, TU Braunschweig, Germany.

Harmathy, T., 1986
A SUGGESTED LOGIC FOR TRADING BETWEEN FIRE-SAFETY MEASURES, *Fire and Materials* (9–12), National Research Council of Canada.

Harmathy, T., and Oleszkiewicz, I., 1987
FIRE DRAINAGE SYSTEM, *Fire Technology,* vol. 23, Feb., pp. 26–48.

Harmathy, T., 1989
WHAT KILLS IN FIRES: SMOKE INHALATION OR BURNS?, *Fire Journal*, May/June.

Harwell Laboratory, 1990
HARWELL-FLOW3D RELEASE 2.3: USER MANUAL, CFD Department, AEA Industrial Technology, Harwell Laboratory, Oxfordshire, U.K., July.

Harwood, B., and Hall, J. R., Jr., 1989
WHAT KILLS IN FIRES: SMOKE INHALATION OR BURNS?, *Fire Journal*, May/June, pp. 29–34.

Hass, R., Quast, U., and Rudolph, K., 1984
STABA-F, A COMPUTER PROGRAM FOR THE DETERMINATION OF LOAD-BEARING AND DEFORMATION BEHAVIOUR OF UNI-AXIAL STRUCTURAL ELEMENTS UNDER FIRE ACTION. Institut für Baustoffe, Massivbau und Brandschutz, TU Braunschweig, Germany.

Hass, R., and Quast, U., 1985
BRANDVERHALTEN VON VERBUNDSTÜTZEN MIT BERÜCKSICHTIGUNG DER UNTERSCHIEDLICHEN STUTZEN/RIEGEL-VERBINDUNGEN, Forschungbericht Projekt 86.2.2, Studiengesellschaft für Anwendungstechnik von Eisen und Stahl e. V.

Heijnen, J. G., and Schreuder, J. A. M., 1982
NUMBER, LOCATION AND CAPACITY OF FIRE STATIONS IN ROTTERDAM (in Dutch), Fire Department Rotterdam, Rotterdam, The Netherlands.

Herschelmann, F., and Rudolph, K., 1983
BERÜCKSICHTIGUNG DER BETONFEUCHTIGKEIT AUF ERWÄRMUNG-SVORGÄNGE IN STAHL-VERBUND-TRÄGERN UNTER BRANDBEANSPRUCHUNG, Arbeitsberichte 1981–1983 des Sonderforschungbereichs 148, Brandverhalten von Bauteilen, TU Braunschweig, Germany.

Home Office, 1935
MANUAL OF SAFETY REQUIREMENTS ON THEATRES AND OTHER PLACES OF PUBLIC ENTERTAINMENT, HMSO, London.

Houghton, E. L., and Carruther, N. B., 1976
WIND FORCE ON BUILDINGS AND STRUCTURES, Wiley, New York.

Ingberg, S. H., 1928
TESTS OF SEVERITY OF BUILDING FIRES, *NFPA Quarterly*, vol. 22, p. 43.

Institution of Structural Engineers/The Concrete Society, 1978
DESIGN AND DETAILING OF CONCRETE STRUCTURES FOR FIRE RESISTANCE, Interim Guidance by a Joint Committee, London.

International Standards Organization, ISO 834, 1975
FIRE RESISTANCE TESTS—ELEMENTS OF BUILDING CONSTRUCTIONS.

Isner, M. S., 1988a
$80 MILLION FIRE IN MONTREAL HIGH-RISE, *Fire Journal*, Jan./Feb., pp. 64–70.

Isner, M. S., 1988b
SMOKY FIRE KILLS FOUR IN NEW YORK HIGH-RISE, *Fire Journal*, Sept./Oct., pp. 72–77.

Jones, W. W., 1985
A MULTICOMPARTMENT MODEL FOR THE SPREAD OF FIRE, SMOKE, AND TOXIC GASES, *Fire Safety Journal*, vol. 9, pp. 55–79.

Journal of Fire Protection Engineering, 1989
THE CAPABILITIES OF SMOKE CONTROL: FUNDAMENTALS AND ZONE SMOKE CONTROL, vol. 1, no. 1, pp. 1–10.

Juillerat, E. E., and Gaudet, R. E., 1967
FIRE AT DALE'S PENTHOUSE RESTAURANT, *Fire Journal*, May, p. 5.

Jungbluth, O., and Bangert, Hahn, J.
FEUERWIDERSTANDSFÄHIGE KONSTRUKTIONEN DURCH PROFILVERBUND. Siehe /6/.

Jungbluth, O., Feyereisen, H., and Oberegge, O., 1980
VERBUNDPROFILKONSTRUKTIONEN MIT ERHÖHTER FEUERWIDESTANDSDAUER, *Bauingenieur*, vol. 55, pp. 371–376.

Jungbluth, O., and Lindhorst, W.
FEUERWIDERSTANDSFÄHIGE SPEZIAL-VERBUNDPROFILE. Siehe /3/.

Kajima Corp., Mitsui Fudousan Co., Ltd., and Science University of Tokyo, Chiba, 1986
LEAKAGE FROM DOORS OF A HIGH-RISE APARTMENT, Grant and Pagni (Eds.), Hemisphere Publishing Corp., New York, pp. 891–900.

Kansai University, Osaka, 1983
INVESTIGATION INTO THE ACTUAL CONDITION OF FOLDING FIRE ESCAPE LADDERS INSTALLED IN THE BALCONIES OF HIGH-RISE APARTMENTS, Fire Research and Safety, Sixth Joint Meeting of the UINR Proceedings (1982), Buildings Research Institute, Tokyo, pp. 191–202.

Kendik, E., 1982
DIE BERECHNUNG DER RÄUMUNGSZEIT IN ABHÄNGIGKEIT DER PROJEK-TIONSFLÄCHE BEI DER EVAKUIERUNG DER VERWALTUNGSHOCH-HÄUSER ÜBER TREPPENRÄUME, VFBD, 6th International Fire Protection Seminar, Karlsruhe, Germany.

Kendik, E., 1983
DETERMINATION OF THE EVACUATION TIME PERTINENT TO THE PRO-JECTED AREA FACTOR IN THE EVENT OF TOTAL EVACUATION OF HIGH-RISE OFFICE BUILDINGS VIA STAIRCASES, Fire Safety Journal, 5.223.

Kendik, E., 1984
DIE BERECHNUNG DER PERSONENSTRÖME ALS GRUNDLAGE FÜR DIE BEMESSUNG VON GEHWEGEN IN GEBÄUDEN UND UM GEBÄUDE, Ph.D. dissertation, Technical University of Vienna, Austria.

Kendik, E., 1985a
ASSESSMENT OF ESCAPE ROUTES IN BUILDINGS AND A DESIGN METHOD FOR CALCULATING PEDESTRIAN MOVEMENT, Paper presented at SFPE's 35th Anniversary Engineering Seminar, Chicago, Ill., May.

Kendik, E., 1985b
METHODS OF DESIGN FOR MEANS OF EGRESS: TOWARDS A QUANTITA-TIVE COMPARISON OF NATIONAL CODE REQUIREMENTS, Paper presented at the First International Symposium on Fire Safety Science, National Bureau of Standards, Gaithersburg, Md., Oct.

Kisko, T. M., and Francis, R. L., 1985
EVACNET + : A COMPUTER PROGRAM TO DETERMINE OPTIMAL BUILDING EVACUATION PLANS, Fire Safety Journal, vol. 9, no. 2, pp. 211–220.

Klem, T. J., 1987
97 DIE IN ARSON FIRE AT DUPONT PLAZA HOTEL, Fire Journal, May/June, pp. 74–83.

Klem, T. J., 1989
LOS ANGELES HIGH-RISE BANK FIRE, Fire Journal, May/June, p. 72.

Klingsch, K., 1984a
ANALYSE DES TRAGVERHALTENS VON GEILINGER-BAUSTÜTZEN BEI NORMAL TEMPERATUR UND BEI BRANDBEANSPRUCHUNG SOWIE DER ZUGEHÖRIGEN BEMESSUNGVERFAHREN, Wuppertal, Germany.

Klingsch, W., Bode, H.-G., and Finsterle, A., 1984b
BRANDVERHALTEN VON VERBUNDSTÜTZEN AUS VOLLSTÄNDIG EIN-BETONIERTEN WALZPROFILEN, Bauingenieur, vol. 59, pp. 427–432.

Klingsch, W., Wurker, K.-G., and Martin-Bullmann, R., 1984c
BRANDVERHALTEN VON HOHLPROFIL-VERBUNDSTÜTZEN, Stahlbau, vol. 53, pp. 300–305.

Klote, J. H., 1980
STAIRWELL PRESSURIZATION, ASHRAE Transactions, vol. 86, pt. I, pp. 604–673.

Klote, J. H., 1982
A COMPUTER PROGRAM FOR ANALYSIS OF SMOKE CONTROL SYSTEMS, NBSIR 82-2512, National Bureau of Standards, Gaithersburg, Md., June.

Klote, J. H., and Fothergill, J. W., 1983
DESIGN OF SMOKE CONTROL SYSTEMS FOR BUILDINGS, National Bureau of Standards, Gaithersburg, Md., Sept.; American Society for Heating, Refrigerating and Air-Conditioning Engineers, Atlanta, Ga.

Klote, J. H., and Tamura, G. T., 1986
SMOKE CONTROL AND FIRE EVACUATION BY ELEVATORS, *ASHRAE Transactions,* vol. 92, no. 1A, pp. 231–245.

Klote, J. H., 1987a
AN OVERVIEW OF SMOKE CONTROL TECHNOLOGY, NBSIR 87-3626, National Bureau of Standards, Gaithersburg, Md., Sept.

Klote, J. H., and Tamura, G. T., 1987b
EXPERIMENTS OF PISTON EFFECT ON ELEVATOR SMOKE CONTROL, *ASHRAE Transactions,* vol. 93, no. 2, pp. 2217–2228.

Knötig, H., 1980
GENERELLES INTER-AKTIONS-SCHEMA, *Humanekologische Blätter,* vol. 9, no.2–3.

Kobayashi, M., 1981
DESIGN STANDARDS OF MEANS OF EGRESS IN JAPAN, International Seminar on Life Safety and Egress, University of Maryland, Md., Nov.

Kordina, K., et al., 1977, 1980, 1983
JAHRESBERICHTE 1975-1977, 1978 AND 1981–1983, Sonderforschungsbereich 148, TU Braunschweig, Germany.

Kordina, K., 1981
BAULICHER BRANDSCHUTZ IN STRASSEN- UND U-BAHN-TUNNELN, *Bauingenieur,* vol. 56.

Kordina, K., and Meyer-Ottens, C., 1981
BETON-BRANDSCHUTZ-HANDBUCH, Beton-Verlag, Düsseldorf, Germany.

Kordina, K., Klingsch, W., and Herschelmann, F., 1982
ZUR FRAGE DES BRANDVERHALTENS VON STAHLVERBUNDKONSTRUK-TIONEN, Sicherheit Brand- und Katastrophenbekämpfung, Notfallrettung; 2. Status-Seminar des Bundesministers für Forschung und Technologie, Deutscher Gemein-deverlag, Kohlhammer.

Kordina, K., and Krampf, L., 1984
EMPFEHLUNGEN FÜR BRANDSCHUTZTECHNISCH RICHTIGES KON-STRUIEREN VON BETONBAUWERKEN, Publ. 352, Deutscher Ausschuss fur Stahlbeton.

Kordina, K., Herschelmann, F., and Richter, E., 1984
PARAMETERSTUDIE FÜR VERBUNDTRÄGER DER FEUERWIDERSTAND-KLASSE F 90 (Versuche zum Erwärmungsverhalten), Forschungbericht Projekt 86.2.5, Studiengesellschaft für Anwendungstechnik von Eisen and Stahl e. V.

Kordina, K., and Klingsch, W., 1984
BRANDVERHALTEN VON STAHLSTÜTZEN IM VERBUND MIT BETON UND VON MASSIVEN STAHLSTÜTZEN OHNE BETON, Forschungsbericht Projekt 35, Studiengesellschaft für Anwendungstechnik von Eisen und Stahl e. V.

Kordina, K., and Hass, R., 1985a
UNTERSUCHUNGSBERICHT Nr. 85636, Amtliche Materialprüfanstalt für das Bauwesen, TU Braunschweig, Germany.

Kordina, K., Wesche, J., and Hoffend, F., 1985b
UNTERSUCHUNGSBERICHT NR. 85833, Amtliche Materialprüfanstalt für das Bauwesen, TU Braunschweig, Germany.

Kordina, K., Richter, E., and Aufmuth, U., 1986
DEHNFUGEN, ANFORDERUNGEN UND KONSTRUKTION UNTER BERÜ-CKSICHTIGUNG DES BRANDFALLS, TEIL II, Institut für Baustoffe, Massivbau und Brandschutz, TU Braunschweig, Germany.

Kruppa, J., 1983
ECHAUFFEMENT DES PLANCHERS BETON A BAC ACIER SOUMIS A L'INCENDIE CONVENTIONAL, CTICM, Paris.

Las Vegas Fire Department, 1982
NOTIFICATION AND ALARM SYSTEMS—THE LAS VEGAS STORY, 1980 Conference on Life Safety and the Handicapped, AIA Research Foundation/National Technical Information Services, pp. 36–38.

Lathrop, J. K., 1975
WORLD TRADE CENTER FIRE, *Fire Journal,* July, pp. 19–24.

Lathrop, J. K., 1976a
TWO FIRES DEMONSTRATE EVACUATION PROBLEMS IN HIGH-RISE BUILDINGS, *Fire Journal,* Jan., pp. 65–68.

Lathrop, J. K., 1976b
TWO FIRES DEMONSTRATE...," *Fire Journal,* Jan., pp. 68–70.

Lathrop, J. K., 1979
ATRIUM FIRE PROVES DIFFICULT TO VENTILATE, *Fire Journal,* Jan., pp. 30–31.

Lathrop, J. K., 1986
LIFE SAFETY CODE HANDBOOK, 3d ed., National Fire Protection Association, Quincy, Mass., p. 140.

Lee, T. R., 1971
PSYCHOLOGY AND ARCHITECTURAL DETERMINISM, pts. 1–3, *Architect's Journal,* vol. 154, pp. 253–262, 475–483, 651–659.

Levin, B. L., 1989
EXITT—A SIMULATION MODEL OF OCCUPANT DECISIONS AND ACTIONS IN RESIDENTIAL FIRES, Proc. of the 2d International Symposium of Fire Safety Science, T. Wakamatsu et al. (Eds.), Hemisphere Publ. Corp., pp. 561–570.

London Transport Board, 1958
SECOND REPORT OF THE OPERATIONAL RESEARCH TEAM ON THE CAPACITY OF FOOTWAYS, Research Report.

Los Angeles City Fire Department, 1988
REPORT ON FIRST INTERSTATE BANK BUILDING FIRE, Executive Summary, Los Angeles, Calif.

LOSS PREVENTION DATA SHEET 1–3, 1983
Factory Mutual Engineering Corp., USA, Mar.

MacDonald, A. J., 1975
WIND LOADING ON BUILDINGS, Wiley, New York.

McGuire, J. H., Tamura, G. T., and Wilson, A. G., 1970
FACTORS IN CONTROLLING SMOKE IN HIGH BUILDINGS, Symposium on Fire Hazards in Buildings, ASHRAE Semi-Annual Meeting, San Francisco, Calif., Jan.

McGuire, J. H., and Tamura, G. T., 1975
SIMPLE ANALYSIS OF SMOKE FLOW PROBLEMS IN HIGH BUILDINGS, *Fire Technology,* vol. 11, Feb., pp. 15–22.

Melinek, S. J., and Booth, S., 1975
AN ANALYSIS OF EVACUATION TIMES AND THE MOVEMENT OF CROWDS IN BUILDINGS, CP 96/75, Building Research Establishment, Fire Research Station, Borehamwood, U.K.

Ministry of Interior, ROC, 1989a
CONSTRUCTION AND PLANNING ADMINISTRATION, A Briefing on the Construction and Planning Administration of Interior ROC, July, p. 14.

Ministry of Interior, ROC, 1989b
STATISTIC ABSTRACT OF INTERIOR OF THE REPUBLIC OF CHINA, A briefing on the Construction and Planning Administration of Interior ROC, July, p. 48.

Ministry of Interior, ROC, 1989c
STATISTICAL DATA BOOK OF THE MINISTRY OF THE INTERIOR, June, p. 142.

Ministry of Municipal Affairs, 1984
SECTION 3.7 HANDBOOK, BUILDING REQUIREMENTS FOR PERSONS WITH DISABILITIES INCLUDING ILLUSTRATIONS AND COMMENTARY, British Columbia, Canada.

Ministry of Works, 1952
FIRE GRADING OF BUILDINGS, MEANS OF ESCAPE, PART 3: PERSONAL SAFETY, Post-War Building Studies No. 29, HMSO, London.

Minne, R., Vandevelde, R., and Odou, M., 1985
FIRE TEST REPORTS 5091–5099, Laboratorium voor Aanwending der Brandstoffen en Warmte-over-dracht, University of Gent, Belgium.

Mitler, H. E., and Rockett, J. A., 1987
USER'S GUIDE TO FIRST, A COMPREHENSIVE SINGLE-ROOM FIRE MODEL, NBSIR 87-3595, National Institute of Standards and Technology, Gaithersburg, Md.

Muess, H., 1978
BRANDVERHALTEN VON BEKLEIDETEN STAHLBAUTEILEN, Stahlbau-Verlag GmbH, Köln, Germany.

Muess, H., and Schaub, W., 1982
NEUBAU EINER ZWEIGSCHOSSIGEN FERTIGUNGSHALLE MIT NEUARTIGER BRANDSCHUTZKONZEPTION, Stahlbau, vol. 51, pp. 225–234.

Müller, W. L., 1966
DIE BEURTEILUNG VON TREPPEN ALS RÜCKZUGSWEG IN MEHRGESCHOSSIGEN GEBÄUDEN, Unser Brandschutz, vol. 16, no. 11, suppl. 4, p. 930

Müller, W. L., 1968
DIE UBERSCHNEIDUNG DER VERKEHRSSTR Ö ME BEI DEM BERECHNEN DER RÄUMUNGSZEIT VON GEBÄUDEN, Unser Brandschutz, vol. 18, no. 11, suppl. 4, p. 87.

Müller, W. L., 1969
DIE DARSTELLUNG DES ZEITLICHEN ABLAUFS BEI DEM RÄUMEN EINES GEBÄUDES, Unser Brandschutz, vol. 19, no. 1, suppl. 4, p. 6.

Nash, P., and Young, R. A., 1978
SPRINKLERS IN HIGH-RISE BUILDINGS, CP 29/78, Building Research Establishment, Fire Research Station, Borehamwood, U.K.

National Bureau of Standards, 1935
DESIGN AND CONSTRUCTION OF BUILDING EXITS, Misc. Publ. M151, National Bureau of Standards, Washington, D.C.

National Fire Protection Association, 1981a
CODE FOR SAFETY TO LIFE FROM FIRE IN BUILDINGS AND STRUCTURES, NFPA 101, Quincy, Mass.

National Fire Protection Association, 1981b
STANDARD FOR THE INSTALLATION OF AIR CONDITIONING AND VENTILATING SYSTEMS, NFPA 90A, Quincy, Mass.

National Fire Protection Association, 1982a
CODE FOR SAFETY TO LIFE FROM FIRE IN BUILDINGS AND STRUCTURES, NFPA 101, vol. 9.

National Fire Protection Association, 1982b
INVESTIGATION REPORT ON THE LAS VEGAS HILTON HOTEL FIRE, Fire Journal, Jan.

National Fire Protection Association, 1985
CODE FOR SAFETY TO LIFE FROM FIRE IN BUILDINGS AND STRUCTURES, 1985 edition, NFPA

National Fire Protection Association, 1988
LIFE SAFETY CODE, NFPA 101, fig. A-15-3.1.1(a) Cell Block Smoke Control Ventilation Curves, sec. A-15-3.1.3, Quincy Mass., p. 219.

National Oceanic and Atmospheric Administration, 1979
TEMPERATURE EXTREMES IN THE UNITED STATES, National Climatic Center, Ashville, N.C.

National Research Council of Canada and Interscience Communications, Ltd., 1989
AUDIBILITY OF FIRE ALARM SYSTEMS IN HIGH-RISE BUILDINGS, International Conference on Fires in Buildings 1989, Extended Abstracts of Papers, Technomic Publishing Co., Lancaster, Pa., pp. 89–93.

Nelson, H. E., 1987
AN ENGINEERING ANALYSIS OF THE EARLY STAGES OF FIRE DEVELOPMENT—THE FIRE AT THE DUPONT PLAZA HOTEL AND CASINO—DECEMBER 31, 1986, NBSIR 87-3560, National Institute of Standards and Technology, Gaithersburg, Md., April.

Nelson, H. E., and MacLennon, H. A., 1988
EMERGENCY MOVEMENT, *SFPE Handbook of Fire Protection Engineering,* National Fire Protection Association, Quincy, Mass., sec. 2, chap. 6, pp. 2/106-2/115.

Nelson, H. E., 1989
SCIENCE IN ACTION, *Fire Journal,* July/Aug., pp. 29–34.

Nelson, H. E., 1990
FPETOOL: FIRE PROTECTION ENGINEERING TOOLS FOR HAZARD ESTIMATION, NISTIR 4380, National Institute of Standards and Technology, Gaithersburg, Md.

Oberg, B. W., 1977
DEVELOPMENT OF AN EVALUATIVE TECHNIQUE FOR THE FIRE PERFORMANCE OF LAMINATE CONSTRUCTION SYSTEMS, M. Arch Thesis, Ohio State University, Columbus, Ohio.

Okabe, T., et al., 1988
ELASTO-PLASTIC CREEP THERMAL DEFORMATION BEHAVIOR OF MULTISTORY STEEL FRAMES DESIGNED BY THE PLASTIC METHODS, Proc. of the 2d International Symposium on Fire Safety Science, Tokyo, June, pp. 739–748.

Ove Arup & Partners, 1977
DESIGN GUIDE FOR FIRE SAFETY OF BARE STRUCTURAL STEEL

Pauls, J. L., 1978
MANAGEMENT AND MOVEMENT OF BUILDING OCCUPANTS IN EMERGENCIES, DBR Paper 78, National Research Council of Canada, Ottawa, Ont., Canada.

Pauls, J. L., 1980a
BUILDING EVACUATION: RESEARCH AND RECOMMENDATIONS, In D. Canter (Ed.), *Fires and Human Behavior,* Wiley, Chichester, pp. 251–276

Pauls, J. L., and Jonas, B., 1980b
BUILDING EVACUATION: RESEARCH METHODS AND CASE STUDIES, in D. Canter (Ed.), *Fires and Human Behaviour,* Wiley, Chichester, chap. 13.

Pauls, J. L., 1982
EFFECTIVE-WIDTH MODEL FOR CROWD EVACUATION, VFDB, 6th International Fire Protection Seminar, Karlsruhe, Germany, Sept.

Pauls, J. L., 1984
DEVELOPMENT OF KNOWLEDGE ABOUT MEANS OF EGRESS, *Fire Technology,* vol. 20, no. 2, May.

Paulsen, R. L., 1981
HUMAN BEHAVIOUR AND FIRE EMERGENCIES: AN ANNOTATED BIBLIOGRAPHY, NBSIR-81-2438. National Bureau of Standards, Washington, D.C.

Pedersen, E., Schleich, J. B., Cajot, L. G., Exner, H., and Andresen, C., 1989
STRUCTURAL FIRE PROTECTION IN MULTI-STOREY OPEN CAR-PARK BUILDINGS, Dansk Brandvaerns Komite, Copenhagen, Denmark.

Peschl, I. A. S. Z., 1971
FLOW CAPACITY OF DOOR OPENING IN PANIC SITUATION, *BOUW,* vol. 26, no. 2, pp. 62–67.

Peterson, C. E., 1989
CONSTRUCTION HELPS LIMIT HIGH-RISE FIRE, *Fire Journal,* Feb., pp. 20–24.

Pettersson, M., and Witteveen, J., 1978/1980
ON THE CRITICAL TEMPERATURES OF STEEL ELEMENTS DERIVED FOR CONVENTIONAL FIRE RESISTANCE TESTS AND FROM CALCULATIONS, *Fire Safety Journal,* vol. 2, Lausanne, Switzerland.

Pigott, B. B., 1979
OUTLINE SPECIFICATION FOR A SYSTEM OF AUTOMATIC FIRE DETECTION OFFERING REDUCED FALSE ALARMS AND ENFORCEABLE MAINTENANCE, Special Paper, Building Research Establishment, Fire Research Station, Borehamwood, U.K.

Powers, W. R., 1971
OFFICE BUILDING FIRE, *Fire Journal,* March, pp. 5–7f.

Predtechenskii, V. M., and Milinski, A. I., 1978
PLANNING FOR FOOT TRAFFIC FLOW IN BUILDINGS, transl. from Russian, Amerind Publishing Co., New Delhi, India.

Public Works Department, 1982
FIRE PRECAUTIONS FOR BUILDINGS, Development and Building Control Division, Singapore.

Purser, D. A., 1988
TOXICITY ASSESSMENT OF COMBUSTION PRODUCTS, *SFPE Handbook of Fire Protection Engineering,* National Fire Protection Association, Quincy, Mass., sec. 1, chap. 14, pp. 1/200–1/245.

Quarantelli, E. L., 1984
ORGANIZATIONAL BEHAVIOR IN DISASTERS AND IMPLICATIONS FOR DISASTER PLANNING, FEMA Monograph Series.

Quast, U.
AUSWIRKUNGEN DER STÜZEN/RIEGEL-VERBINDUNG AUF DAS BRANDVERHALTEN VON VERBUNDSTÜTZEN. Siehe /3/.

Quast, U., 1983
VERBUNDSTÜTZEN UNTER BRANDBEANSPRUCHUNG, Brandverhalten von Stahlund Stahlverbundkonstruktionen, Status-Seminar, Studiengesellschaft für Anwendungstechnik von Eisen und Stahl e. V., Verlag TOV Rheinland, Köln, Germany.

Quast, U., and Rudolph K., 1985
BEMESSUNGSHILFEN FÜR STAHLVERBUNDSTÜTZEN MIT DEFINIERTEN FEUERWIDERSTANDSKLASSEN, Forschungsbericht Projekt 86.2.3, Studiengesellschaft für Anwendungstechnik von Eisen und Stahl e. V.

Ravers, W. R., 1971
NEW YORK OFFICE BUILDING FIRE, *Fire Journal,* Jan., pp. 18–23f.

Read, R. E. H., to be published
MEANS OF ESCAPE IN CASE OF FIRE: THE DEVELOPMENT OF LEGISLATION AND STANDARDS IN GREAT BRITAIN.

Read, R. E. H., and Shipp, M. P., 1979
AN INVESTIGATION OF FIRE DOOR CLOSER FORCES, Building Research Establishment, Fire Research Station, Borehamwood, U.K.

Richter, C. F., 1958
ELEMENTARY SEISMOLOGY, Freeman, San Francisco, Calif., 768 pp.

Rilling, J., 1978
SMOKE STUDY, 3RD PHASE, METHOD OF CALCULATING THE SMOKE MOVEMENT BETWEEN BUILDING SPACES, Centre Scientifique et Technique du Bâtiment (CSTB), Champs-sur-Marne, France, Sept.

Sachs, P., 1972
WIND FORCES IN ENGINEERING, Pergamon, New York.

Said, M., and Nady, A., 1988
A REVIEW OF SMOKE CONTROL MODELS, *ASHRAE Journal,* vol. 30, April, pp. 36–40.

Salse, E. A. B., and Gustaferro, A. H., 1971
STRUCTURAL CAPACITY OF CONCRETE BEAMS DURING FIRES AND AF-FECTED BY RESTRAINT AND CONTINUITY, Proceedings of the Fifth CIB Congress, Paris.

Sander, D. M., and Tamura, G. T., 1973
FORTRAN IV PROGRAM TO SIMULATE AIR MOVEMENT IN MULTI-STORY BUILDINGS, DBR Computer Program no. 35, National Research Council of Canada, March.

Sander D. M., 1974
FORTRAN IV PROGRAM TO CALCULATE AIR INFILTRATION IN BUILDINGS, DBR Computer Program no. 37, National Research Council of Canada, May.

Scawthorn, C., 1987
FIRE FOLLOWING EARTHQUAKE, ESTIMATE OF THE CONFLAGRATION RISK TO INSURED PROPERTY IN GREATER LOS ANGELES AND SAN FRANCISCO, All-Industry Research Advisory Council (AIRAC), Oak Brook, Ill.

Schleich, J. B., 1987a
CEFICOSS: A COMPUTER PROGRAM FOR THE FIRE ENGINEERING OF STEEL STRUCTURES, International Conference on Mathematical Models for Metals and Materials Applications, Institute of Metals, London, Oct.

Schleich, J. B., 1987b
L'ACIER FACE AUX INCENDIES, Congrès National du Syndicat National des Architectes Agrées et des Maîtres d'Oeuvre en Bâtiment, Angers, France, May.

Schleich, J. B., 1987c
NUMERICAL SIMULATIONS, THE FORTHCOMING APPROACH IN FIRE ENGI-NEERING DESIGN OF STEEL STRUCTURES, *Fachberichte Hüttenpraxis Metall-weiterverarbeitung,* vol. 25, no. 4.

Schleich, J. B., 1987d
NUMERISCHE SIMULATION—ZUKUNFTSORIENTIERTE VORGEHENSWEISE ZURFEUERSICHERHEITSBEURTEILUNGVONSTAHLBAUTEN,*DerMaschinen-schaden,* vol. 60, no. 4.

Schleich, J. B., 1987e
REFAO-CAFIR, COMPUTER ASSISTED ANALYSIS OF THE FIRE RESISTANCE OF STEEL AND COMPOSITE CONCRETE-STEEL STRUCTURES, C.E.C. Re-search 7210-SA/502, Final Report EUR 10828 En, ARBED Research Centre, Luxembourg.

Schleich, J. B., 1987–1989a
GLOBAL BEHAVIOUR OF STEEL STRUCTURES UNDER LOCAL FIRES, R.P.S. Working Documents, ARBED Research Centre, Luxembourg.

Schleich J. B., 1987–1989b
PRACTICAL DESIGN TOOLS FOR COMPOSITE STEEL-CONCRETE CON-STRUCTION ELEMENTS, C.E.C. Research 7210-SA/504, ARBED Research Centre, Luxembourg.

Schleich, J. B., 1987–1989c
PRACTICAL DESIGN TOOLS FOR UNPROTECTED STEEL COLUMNS, Research Reports of C.E.C. Research 7210-SA/505, ARBED Research Centre, Luxembourg.

Schleich, J. B., Lahoda, J. P., and Hutmacher, H., 1983
GARANTIERTER FEUERWIDERSTAND IN STAHLBAU, EINE NEUE TECH-NOLOGIE, *Acier-Stahl-Steel.*

Schmidt, H., 1984
STAHLTRAPEZPROFILDECKEN—BEMESSUNG UND BRANDSCHUTZ, *Stahl-bau,* vol. 53, pp. 295–299.

Schmidt, H., and Lehmann, R.
STAHLPROFILDECKEN MIT AUFBETON—BEMESSUNG UND FEUERWIDER-STAND. Siehe /6/.

Schreuder, J. A. M., 1981
APPLICATION OF A LOCATION MODEL TO FIRE STATIONS IN ROTTERDAM, *European Journal of Operational Research,* vol. 6, pp. 212–219.

Schreuder, J. A. M., 1984
RISK-COVERING FIRE STATIONS IN ROTTERDAM, Contract-Research TH Twente-TW Enschede, in Dutch.

Schreuder, J. A. M., Haken, J. H., and W. Brinkhuis, 1984
RISK-COVERING FIRE DEPARTMENT HENGELO (in Dutch), Contract Research, TH Twente-TW Enschede.

Seeger, P., and John, R., 1978
UNTERSUCHUNG DER RÄUMUNGSABLÄUFE IN GEBÄUDEN ALS GRUNDLAGE FÜR DIE AUSBILDUNG VON RETTUNGSWEGEN, TEIL III: REALE RÄUMUNGSVERSUCHE, Informationszentrum für Raum and Bau der FgG, Stuttgart, Germany, p. 395.

Shannon, C., and Weaver, W., 1949
THE MATHEMATICAL THEORY OF COMMUNICATION, University of Illinois Press, Urbana.

Sharry, J. A., 1974
SOUTH AMERICAN BURNING, *Fire Journal,* July, pp. 23–28.

Sharry, J. A., 1975
HIGH-RISE HOTEL FIRE, *Fire Journal,* Jan., pp. 20–22.

Shaw, B. H., and Whyte, W., 1974
AIR MOVEMENT THROUGH DOORWAYS—THE INFLUENCE OF TEMPERA-TURE AND ITS CONTROL BY FORCED AIR FLOW, *Building Research Services Engineer,* vol. 42, Dec., pp. 210–218.

Sime, J. D., 1983
AFFILIATIVE BEHAVIOUR DURING ESCAPE TO BUILDING EXITS, *Journal of Environmental Psychology,* vol. 1, March, pp. 21–41.

Simiu, E., and Scanlan, R. H., 1978
WIND EFFECTS ON STRUCTURES: AN INTRODUCTION TO WIND ENGINEER-ING, Wiley, New York.

Simpson, S., 1982
THE LIFT AS A MEANS OF ESCAPE FOR HANDICAPPED EMPLOYEES, Fire Prevention 135, Fire Protection Association, U.K., Dec.

Stahl, F. I., and Archea, J., 1977
AN ASSESSMENT OF THE TECHNICAL LITERATURE ON EMERGENCY EGRESS FROM BUILDINGS, NBSIR 77-1313, National Bureau of Standards, Wash-ington, D.C.

Stahl, F. I., 1982a
BFIRES II—A BEHAVIOR-BASED COMPUTER SIMULATION OF EMERGENCY EGRESS DURING FIRES, *Fire Technology,* vol. 18, Feb., pp. 49–65.

Stahl, F. I., 1982b
TIME BASED CAPABILITIES OF OCCUPANTS TO ESCAPE FIRES IN PUBLIC BUILDINGS: A REVIEW OF CODE PROVISIONS AND TECHNICAL LITERA-TURE, NBSIR 82-2480. National Bureau of Standards, Washington, D.C., April.

Tamura, G. T., and Wilson, A. G., 1966
PRESSURE DIFFERENCES FOR A 9-STORY BUILDING AS A RESULT OF CHIM-NEY EFFECT AND VENTILATION SYSTEM OPERATION, *ASHRAE Transac-tions,* vol. 72, pt. 1, pp. 122–134.

Tamura, G. T., and Shaw, C. Y., 1976a
AIR LEAKAGE DATA FOR THE DESIGN OF ELEVATOR AND STAIR SHAFT PRESSURIZATION SYSTEMS, *ASHRAE Transactions,* vol. 83, pt. 2, pp. 179–190.

Tamura, G. T., and Shaw, C. Y., 1976b
STUDIES ON EXTERIOR WALL AIR TIGHTNESS AND AIR INFILTRATION OF TALL BUILDINGS, *ASHRAE Transaction,* vol. 83, pt. 1, pp. 122–134.

Tamura, G. T., and Shaw, C. Y., 1978
EXPERIMENTAL STUDIES OF MECHANICAL VENTING FOR SMOKE CONTROL IN TALL OFFICE BUILDINGS, *ASHRAE Transactions,* vol. 86, pt. 1, pp. 54–71.

Tamura, G. T., and Klote, J. H., 1987
EXPERIMENTAL FIRE TOWER STUDIES OF ELEVATOR PRESSURIZATION SYSTEMS FOR SMOKE CONTROL, *ASHRAE Transactions,* vol. 93, no. 2, pp. 2235–2256.

Tanaka, T., 1983
A MODEL OF MULTIROOM FIRE SPREAD, NBSIR 83-2718, National Institute of Standards and Technology, Gaithersburg, Md.

Templer, J. A., 1975
STAIR SHAPE AND HUMAN MOVEMENT, Ph.D. dissertation, Columbia University, New York.

Thomas, P. H., 1970
MOVEMENT OF SMOKE IN HORIZONTAL CORRIDORS AGAINST AN AIR FLOW, *Institution of Fire Engineers Quarterly,* vol. 30, no. 77, pp. 45–53.

Thomas, P. H., 1986
THE ROLE OF FLAMMABLE LININGS IN FIRE SPREAD, in P. H. Thomas—Fire Research Station 1951–1986 Selected Papers, Paper 17, Garston, Watford, Building Research Establishment, Fire Research Station, Borehamwood, U.K.

Tidey, J., 1983
Greater London Council, private communication.

Ueno, T., Maeda, J., Yoshida, T., and Suzuki, S., 1986
CONSTRUCTION ROBOTS FOR SITE AUTOMATION, Proc. of the Conference on CAD and Robotics in Architecture and Construction, Marseilles, France.

Walton, D., 1985
ASET-B: A ROOM FIRE PROGRAM FOR PERSONAL COMPUTERS, NBSIR 85-3144-1, National Institute of Standards and Technology, Gaithersburg, Md.

Wakamatsu, T., 1977
CALCULATION METHODS FOR PREDICTING SMOKE MOVEMENT IN BUILDING FIRES AND DESIGNING SMOKE CONTROL SYSTEMS, Fire Standards and Safety, ASTM STP 614, A. F. Robertson (Ed.), American Society for Testing and Materials, Philadelphia, Pa., pp. 168–193.

Wakamatsu, T., 1988
DEVELOPMENT OF DESIGN SYSTEM FOR BUILDING FIRE SAFETY, Proc. of the 2d International Symposium on Fire Safety Science, Tokyo, June, pp. 881–895.

Walter, R., 1981
PARTIELL BRANDBEANSPRUCHTE STAHLBETONDECKEN—BERECHNUNG DES INNEREN ZWANGES MIT EINEM SCHEIBENMODELL, Thesis, TU Braunschweig, Germany.

Watrous, L. D., 1969
FIRE IN A HIGH-RISE APARTMENT BUILDING, *Fire Journal,* May, pp. 5–11.

Watrous, L. D., 1971
28 DIE IN PIONEER HOTEL, TUCSON, ARIZONA, *Fire Journal,* May, pp. 21–27.

Watrous, L. D., 1972a
FATAL HOTEL FIRE, *Fire Journal,* Jan., pp. 5–8.

Watrous, L. D., 1972b
FOUR DIE IN NEW YORK YMCA FIRE, *Fire Journal,* Nov., pp. 32ff.

Watrous, L. D., 1973
HIGH-RISE FIRE IN NEW ORLEANS, *Fire Journal,* May, pp. 6–9.

Weinroth, J., 1988
   EXITT: A SIMULATION MODEL OF OCCUPANT DECISIONS AND ACTIONS IN
   RESIDENTIAL FIRES, *Fire Technology,* Aug., pp. 273–278.

Wesche, J., 1985
   TRAGVERHALTEN VON STAHLBETONPLATTEN IM BAUPRAKTISCHEN
   EINBAUZUSTAND BEI BRANDBEANSPRUCHUNG, Thesis, TU Braunschweig,
   Germany.

Wickstrom, U., 1979
   TASEF, A COMPUTER PROGRAM FOR TEMPERATURE ANALYSIS OF STRUC-
   TURES EXPOSED TO FIRE, Lund, Sweden.

Willey, A. E., 1972a
   TAE YON KAK HOTEL FIRE, *Fire Journal,* May, pp. 5–10.

Willey, A. E., 1972b
   HIGH-RISE BUILDING FIRE, *Fire Journal,* July, pp. 7–13.

Williamson, R. B., and Ling, W. C., 1980
   THE MODELLING OF FIRE THROUGH PROBALISTIC NETWORKS, Proc. of System Methodologies and Some Applications Symposium, University of Maryland, Feb.
   27–29, Society of Fire Protection Engineers, Boston.

Williamson, R. B., 1981
   COUPLING DETERMINISTIC AND STOCHASTIC MODELING TO UNWANTED
   FIRE, *Fire Safety Journal,* vol. 3, no. 4, pp. 243–259.

Williamson, R. B., and Ling, W. C., 1985
   MODELING OF FIRE SPREAD THROUGH PROBALISTIC NETWORKS, *Fire
   Safety Journal,* vol. 9, no. 3, pp. 287–300.

Williamson, R. B., and Ling, W. C., 1986
   USE OF PROBALISTIC NETWORKS FOR ANALYSIS OF SMOKE SPREAD AND
   THE EGRESS OF PEOPLE IN BUILDINGS, Proc. of the 1st International Sympo-
   sium on Fire Safety Science, Grant and Pagni (Eds.), Hemisphere Publishing Corp.,
   New York, pp. 953–962.

Wilson, R., 1962
   T-I-M-E!: THE YARDSTICK OF FIRE CONTROL, *NFPA Firemen,* Sept./Nov./Dec.

Wolpert, J., and Zillman, D., 1969
   THE SEQUENTIAL EXPANSION OF A DECISION MODEL IN SPATIAL CON-
   TEXT, *Environment and Planning,* vol. 1, pp. 91–104.

Wood, P. G., 1972
   THE BEHAVIOUR OF PEOPLE IN FIRES, Fire Research Note 953, Building Research Establishment, Fire Research Station, Borehamwood, U.K.

Wright, J. C., 1989
   FIRE PROTECTION ENGINEERING IN EUROPE (Editorial), *FPC Fire Protection
   Newsletter,* vol. 2, no. 4, Nov.

Yamaguchi, T., et al., 1990
   FULL SCALE FIRE TEST OF STEEL COLUMN WITH EXCELLENT MECHANI-
   CAL PROPERTIES AT ELEVATED TEMPERATURE, General Building Research
   Center, vol. 57, Jan., pp. 9–16.

Yoshida, H., Shaw, C. Y., and Tamura, G. T., 1979
   A FORTAN IV PROGRAM TO CALCULATE SMOKE CONCENTRATIONS IN A
   MULTI-STORY BUILDING, DBR Computer Program no. 45, National Research
   Council of Canada, June.

Yoshida, H., Ueno, T., Nonaka, M., and Yamazaki, S., 1984
   DEVELOPMENT OF SPRAYROBOT FOR FIREPROOF COVER WORK, Proc. of
   Workshop Conference on Robotics in Construction, Carnegie-Mellon University, Pitts-
   burgh, Pa.

Contributors

The following is a list of those who have contributed manuscripts for this volume. The names, affiliations, cities, and countries of each contributor are given.

Cliff Barnett, MacDonald Barnett Partners, Ltd., Auckland, New Zealand

Richard W. Bletzacker, Richard W. Bletzacker & Assoc., Inc., Columbus, Ohio, USA

Christopher K. Booker, University of Surrey, Surrey, United Kingdom

R. Bossart, Getlinger AG, Winterthur, Switzerland

David Canter, University of Surrey, Surrey, United Kingdom

K. P. Cheung, Hong Kong Polytech, Hung Hom, Hong Kong

G. M. E. Cooke, Fire Research Station, Borehamwood, England

Leonard Y. Cooper, National Institute of Standards and Technology, Gaithersburg, Maryland, USA

Yuh-Chyurn Ding, Ministry of Interior, Republic of China

M. David Egan, Clemson University, Clemson, South Carolina, USA

Charles M. Fleischman, University of California, Berkeley, California, USA

R. Hass, Institut für Baustoffe, Braunschweig, Germany

Yasao Kajioka, Shimizu Corporation, Tokyo, Japan

K. Keira, Nippon Steel Corporation, Tokyo, Japan

Ezel Kendik, COBAU Consult, Vienna, Austria

John H. Klote, National Institute of Standards and Technology, Gaithersburg, Maryland, USA

Karl Kordina, Technical University of Braunschweig, Braunschweig, Germany

L. Krampf, Technical University of Braunschweig, Braunschweig, Germany

Patrick R. Kroos, J. Roger Preston & Partners, Hong Kong, Hong Kong

J. Kruppa, CTICM, Chevreuse, France

Junichiro Maeda, Shimizu Corporation, Tokyo, Japan

Harold E. Nelson, National Institute of Standards and Technology, Gaithersburg, Maryland, USA

M. Okamatsu, Nippon Steel Corporation, Tokyo, Japan

Nobuhiro Okuyama, Shimizu Corporation, Tokyo, Japan

M. Oohashi, Nippon Steel Corporation, Tokyo, Japan

Jille Powell, University of Surrey, Surrey, United Kingdom

Ulrich Quast, Institut für Baustoffe, Braunschweig, Germany

E. Richter, Institut für Baustoffe, Braunschweig, Germany

K. Rudolph, Institut für Baustoffe, Braunschweig, Germany

Y. Sakamoto, Nippon Steel Corporation, Tokyo, Japan

Hitoshi Sato, Shimizu Corporation, Tokyo, Japan

Charles S. E. Scawthorn, EQE Engineering, San Francisco, California, USA

J. B. Schleich, ARBED Research Centre, Luxembourg, Luxembourg

Jan A. M. Schreuder, Twente University of Technology, Enschede, The Netherlands

Duiliu Sfintesco, Vanves, France

L. Twilt, Institut TNO IBBC, Delft, The Netherlands

Takatoshi Ueno, Nippon Steel Corporation, Tokyo, Japan

Robert B. Williamson, University of California, Berkeley, California, USA

Joseph Zicherman, IFT Technical Services, Inc., Berkeley, California, USA

Building Index

The following index enables the reader to identify the page numbers on which a particular building is mentioned. Numbers in italics that follow cities and buildings refer to photographic views.

Name Index

The following list cites the page numbers on which the indicated names are mentioned. The list includes authors as well as other individuals or organizations named in the text. Names followed by years refer to bibliographic citations that are included in the appendix entitled "References/Bibliography."

339

Subject Index